长江水生生物多样性保护系列丛书

Series of Aquatic Biodiversity Conservation in the Yangtze River

海洋馆中的中华鲟保护研究

Conservation Research of Chinese Sturgeon in Aquarium

危起伟　主编

Edited by Wei Qiwei

科学出版社

北京

内 容 简 介

中华鲟在水族馆内的养殖、保护与展示具有很大的应用前景。本书总结了中华鲟养殖与保护工作中与海洋馆相结合的养殖、展示、保护等方面的研究成果，具有较高的学术价值；主要收录了与水族馆养殖中华鲟相关的论文，包括繁殖学、行为学、疾病学、营养学、生理生化等方面的研究成果。

本书主要供中华鲟保护相关的人士、科研人员、学生、水族馆技术人员等参考使用。

图书在版编目（CIP）数据

海洋馆中的中华鲟保护研究/危起伟主编. —北京：科学出版社，2016.9

（长江水生生物多样性保护系列丛书）

ISBN 978-7-03-049889-2

Ⅰ.①海… Ⅱ.①危… Ⅲ.①中华鲟－保护－研究 Ⅳ.①Q959.46

中国版本图书馆 CIP 数据核字(2016)第 217222 号

责任编辑：张颖兵/责任校对：王 莹

责任印制：彭 超/封面设计：苏 波

科学出版社 出版

北京东黄城根北街 16 号

邮政编码：100717

http://www.sciencep.com

武汉中远印务有限公司印刷

科学出版社发行 各地新华书店经销

*

开本：889×1194 1/16

2016 年 9 月第 一 版 印张：14 1/2

2016 年 9 月第一次印刷 字数：400 000

定价：90.00 元

（如有印装质量问题，我社负责调换）

长江水生生物多样性保护系列丛书

Series of Aquatic Biodiversity Conservation in the Yangtze River

《海洋馆中的中华鲟保护研究》编委会

Editorial Board of "Conservation Research of Chinese Sturgeon in Aquarium"

主编(Chief editor)

危起伟　Wei Qiwei

副主编(Associate editor)

杜　浩　Du Hao
张　辉　Zhang Hui
吴金明　Wu Jinming
王成友　Wang Chengyou
厉　萍　Li Ping
胡维勇　Hu Weiyong
徐海明　Xu Haiming
张晓雁　Zhang Xiaoyan
张军英　Zhang Junying
张书环　Zhang Shuhuan

编委(Editor)

杜　浩　Du Hao
胡维勇　Hu Weiyong
厉　萍　Li Ping
沈　丽　Shen Li
王海燕　Wang Haiyan
吴金明　Wu Jinming
杨月媚　Yang Yuemei
张军英　Zhang Junying
张晓雁　Zhang Xiaoyan
郭　霞　Guo Xia
冷小茜　Leng Xiaoqian
刘志刚　Liu Zhigang
王成友　Wang Chengyou
危起伟　Wei Qiwei
徐海明　Xu Haiming
张　辉　Zhang Hui
张书环　Zhang Shuhuan
张艳珍　Zhang Yanzhen

序

（PREFACE）

中华鲟为国家一级重点保护野生动物，是长江的旗舰保护物种，受到国际社会和我国政府的高度关注。中华鲟生长在我国东部沿海，在长江、珠江产卵繁殖，其自然种群兴衰与水域环境变化息息相关。随着我国经济的快速发展，其自然种群日趋衰退，并有灭绝风险。易地保护是中华鲟保护的重要手段，在目前长江经济带建设、“共抓大保护，不搞大开发”的背景下，积极探索中华鲟易地保护的技术与方法显得尤为重要。

中国水产科学研究院长江水产研究所长期致力于中华鲟的保护研究工作，取得了丰硕的成果。2007 年“中华鲟物种保护技术研究”获得了国家科技进步二等奖。在易地保护和人工种群建设方面，先后突破了中华鲟野生个体的人工繁殖、苗种培育、子一代亲本培育等技术瓶颈，实现了中华鲟亲本的规模化驯养与保种，2012 年突破了中华鲟的全人工繁殖，从而取得了中华鲟在全人工环境中的保种成功，为实施大规模中华鲟增殖养护提供了基础。

我在中国水产科学研究院及长江水产研究所任职期间，有幸见证并直接参与了一些中华鲟的保护研究工作。2005 年 1 月，长江水产研究所与北京海洋馆达成协议，启动科企联合方式开展中华鲟驯养展示及保护相关的科学研究。利用长江水产研究所的技术优势和北京海洋馆的设施条件与展示平台，最大程度上实现了国宝中华鲟的保育宣传，提高了公众认知和环保意识。经过 12 年持续合作与研究，先后突破了野生中华鲟的开口摄食、产后亲本的康复和再发育、子一代中华鲟的批量性成熟、子一代亲本的产后再发育等技术难题；建立了中华鲟驯养技术规范；初步建立了中华鲟健康养殖生理生化指标监测体系等。中华鲟在北京海洋馆中的养殖与保护效果证明，通过科企合作开展中华鲟的养殖与科研，是实施中华鲟保护的有效途径。

本书是海洋馆养殖条件下中华鲟易地保护相关研究的系统总结，不仅具有较高的科学意义，而且可为中华鲟在其他水族馆中的驯养、展示与保护工作提供参考借鉴。

农业部渔业渔政管理局　局长　张显良

2016 年 9 月 15 日

前　　言

（FOREWORD）

中华鲟作为国家一级保护动物和大型洄游性鱼类，一直备受科研工作者和社会公众关注。然而，在长江及近海水域生态环境持续衰退的背景下，其自然种群数量急剧下降。在自然种群的保护尚未取得显著效果的同时，易地保护成为物种保护的有效手段。2005 年 4 月 22 日，首批 26 尾中华鲟从湖北荆州踏上旅程到达北京海洋馆。十多年来，北京海洋馆、香港海洋公园与长江水产研究所进行了多次中华鲟的交换、养殖和中华鲟在海洋馆的保护性公众展示。海洋馆在中华鲟的养殖与展示方面投入了大量的物力与精力，取得了良好的社会宣传效果。同时，海洋馆在保护性养殖方面也取得了丰硕的成果，开启了中华鲟易地保护的新篇章。

中华鲟在水族馆养殖的成功，离不开一系列研究工作的支撑。本文集收录了 2006 年以来公开发表或在学术会议上宣读的 28 篇论文。文集中，既有海洋馆养殖系统及设施的设计，也有鱼体在海洋馆环境下的生物学、生态学、行为学等方面的观察与研究；既有中华鲟成体的养殖与繁殖技术的研究，也有对幼鱼培育方法的探索。因此，本文集是目前为止中华鲟海洋馆养殖相关成果的最新、最全的总结，希望文集的出版能为中华鲟在各地海洋馆中的养殖提供技术指导，同时也为中华鲟的易地保护提供参考和借鉴。

由于我们水平有限，书中难免存在一些不妥之处，敬请广大读者批评指正。

编　者

目　　录

(CONTENTS)

中华鲟水族馆迁地驯养的适应性研究

危起伟[1,3,*]　张晓雁[1,2,4]　张先锋[1,**]　刘鉴毅[3]　杨道明[4]
朱永久[3]　陈细化[3]　吴香香[4]　李罗新[3]　刘建伟[5]

(1. 中国科学院水生生物研究所　湖北武汉　430072
2. 中国科学院研究生院　北京　100049
3. 农业部淡水生物多样性保护重点开放实验室 中国水产科学研究院长江水产研究所　湖北武汉　430223
4. 北京信沃达海洋科技有限公司(北京海洋馆)　北京　100081
5. 中国水产科学研究院　北京　100141)

摘　要　于2005年4月到2007年8月,观测了5尾野生中华鲟(W)和323尾子一代中华鲟(F_1)在水族馆中的存活、生长发育以及行为。W全部存活,F_1存活超过86%。W的体质量月增长率介于0.55%~10%,1尾雌性W在驯养28个月后性腺再次发育至IV期初期。子一代中华鲟体质量和全长月最低增长率为1.39%(F_1-1997)和4.28%(F_1-2001),最高为35.70%和29.35%(F_1-2005)。W主动摄食前后游泳行为变化明显,对食物具选择性,喜摄食鲜活饵料,对5种鲜饵有较好选择性,对其余8种选择性差。中华鲟可以适应水族馆环境,获得较好的生长发育效果表明,水族馆可作为中华鲟易地保护的栖息地之一。

关键词　中华鲟　行为　食物选择　生长　性腺发育　适应　水族馆

Acclimating and Maintaining Chinese Sturgeons (*Acipenser sinensis*) in the Captive Environment of a Large Aquarium

Wei Qiwei[1,3,*]　Zhang Xiaoyan[1,2,4]　Zhang Xianfeng[1,**]　Liu Jianyi[3]
Yang Daoming[4]　Zhu Yongjiu[3]　Chen Xihua[3]
Wu Xiangxiang[4]　Li Luoxin[3]　Liu Jianwei[5]

(1. Institute of Hydrobiology, Chinese Academy of Sciences, Wuhan 430072, China;
2. Graduate University of Chinese Academy of Science, Beijing 100049, China;
3. Key Laboratory of Freshwater Biodiversity Conservation and Utilization, Ministry of Agriculture; Yangtze River Fisheries Research Institute, Chinese Academy of Fisheries Sciences, Jingzhou 434000, Hubei Province, China;
4. Beijing Aquarium, Beijing, 100081, China;
5. Chinese Academy of Fisheries Science, Beijing 100141, China)

Abstract　In this study, the 5 wild Chinese sturgeons which were captured from the Yichang section of the Yangtze River, and the 323 offsprings of the Chinese sturgeons, were studied as to their survival, development and behaviour under aquarium conditions between 2005 and 2007 in the captive environment of an aquarium. All 5 wild caught females survived, with weight gains per month raging from 0.55%(Specimen W1#: Arrival: 220.00 kg, Final: 252.50 kg; across 27 months) to 10.00% of original body weight (Specimen W32#: Onset:

* 通讯作者(Corresponding author), E-mail: weiqw@ yyi. ac. cn
** 通讯作者(Corresponding author), E-mail: zhangX@ ihb. ac. cn

125.00 kg, Final: 175.00 kg; across 4 months). In addition, the gonads of one captured fish redeveloped to the stage IV by August 2007. The changes of the depth in the tank were observed before, during and after active food intake. Furthermore, the fish formed strong feeding associations with live food items, such as *Hypophthalmichthys molitrix*, *Cyprinus carpio* and *Paralichthys lethostigma*, while rarely feeding on *Hexagrammos otakii* and *Epinephelus awoara*, occasionally on *Monopterus albus*, *Penaeus vannamei*, *Leiocassis longirostris*, and never on *Eriocheir sinensis* H. Milne-Edwards, *Misgurnus anguillicaudatus*, *Silurus asotus* Linnaeus, *Octopus Variabilis*, *Anadara uropygimelana*. The development of food preferences in the captive environment suggest that theses Chinese sturgeon successfully acclimated to the large aquarium, in addition, the survival rate of the filial generation of offspring was above 86 %. Across the study period, for the older captive bred offspring (1997 and 2001 stocks) minimum weight increase per month was 4.28 % (F_1-2001: Onset: (12.24 ± 1.11) kg, Final: (25.35 ± 2.10) kg; over 25 months) and the minimum lengthwise growth rate per month was 1.39 % (F_1-1997: Onset: (187.32 ± 1.32) cm, Final: 260.00 cm; across 28 months). However, the youngest (the 2005 offspring), had the highest weight gain of 35.70 % per month (Onset: (16.50 ± 1.92) g, Final: (8733.33 ± 739.80) g; across 18 months) and lengthwise growth rate 29.35 % per month (Onset: (15.53 ± 0.77) cm, Final: (115.33 ± 4.14) cm; across 18 months) respectively.

Key words *Acipenser sinensis* Gray Behaviour Food predisposed Growth Gonad development Adaptability Aquarium

1 Introduction

The Chinese sturgeon (*Acipenser sinensis* Gray) is a typical anadromous fish that matures in the East China Sea and Yellow Sea, and spawns in the Zhu River and Yangtze River (Wei et al., 1997). However, the construction of the Gezhouba Dam in the 1980s and the Three Gorges Dam in 2000s in the Yangtze River block the migration of the sturgeons to their natural spawning grounds, contributing to the decline and changes in stock structure of this critically endangered population (Wei et al., 2005) which is further exacerbated by females only breeding three to four times during the course of a life cycle, with < 1 % of offspring surviving due to high predation rates (Chen, 2007). It has been calculated that juveniles were further reduced by 80 %-90 % between 1981 and 1991, despite already being at critical levels (Ke, 1999). In response to the anthropogenic changes to the river structure, some breeding cohorts that are still functional have formed new spawning grounds below the Gezhouba Dam (Wei et al., 2004).

Due to the loss of natural habitats which are critical for reproduction, *ex-situ* conservation studies (i. e. outside of the natural environment) of Chinese sturgeon are required to safeguard population numbers and prevent the genetic diversity from becoming extinct. One of the primary problems of *ex-situ* conservation is the construction of suitable artificial environments that replicate sufficiently the natural environmental features to foster feeding and breeding activity by wild caught fish (Li et al., 1994). Due to the large size of sturgeon (with adults reaching up to 4 m in total length), Beijing aquarium is considered to be one of the few units in China that are equipped with advanced technical skills and large enough aquaria tanks suitable to identify the conditions required for the acclimation of sturgeons while serving species protection needs (Liu et al., 2006). In the wild, adult sturgeon usually stop food intake on initiation of the upstream migration to spawning grounds. Fish will start feeding again only after returning to the sea following spawning (The Yangtze Aquatic Resources Survey Group, 1988). Since the 1970s, several captive breeding programmes have been initiated to boost population numbers. However, the reintroduction of captive bred sturgeon to the Yangtze River, following spawning, has so far resulted in extremely low survival rates (Wei et al., 2004). Therefore, to improve both *ex-situ* captive conditions and reintroduction success, it is necessary to develop an understanding of the feeding behaviour, and understand recovery needs after spawning, while also identifying the conditions required for gonad redevelopment of wild sturgeon that are housed in controlled environments. Such data are urgently needed to safeguard the critically endangered Chinese sturgeon.

Based on these requirements, the current study investigated the behaviour, growth and food preference of wild caught adult female Chinese sturgeon and their offsprings derived from three different captive stocks. The capacity of the wild caught sturgeon to acclimatize to the aquarium environment was also studied to identify parameters, which may be used to indicate performance and hence optimal conditions for welfare. Furthermore, the changes in the behaviour and morphology of one adult female were recorded during gonad re-development in captivity. We apply the findings of this study to identify parameters which can contribute to the successful acclimation of wild caught Chinese sturgeon. The implications of these findings will assist in the long-term captive breeding and re-introduction programmes.

2 Materials and methods

2.1 Fish stock

Between 2005 and 2006, five wild gravid female Chinese sturgeon were captured from the Yichang section of the Yangtze River, due to breed. The sturgeon were initially held at the Yangtze River Fisheries Research Institute of the Chinese Academy of Fisheries Sciences in Jingzhou city of Hubei Province, China. After 1 or 2 weeks in captivity the fish produced filial sturgeons and were held for a further 2 months at this site before transfer by ambulance to Beijing Aquarium. The origin of the wild caught fish, and accompanying basic measurement data are given in Table 1.

Table 1 Parameters of the wild caught adult female Chinese sturgeon that were housed at the Beijing Aquarium

Specimen identification	*N*	Origin of the fish	Age at arrival	Starting date (yyyy/mm/dd)
W1#	1	Capture in Yichang section of Yangtze River on 12th Oct, 2004, post mature, not propagated	Not determined	2005/04/03
W28#	1	Capture in Yichang section of Yangtze River on 5th Nov, 2005, post mature, not propagated	Not determined	2005/12/29
W29#	1	Capture in Yichang section of Yangtze River on 9th Nov, 2005, artificial propagation	Not determined	2006/01/13
W31#	1	Capture in Yichang section of Yangtze river on 30th Oct, 2006, artificial propagation	Not determined	2006/12/15
W32#	1	Capture in Yichang section of Yangtze river on 30th Oct, 2006, artificial propagation	Not determined	2006/12/19
F1-1997	2	Controlled propagation in 1997	7 +	2005/04/16
F1-2001	23	Controlled propagation in 2001	3 +	2005/03/25
F1-2005	298	Controlled propagation in 2005	0 +	2005/12/29

N, number of fish.

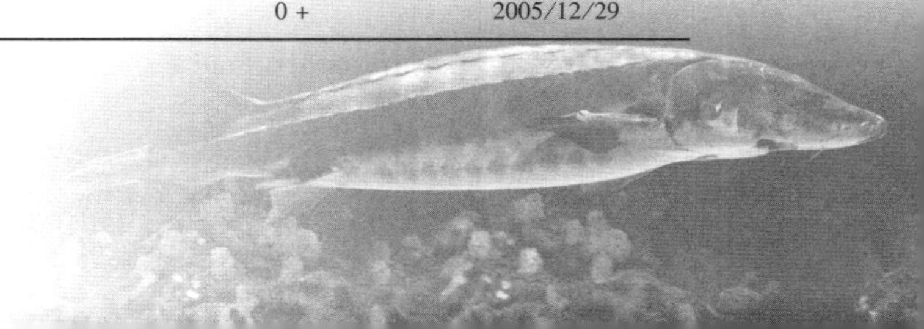

In addition, filial sturgeons from captive bred Chinese sturgeon stocks were assessed. These comprised two fish that were produced in 1997(F1-1997), 23 fish that were produced in 2001(F1-2001), and 298 fish that originated from the 2005(F1-2005) reproduction. The F1-2005 sturgeons came from one of the experimental wild females (W29#), however, the F1-1997 and F1-2001 sturgeons did not originate from any of these females.

2.2 Housing conditions

The five wild caught adult females were accommodated in a square tank together with captive bred juveniles from the filial generation stock that were >1 m in total length(juvenile length ranged from 148 to 200 cm). The tank size was 29.0 m(length) ×11.0 m(width) ×4.4 m(height), with an exhibition-window on one side of the tank(20 m in length; 3.0 m height). The tank held a maximum water volume of 1 200 m^3. The stocking density was 2.2 kg/m^3.

F1-2005 generation from the wild caught females(also <1 m) were housed in two connected circular tanks, with two observation-windows(20 cm ×20 cm). The diameter and height of each tank were 3.5 and 1.4 m, respectively, which when combined held 25 m^3 water volume. The stocking density was maintained below 2.2 kg/m^3.

All tanks were equipped with advanced life support systems, which automatically control water quality to set specifications. For example, several key water parameters are controlled continuously, including water temperature at 21-24℃, oxygen content to be between 7 and 10 mg/L, while pH values were maintained at 7.5-8.0, and ionized ammonia [NH_4^+] concentrations leveled at 0.01-0.05 mg/L, and [NO_2^-] concentrations remained low (0-0.1 mg/L). The main water quality parameters were determined daily, and tanks were cleaned at fixed times twice weekly.

2.3 Feeding information

Both natural food(i.e. fish) and artificial dry pelleted commercial sturgeon diets were used for all four fish groups. The natural food was purchased from the market, and included a range of fish species of size range 5-40 cm; *Hypophthalmichthys molitrix*, *Cyprinus carpio*, *Paralichthys lethostigma*, *Hexagrammos otakii*, *Epinephelus awoar*, *Monopterus albus*, *Leiocassis longirostris*, *Penaeus vannamei*, *Eriocheir sinensis* H. Milne-Edwards, *Misgurnus anguillicaudatus*, *Silurus asotus* Linnaeus, *Octopus variabilis* and *Anadara uropygimelana*. The natural food was offered in three different forms; including frozen, fresh but dead, and live.

In the first tank, containing the wild caught sturgeon and F1-1997 and F1-2001 generation fish of >1 m, the fish were individually hand-fed by a diver. Hence all food intake was documented. Generally, after 2 weeks, force feeding of wild fish that had been recently caught except for W1# was required on arrival, and continued until active feeding was initiated, which individually varied between 20 days and 10 months(Zhang et al., 2007). Fish were fed once a day at fixed location. In the second tank, containing F1-2005 stock of <1 m, prepared commercial feed and natural food were offered by hand at the same location(tank surface) once daily.

2.4 Data collection

Growth measurements. Four different growth measurements were obtained during the course of the study: (i) body circumference, (ii) waist girth, (iii) total body length (head to the end of vertebra), and (iv) body weight. The 'body circumference' was measured at the highest point of the third dorsal bony plate. The 'waist girth' was measured at the highest point of the fifth dorsal bony plate. For the wild caught sturgeon, these two measurements were obtained monthly by a diver using a tape measure under water while swimming with the fish. The total body length was measured using a tape measure, and the body weight was recorded after the fish removed from the water. For juvenile stock of >1 m the same procedure was followed. For juvenile stock of <1 m the fish were removed from the water to acquire all measurements every 3 months. Each measurement was repeated three times on each occasion for each fish. The means of these three measurements were used in the results.

Measurements were taken underwater due to the large size of the fish, a validation test was run to determine the degree of error that might exist by the same individual and by different individuals measuring the same fish. A total of three people made measurements on separate occasions during the course of 1 day. We found that measurement error was on average 0.5 cm (range 0-1; (0.5 ±0.32) cm) for repeated measurements by one person, and on average 1.5 (range 0-2.5 cm; (1.5 ±0.41) cm) for repeated measurements by different people. Based on these findings, during the main experiment only one person conducted all measurements to keep measurement error to a minimum.

Gonad development. The oocytes of one wild caught adult Chinese sturgeon (W1#) were removed by a veterinarian by biopsy and photographed on 4 August 2007. The diameter of the most of the oocytes was measured.

Behavior observations. Opportunistic sampling of the behavior of the wild caught sturgeon and the three different juvenile age groups was conducted. Behaviour was observed and documented for three times per day between 8:30 and 17:30 (total 90 min of observation period). In addition, feeding behavior was observed and documented daily. Food preference calculations followed the methodology of Liang (1995). The wild sturgeons feeding behaviours of 13 live baits were recorded and observed lasting for 1 month. Calculations on feeding behaviour was referred to the rate of special behaviour after Liang (1995). Each bait was given 10 days, three times per day, and 6-7 baits were given simultaneously and randomly in the 10 days, and the response of fish to the food item was recorded. The results are presented as percentage of observations.

2.5 Data processing

The selection of different food items by the wild caught sturgeon were analyzed for differences using the Kruskal-Wallis Test and Mann-Whitney Test.

3 Results

3.1 Survival

All five wild caught Chinese sturgeon survived the assessment period. After the study, three of the individuals (W1#, W29#, W31#) remained at Beijing Aquarium, while the other two (W28#, W32#) were released back into the wild in the Jinzhou region of the Yangtze River on 22 April 2007. The F1-1997 and F1-2001 captive bred filial generation juvenile stock survived the assessment period, and to date remain housed at Beijing Aquarium. Out of the 298 offspring in F1-2005 stock, 257 survived the assessment period, which is an 86 % survival rate. Of these, 148 were released into the Jinzhou region of the Yangtze River on 22 April 2006, while the rest of the fish remained housed at Beijing Aquarium.

3.2 Growth and development

Wild Chinese sturgeon. The first wild caught female (W1 #) was initially captured on 12 October 2004 and admitted to the aquarium on 3 April 2005. The fish initiated active food intake from 7 April 2005 until suddenly stopping feeding on 20 August 2006. As a result, an increase followed by a declining trend in body circumference was recorded, as shown in Fig. 1. On the removal of gonad tissue by biopsy indicated the presence of brown and elliptic oocytes, with an average diameter of 3.2 mm. The polar spot was vague on the animal pole of oocyte.

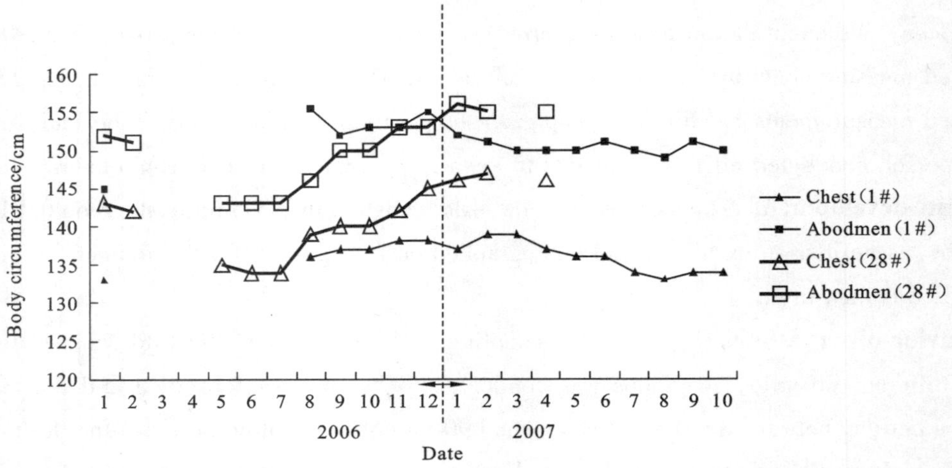

Fig. 1 Change in body circumference of W1#(female) and W28#(female) wild caught Chinese sturgeon during 2006-2007

The second female (W28#) was initially captured on 5 November 2005 and admitted to the aquarium on 29 December, 2005. The body circumference parameters declined from arrival until August 2006. At this point force feeding was initiated, and after 21 days the fish began active feeding, at which point body circumference measurements showed a steady increase (Fig. 1).

The third female (W29#) was initially captured on 9 November 2005 and admitted to the aquarium on 13 January 2006. The body circumference measurements declined after arrival. Force feeding was initiated on 29 May 2006 and began active feeding on 17 June 2006, after which the body circumference measurements steadily increased (Fig. 2).

The fourth female (W31#) was initially captured on 30 October 2006 and admitted to the aquarium on 15 December 2006. Force feeding was initiated on 18 June 2007, and at the end of the study period (October 2007) active feeding had not been initiated. The body circumference measurements are shown in Fig. 3.

The fifth female (W32#) was initially captured on 30 October 2006 and admitted to the aquarium on 19 December 2006. Force feeding was initiated on 9 January 2007, while active feeding commenced on 1 February 2007. Subsequently, body circumference measurements increased steadily (Fig. 3).

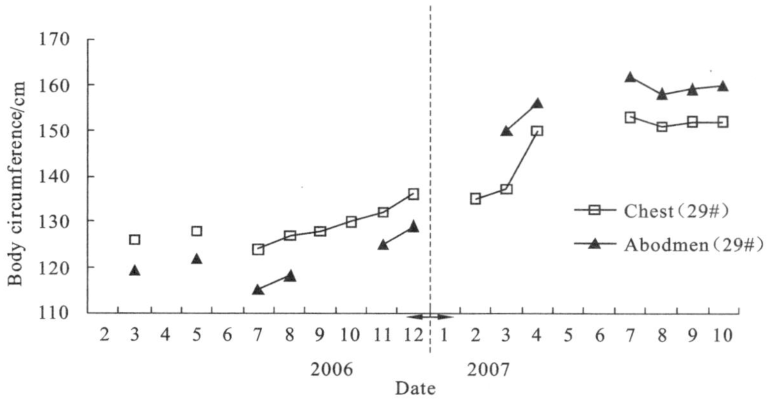

Fig. 2 Change in body circumference of W29#(female) wild caught Chinese sturgeon during 2006-2007

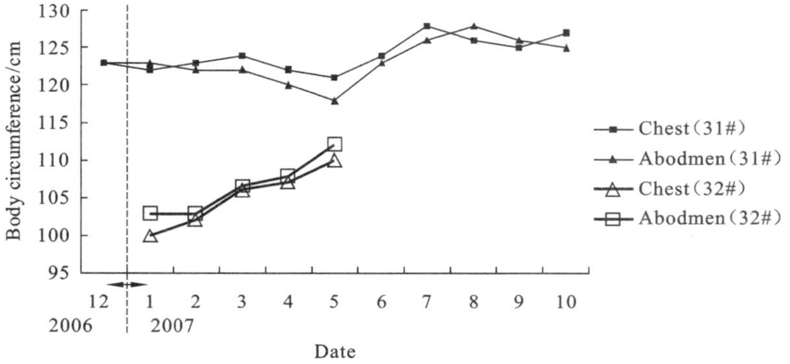

Fig. 3 Change in body circumference of W31#(female) and W32#(female) wild caught Chinese sturgeon during 2006-2007

Figures 4 and 5 show the variation of total length and body weight respectively for the five wild captured female Chinese sturgeon.

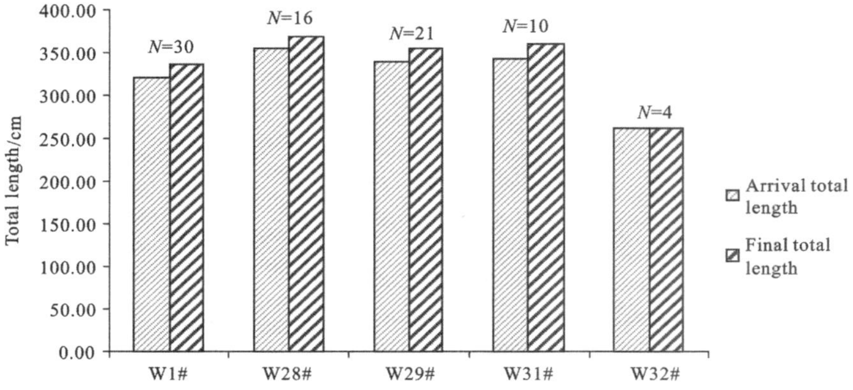

Fig. 4 Variation in total length measurements of wild Chinese sturgeon($n = 5$) during the study period in captivity(N = number of months in captivity)

First generation Chinese sturgeon. For the three different age groups of juvenile stock(F1-

1997, F1-2001, and F1-2005), Table 2 shows the measurement data for total body weight and total body length. All age groups of juvenile sturgeon grew with respect to both body weight and body length.

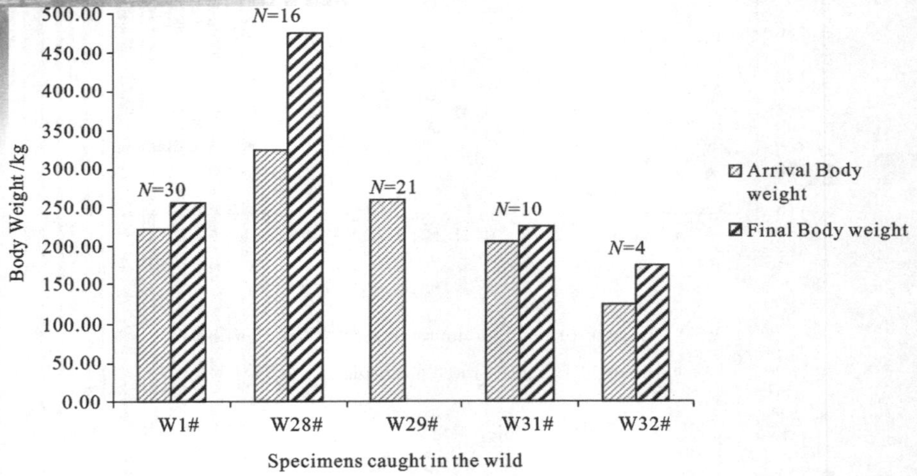

Fig. 5 Variation in total length measurements of wild Chinese sturgeon ($n=5$) during the study period of captivity (N = number of months in captivity)

Table 2 Growth performance of F1-2005, F1-2001 and F1-1997 captive bred Chinese sturgeon

Indicated group	Sampling date	N	Total length/cm		Body weight/kg	
			Mean ± SE	CV/%	Mean ± SE	CV/%
F1-2005	Jan-06	15	15.53 ±0.77	4.96	0.02 ±0.00	11.64
	Feb-06	15	27.27 ±1.14	4.18	0.10 ±0.01	7.94
	Mar-06	15	40.30 ±2.88	7.15	0.24 ±0.03	15.74
	May-06	15	39.07 ±2.75	7.04	0.26 ±0.05	18.57
	Jun-06	15	53.90 ±1.97	3.65	0.58 ±0.06	10.00
	Jul-07	15	115.33 ±4.14	3.59	8.73 ±0.74	8.47
F1-2001	Apr-05	23	119.11 ±2.57	2.16	12.24 ±1.11	9.07
	Jul-05	20	129.20 ±2.28	1.76	13.80 ±0.71	5.14
	Oct-05	23	137.70 ±1.52	1.10	—	—
	Feb-06	23	145.78 ±2.13	1.46	—	—
	May-07	23	160.72 ±3.41	2.12	25.35 ±2.10	8.28
F1-1997	Apr-05	2	187.32 ±1.32	0.70	35.00 ±0.00	0.00
	Jun-05	2	187.32 ±1.32	0.70	40.00	—
	Aug-07	1	260.00	—	99.50	—

N, number of fish.

For example, the F1-1997 stock exhibited a mean increase in body length and weight of 1.39 % per month ($n=2$; Onset: (187.32 ± 1.32) cm, Final: 260.00 cm; across 28 months;) and 6.58 % per month ($n=2$; Onset: 35 kg, Final: 99.5 kg; over 28 months), respectively. The F1-2001 stock exhibited a mean monthly increase in body length and weight of 1.40 % per month ($n=23$; Onset: (119.11 ±2.57) cm, Final: (160.72 ±3.41) cm; across 25 months) and 4.28 % per month ($n=$

23; Onset: (12.24 ± 1.11) kg, Final: (25.35 ± 2.10) kg; over 25 months), respectively. The F1-2005 stock exhibited a mean monthly increase in body length and weight of 29.35 % per month (n = 15; Onset: (15.53 ± 0.77) cm, Final: (115.33 ± 4.14) cm; across 18 months) and 35.70 % per month (n = 15; Onset: (16.50 ± 1.92) g, Final: (8 733.33 ± 739.80) g; across 18 months). In other words, as the fish became older the growth rate became slower and stabilized. These observations were supported by statistical analyses, in which it was confirmed that with increasing age the coefficient of variation declined with respect to total body length and body weight, respectively.

An alternative presentation of the actual total body weight and total body length, indicated that over a 19-28 months period for Chinese sturgeon of: (i) 1 year old (F1-2005) these measurements were 80.91 cm and 3.20 kg; (ii) for those of 4 years old (F1-2001) these measurements were 138.00 cm and 25.53 kg; (iii) for those of 5 years old (F1-2001) these measurements were 154.05 cm and 21.74 kg; and (iv) for those of 8 years old (F1-1997) these measurement were 193.54 cm and 45.09 kg respectively.

3.3 Behaviour

Preferred swimming depth. A total of 45 h behavioural observations were made during the study period, comprising 90 min observations over 30 days. It was found that the five wild caught Chinese sturgeon primarily occupied the upper (i. e. 2.5-4.0 m depth range) and middle (i. e. 1.5-2.5 m depth range) depths of the aquarium prior to the active taking of food and did not changed their position during the period of feeding which lasted 40 min on average (range 30-60 min; 40 ± 7.1). After the completion of active food intake, the sturgeon primarily swam at medium depth for other period of time. The exception was active swimming near the bottom (i. e. 0-1 m depth range) of the aquarium during the period of feeding. Summary information about depth use by the individual sturgeon is presented in Table 3.

Table 3 The swimming behaviour of wild caught Chinese sturgeon (n = 5) in a large public aquarium with respect to depth positioning

Indicated fish	Ordinary depth/m (distance from tank bottom)	Feeding depth/m (distance from tank bottom)
W1#	2.0-4.0	0.1-1.0, ascending to upper depth level after tanking some food
W28#	2.0-2.5	0.1-0.3
W29#	3.0-4.0	0.1-0.3
W31#	1.5-2.0	0.1-0.3
W32#	2.0-2.5	0.1-0.3

Behavioural observations of the two F1-1997 sturgeon showed a primary depth use of 1.5-2.0 m (85.3 % of total observations), but descended to a depth of 0.5-1.0 m during feeding (i. e. for about 30-50 min), followed by a return to the original depth afterwards. The F1-2001 and F1-2005 stocks exhibited a similar pattern of depth use, in which the lower and bottom parts of the aquarium were primarily used.

Food intake by wild caught specimens. The acclimation to captive feeding by the five wild caught sturgeon was assessed. One fish (W1#) took food actively without the requirement of force feeding, probably due to it having been in captivity since 2004 and having already undergone the acclimation process and force feeding prior to the onset of the study. Three fish, which were captured within the timeframe of the study period, began active feeding after about 20 days of force feeding (W28# and W29# after 21 days; W32# after 19 days). However, one fish (W31#) required 10 months of force feeding before the initiation of active food intake.

During the active feeding process, the wild caught sturgeon required 1-5 min to occupy lower water depths before entering the feeding areas. The fish then reduced water depth gradually while feeding, and finally returned to normal depth use within 15 min after the feeding frenzy. When feeding, the fish remained close to the diver handing out food, and maintained a distance of about 0.5-1.0 m from adjacent fish, at which distance no reaction from or to the other fish was noted.

Prey preference. The wild caught sturgeon showed certain selectivity towards different food items when offered a range of 13 different species. The sturgeon preferred live fish as prey, Fig. 6 shows the differences in food preference for the 13 live species that were offered. Following acclimation, the wild caught sturgeon appeared to have strong food preferences to live food items, selecting *Hypophthalmichthys molitrix*, *Cyprinus carpio* and *Paralichthys lethostigma* on over 80 % of occasions that they were offered. The following species were only selected between 15 and 60 % of occasions that they were offered; *Hexagrammos otakii* and *Epinephelus awoara*, *Monopterus albus*, *Penaeus vannamei*, *Leiocassis longirostris*. The sturgeon did not take the following fish on any occasion that it was offered; *Eriocheir sinensis* H. Milne-Edwards, *Misgurnus anguillicaudatus*, *Silurus asotus* Linnaeus, *Octopus variabilis*, *Anadara uropygimelana*. Occasionally they preferred dead *Paralichthys lethostigma*, *Hypophthalmichthys molitrix*, *Cyprinus carpio*, *Hexagrammos otakii*, *Eriocheir sinensis* H. Milne-Edwards, *Sciaenops ocellatus*.

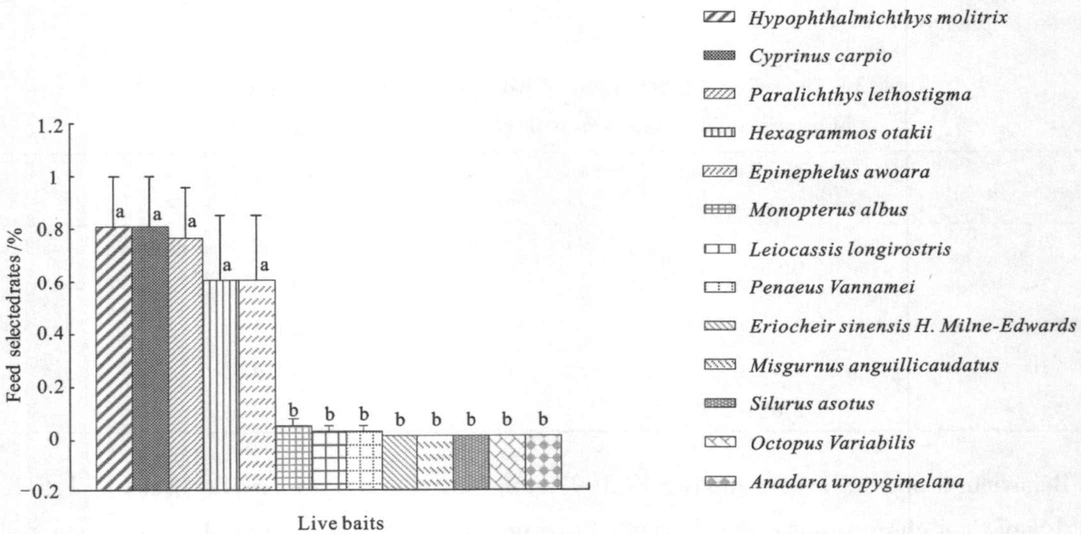

Fig. 6 The feed selected rates of the five wild caught female Chinese sturgeon ($n=5$) with respect to 13 different bait types that were offered

When the diver offered food to the wild sturgeon, the fish elongated their snout in front of the items(fresh, frozen and artificial bait), appearing to sense the item. Subsequent to this action, the following behavior patterns were observed: (i) swallow the food item directly; (ii) swallow the food item after chewing; (iii) spit out the food item after chewing; and (iv) spit out the food item immediately without chewing. The F1-1997, F1-2001 and F1-2005 juvenile stocks ate both natural food and artificial granular diets, exhibiting no food preference behaviour.

4 Discussion

4.1 Development and growth of Chinese sturgeon

In our study, all age groups of the first generation offspring exhibited growth measurements that were much higher than those theoretically calculated for sturgeon in the wild (Anonymous, 1988). For example, sturgeon of 1 year in age are predicted to weigh 1.55 in the wild vs 3.20 kg in our study; at 4 years this figure is 10.35 vs 25.53 kg, respectively; and at 8 years this figure is 34.75 vs 49.09 kg. However, the values obtained in this study were similar to those reported in controlled aquacultures for Chinese sturgeon of about 1 year in age in the Chengdu region (total body length 97.5 cm; total body weight 3.30 kg; Zhao et al., 1997) and slightly larger than the sizes recorded (71.77 cm and 1.96 kg, respectively) by Xiao et al. (1994). It should be noted that the body weight of the Chinese sturgeon that were caught in the wild and transferred directly to Beijing Aquarium increased significantly after the onset of active feeding. In contrast, total body length of the mature females increased slowly, which may be related to the slowing of growth with aging, as is recorded in many species. Water temperature was an important environmental parameter, particularly because fish are poikilothermic. The maintenance of optimal water temperatures in the aquarium (in contrast to replicating seasonal changes and thereby low winter water temperatures) boosted fish growth of the fish, while body weight increases were directly related to both the environment and the high nutrition levels of the food being offered (Yin, 2003).

In addition, the wild caught mature fish showed positive growth trends, for one fish (W1#) the maturation of the ovaries to stage IV (Hu et al., 2007) was recorded, which was recognized by the swelling of the belly and the cessation of food intake. However, subsequent regression of gonad status to stage II (Webb et al., 1999) and complete degeneration was recorded. Several studies have shown that reproductively viable wild sturgeon reach the bayou of Yangtze River between July and August, then migrate for 16 months. In addition, Doroshov et al. (1997) found that subsequent spawning events by *Acipenser transmontanus* occurred at almost identical times of the season. During this period the sturgeon cease all food intake (Wei et al., 1997; Anonymous, 1988). Hence, the captive female (W1#) followed a consistent seasonal pattern with respect to the onset of ovarian development (i. e. in August) and fasting behaviour. A number of factors may modify the functioning of reproductive growth hormones, in turn affecting energy storage, such as nutrition, metabolic rate and hormone actions (Lin, 1996). For example, the hypothalamus-pituitary-gonad axis adjusts the reproductive endocrine system of the sturgeon. Hence, age, body size, and seasonality may contribute towards inducing the hypothalamus to release GnRH. Subsequently, GnRH induces the pituitary release hormone GtH, which in turn boosts the development of the ovaries (Doroshov et al., 1997).

In the current study, the female (W1#) was housed in an artificial environment, in which a constant water temperature (21-24 ℃) was maintained, hence resulting in the growth compared to the

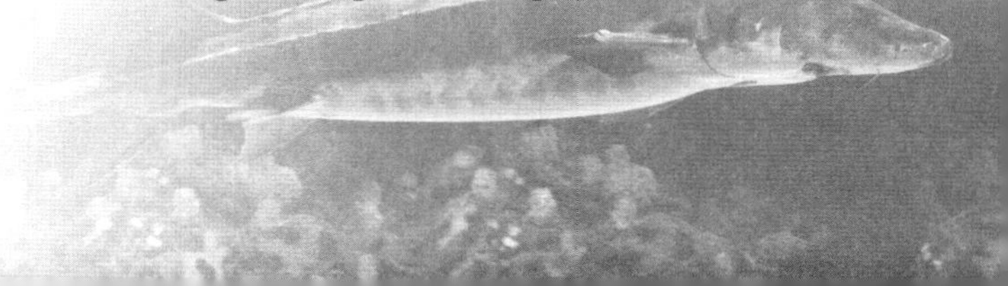

conditions in the natural environment, in which the sturgeon would experience overall annual marine/freshwater fluctuations of between 16 and 21℃. Furthermore, the aquarium was equipped with a metal halogen lamp that provided constant light, which may also have caused the internal body rhythm of W1# to act as a trigger to induce gonad re-development. In fact, under natural conditions, the spawning cycle of sturgeon generally occurs at approximately 1-10 year intervals (Holcik, 1989), with wild female Chinese sturgeon spawning having been documented at 5-7 year intervals (Chen, 2007). In contrast, sturgeon raised under controlled conditions have been found to require 2 years to regain optimal body condition for reproduction (Doroshov et al., 1997, 1994; Williot et al., 1991), which was supported by our observations of W1# (requiring <2 years) in the current study. The difference in the time required to reach optimal body condition is probably influenced by a combination of external environmental factors, genetic inheritance and physical factors (Lin, 2004; Doroshov et al., 1997). In fact, our study clearly demonstrated that the presence of an abundant food resource and stable environmental conditions were integral towards reducing the period of time required for gonad re-development. Basically, high water temperatures accelerated the rate of the early phase of gonad development, but also accelerated the onset of regression in the later stage (Webb et al., 1999), or possibly degraded gonad quality and fertility rate. More research is required to understand how different temperatures affect gonad maturation.

4.2 Depth distribution of Chinese sturgeon

In the present study, differences were found in the depth utilization of individual sturgeon occupying the large aquarium. The Chinese sturgeon is a large fish (up to 4 m in length) that under natural conditions is adept at swimming along the river bed or seabed (Wei et al., 1998). Hence, the limited living space that is available in an artificial landscape may restrict the free space required for swimming, consequently altering the pattern of distribution of individuals that was observed in the aquarium in the current study.

4.3 Acclimation of Chinese sturgeon feeding habits

Based on the behavioural observations of this study, it was found that the captive bred Chinese sturgeon juvenile stock adjusted better to the captive environment and hand feeding regimes, than the wild caught Chinese sturgeon. The wild caught fish exhibited higher levels of food selectivity than the captive bred fish reared from hatching. In fact, a number of studies exist demonstrating that the degree of prey selectivity exhibited by fish is influenced by parameters of prey such as size (or age), morphology and activity (Yin, 2003; Magnhagen et al., 1982; Coates, 1980). In the current study, the wild caught Chinese sturgeon were observed to ‘sense’ food that was present in front of the snout by possibly taste or smell, with varied results. From the existing literature, Chinese sturgeon suck food based on electric feeling, whose barbells have little taste buds, which are known to play less role in foraging (Liang, 1996). Hence, Chinese sturgeon could suck any food. However, as food was also observed to be taken into the oral cavity and accepted or rejected, the chemical sense organs in the oropharyngeal cavity may be ultimately responsible for food selection. Establishing food preferences is not easy and requires long-term observations. Hence, the identification of suitable bait that reflects the preferences of each individual fish under natural conditions is a key component in the acclimating these wild caught Chinese sturgeon to the captive environment. In the current study, offering food by divers may have increased the accessibility of the food to the fish which may have also influenced prey choice by the wild caught Chinese sturgeon. In existing studies, age (assumed to be indicated by total body length) has been considered to alter food choice composition (Yin, 2003). In our study, we found that the wild caught Chinese sturgeon almost completely avoided *Penaeus vannamei*, crayfish and mollusks, which differed from the foods selected by younger sturgeon in bayou region (Liu et al., 2006).

In the current study, the wild caught sturgeon required a short time of force feeding before the initiation of active feeding, with the exception of one fish. In general, by the fifth day of force feeding, the wild caught sturgeon occupied the bottom of the tank during forced feeding. This shift in positioning, may indicate improved appetite, however further studies with a larger sample size are required.

In addition, one mature sturgeon (W31#) appeared to exhibit an affinity to the diver, while a slow reaction to the stimulus of diver was observed. The training of individuals plays an important role in the establishment of behavioral acclimation of fish (He, 1996; Liang, 1996). Hence, in the current study, the sturgeon may have developed a positive conditional reflex to force feeding by the diver over time, or just became habituated to the force feeding procedure. On the other hand, the sturgeon exhibited a weak ability to adjust to the wide range of food that was on offer. We strongly suggest that the variability in the behaviour of individuals that are transferred from the wild to captivity requires more detailed study, particularly for application to both conservation and aquaculture.

5 Conclusions

The current study on the long-term welfare and acclimation of sturgeon originating from the wild and aquaculture, led to the documentation of gonad development and variability of growth parameters in the captive environment of Beijing Aquarium. The carefully regulated environment proved to be a stable and viable habitat, which facilitated the adjustment of wild caught fish to captive conditions, due to the consideration of parameters, such as fish preference, that may contribute to the transition process. In conclusion, our study provides strong support for the ability of wild caught Chinese sturgeon to be held in captivity, which may in time contribute to the repopulation of the species through reintroduction and captive breeding programs.

Acknowledgements

We extend special thanks to Mr. Hu Weiyong, the general manager of Beijing Aquarium and technical team of the aquarium, to Professor Zhang Xianliang, the president of Chinese Academy of Fishery Sciences, and to other members of Conservation of Endangered Fish, Yangtze River Fisheries Research Institute, for there sustainable supports. This study was supported by the Essential Operation Fund of Freshwater Fisheries Research Center, Chinese Academy of Fishery Sciences.

References

Anonymous. 1988. The biology of the sturgeon in Yangtze and their artifical propagation. The Yangtze Aquatic Resources Survey Group. Sichuan Scientific and Technical Publishing House, Sichuan, 88.

Chen X. 2007. Biology and resources status of Acipenseriformes. Beijing: Ocean press.

Coates D. 1980. Prey-size intake in humbug damselfish, Dascyllus aruanus (Pisces, Pomacentridae) living with social

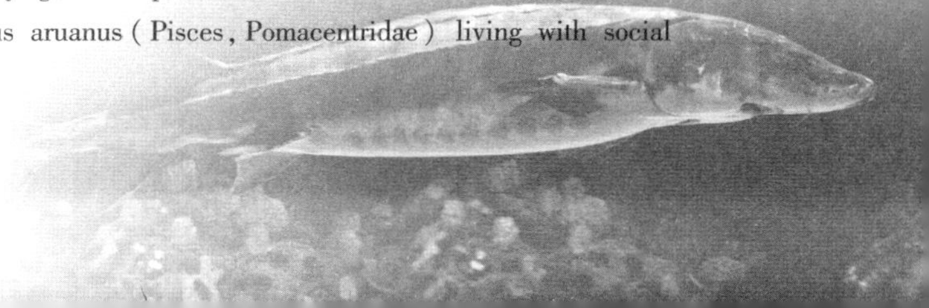

groups. J. Anim. Ecol, 49: 335-340.

Doroshov S I, van Eenennaam J P, Moberg G P, et al. 1994. Reproductive management of cultured white sturgeon (*Acipenser transmontanus*) High performance fish: Proceedings of an International Fish Physiology Symposium held at the University of British Columbia in Vancouver, Canada, July, 16-11: 156-161.

Doroshov S I, Moberg G P, Van Eenennaam J P. 1997. Observations on the reproductive cycle of cultured white sturgeon, Acipenser transmontanus. Environ. Biol. Fish 48: 265-278.

He D. 1996. Studies on behaviour and sensory systems of fishes in Russia (III). J. Ocean. Taiw. Str, 15: 416-421.

Holcik J. 1989. The freshwater fishes of Europe, Vol. 1, Part II. Acipenseriformes. Wiesbaden: AULA-Verlag.

Hu H, Liu X, Zhu H, Lin H. 2007. Gonadal development, artificial reproduction in cultured *Acipenser gueldenstaedti*. Acta Scient. Natural. Univers. Suny, 46: 81-85.

Ke, F., 1999: Conservation and exploitation of *Acipenser sinensis*. Freshw. Fish, 29: 4-7.

Li F, Liu B, Shi S, et al. 1994. Study on *ex-situ* conservation in crested ibis (*Nipponia nippon*). Biodiv. Sci, 2: 24-28.

Liang X. 1995. On the feeding behaviour of mandarin fish (*Siniperca chuatsi*). Oceanol. Limnol. Sin, 26 (Suppl): 116-125.

Liang X. 1996. The structure and function of the ampullae on the ventral part of the snout of Chinese sturgeon (*Acipenser sinensis* Gray). Oceanol. Limnol. Sin, 27 (Suppl): 1-5.

Lin H. 1996. The regulation of growth and growth hormone secretion in fish. Acta. Zool. Sin, 42: 69-79.

Lin H. 2004. The physiology of fishes. Guangzhou: Guangdong Higher Education Press: 146-148.

Liu J, Zhang X, Wei Q. 2006. Observation on domestication of wild Chinese sturgeon in an Aquarium. Chin. J. Zool, 41: 48-53.

Magnhagen C, Wiederholm A M. 1982. Food selectivity versus prey availability: a study using the marine fish Pomatoschistus microps. Oecologia, 55: 311-315.

Webb M A H, van Eenennaam J P, Doroshov S I, et al. 1999. Preliminary observations on the effects of holding temperature on reproductive performance of female white sturgeon, *Acipenser transmontanus* Richardson. Aquaculture, 176: 315-329.

Wei Qiwei, Ke F E, Zhang J M, et al. 1997. Biology, fisheries, and conservation of sturgeons and paddlefish in China. Environ. Biol. Fishes, 48: 241-256.

Wei Qiwai, Yang D, Ke F. 1998. Technique of ultrasonic telemetry for Chinese sturgeon, *Acipenser sinensis*, in Yangtze River. J. Fish. Chin, 22: 211-217.

Wei Qiwei, He J, Yang D, et al. 2004. Status of sturgeon aquaculture and sturgeon trade in China: a review based on two recent nationwide surveys. J. Appl. Ichthyol, 20: 321-332.

Wei Qiwei, Chen X, Yang D, et al. 2005. Variations in spawning stock structure of *Acipenser sinensis* within 24 years since damming of Gezhouba Dam. J. Fish. Sci. Chin, 12: 452-457.

Williot P R, Brun R, Rouault T, et al. 1991. Management of female spawners of the siberian sturgeon, *Acipenser baeri* Brandt: first results//Williot P. Acipenser. Bordeaux: CEMAGREF Publ: 365-380.

Xiao H, Li S. 1994. Study on growth of *Acipenser sinensis* of one year old. Freshw. Fish, 24: 6-9.

Yin M. 2003. Fish ecology. Beijing: China Agriculture Press: 55-56.

Zhang X, Liu J, Wei Q. 2007. Recovery of foraging ability of wild female Chinese sturgeon after artificial propagation in aquarium by different feeding methods. Chin. J. Zool, 42: 142-146.

Zhao Y, Du J, Zhang Y, et al. 1997. Experimentation on culturing *Acipenser sinensis* in pond in Chengdu area. Chin. J. Zool, 32: 41-43.

中国水族馆行业的发展与政府关注：北京海洋馆的动物保护与发展之路*

胡维勇[1,**]　樊祥国[2]　郭志杰[2]　王元群[1]　张军英[1]

（1. 北京海洋馆　北京　100081　　2. 农业部渔政指挥中心　北京　100125）

摘　要　中国水族馆已走过80年的历程，水族馆发展能否与水生动物保护工作结合？怎样结合？北京海洋馆对2004年以来所做的保护工作，如开展野生中华鲟易地保护及展示；推动、参与水生动物保护技术标准、职业标准的编制，促进水族馆行业水生动物保护的规范化；拓展、深化水生动物保护等工作进行了深入的探讨。结论是，海洋馆发展与水生动物保护工作直接相关，北京海洋馆开展的保护工作，是整体水生动物保护的组成部分；开展保护工作既有利于动物保护，体现海洋馆的社会价值，也可提升企业的经济价值，拓宽自身发展渠道；海洋馆馆在水生动物保护方面可以有更大作为。

关键词　海洋馆　保护　发展

1　中国水族馆发展状况及政府的关注

中国水族馆已有80年的历史，随着中国改革开放的大潮，水族馆获得很大发展。

据农业部2011年初对全国水族馆发展状况进行的调查显示，截至2010年年底，全国共有水族馆单位80余家，建筑总面积超过70万平方米。养殖水体总量40余万吨。饲养各种水生物近千种，其中，水生哺乳动物有20多种，总量超过1 200头。水族界直接从业人员近万人。作为新兴人造景观，水族馆受到广大游客的喜爱。

在农业部的指导下，全国水生野生动物保护分会于2000年成立，协会组织开展了大量水生野生动物保护、宣传与管理工作，对珍稀和特有物种资源给予了关注和有效保护。

2011年初，农业部组织了全国水族馆行业发展状况调查和评估，对推动水生野生动物驯养、繁殖和合理开发利用起到了规范作用。

2　时代呼唤我们对水生动物进行保护

改革开放以来，中国经济体制发生了很大的变化，逐步形成了社会主义计划经济与市场经济并存的局面。一方面经济快速提升，带来发展新气象，一方面也出现大量生态与环境问题。据不完全统计，中国有各种水生动物约2万种，约占世界水生动物的10％，其中海洋水生动物约16 200种，淡水水生动物约3 300种。

随着人类活动的增加，有相当数量的水生动物已经灭绝或濒临灭绝。保护生物多样性、拯救濒危水生物种，是我们面临的重要问题。国家为此颁布了《野生动物保护法》，编制了《国家重点保护水生野生动植物名录》。这些法律法规和规范的制定，对水生野生动物的保护和管理工作起到了很好的支撑作用。

---

* 本文在2012年南非世界水族大会上宣读。

** 第一作者，高级经济师，北京海洋馆总经理。

中国的水族馆在保护水生动物方面也做出了自己的努力，在向游客传播海洋知识、水生生物知识方面发挥着愈来愈重要的作用。面对水生生物保护的现状，我们已经感到：时代正呼唤水族馆有更大的作为。

3 北京海洋馆参与水生动物保护工作的尝试

北京海洋馆1999年开业，建筑面积4.2万平方米，她集观赏、科普教育和娱乐为一体，有先进的维生系统设备，水体总量超过1.8万吨，养殖水生动物近千种、数万尾。开业以来，累计接待中外游客超过1 800万人次，其中，2011年游客数量达到200万人次。

北京海洋馆的企业宗旨是，陶怡大众、教益学生、维系生态。北京海洋馆的企业口号是，关爱水生动物、保护地球家园。

在利用自身优势条件参与水生动物保护方面，北京海洋馆就保护工作与水族馆发展的关系，做了一些有益的探索。

3.1 实施野生中华鲟的易地保护

中华鲟是我国一级保护的野生动物。2005年年初，在中国农业部的支持、指导下，北京海洋馆同中国水产科学研究院合作，共同搭建了野生中华鲟易地保护的平台。通过三方创造性的合作，实现了多项阶段性重要成果。

(1) 北京海洋馆推出了一个驯养、展示、研究野生中华鲟的大型展馆，建立了子一代中华鲟在水族馆的蓄养群体。

(2) 中华鲟在水族馆的保育技术得到大幅提升；馆养中华鲟性腺再次发育；2007年康复后的2尾大型野生中华鲟成功放归海洋，一年后其中1尾再次洄游长江参加繁殖。

(3) 中华鲟保护宣传效果空前，七年间有超过1 000万人次的中外游客在北京海洋馆观看了中华鲟的保护展示，媒体曝光率达到1 000次/年以上 。

(4) 搭建了中华鲟保护平台，建立了政府与科企合作的模式与运行机制。

北京海洋馆2005年开展的中华鲟易地保护项目，是企业具有里程碑意义的一件大事，使我们成为国家中华鲟保护体系中的一个部分，成为整个保护链条中的一个环节。

实践表明，海洋馆与政府、与科研机构合作是实施珍稀濒危水生动物保护的一条有效途径。

3.2 做水生动物保护技术标准

水族馆行业的规范化得到了政府相关部门的大力支持，也得到各相关单位领导高度重视和积极参与。

2010年4月，中国水族馆行业在全国水生野生动物保护分会的领导下，组织27家水族馆共同开展了水族馆技术标准的编写。该标准已通过专家评审。五项技术标准是《水族馆术语》《水生哺乳动物饲养设施要求》《水族馆水生哺乳动物饲养水质》《水生哺乳动物谱系记录要求》《水族馆水生哺乳动物驯养技术等级划分要求》。

标准主要规范的是保护类水生哺乳动物的驯养，重点是与行业发展、水族馆建设、动物驯养、保护直接相关的管理及技术层面。标准秉持了保护动物与行业发展兼顾，有一定先进性且不保护落后的思想。自此，中国有了自己的水族馆技术标准。

北京海洋馆在标准化建设中，承担了《水生哺乳动物驯养技术等级划分要求》标准的首席责任，还承担了协助全国水生野生动物保护分会标准化建设的日常组织、协调工作。

2004 年，水族馆行业还组织过《水生哺乳动物驯养师国家职业标准》的制定，有十家单位共同携手。北京海洋馆承担了执行主编职责。该职业标准已于 2005 年由原劳动和社会保障部批准并发布实施。目前，我国水生哺乳动物驯养师职业培训与职业技能鉴定工作已经依此展开。

为了水族馆行业的健康、可持续发展，北京海洋馆会继续推动和参与本行业的规范化建设。

3.3 广泛参与水生动物保护活动

北京海洋馆广泛参与了各种形式的水生动物救助、保护、放流、科普、宣传活动。

2006 年年初，一条受伤搁浅于渤海湾的江豚需要救助。在接到救助信息后，北京海洋馆派员驱车数百公里，对搁浅受伤江豚实施远距离救助，易地至北京海洋馆医疗保护。两个月后，治愈的江豚放归渤海。

2012 年，北京海洋馆参与了农业部“长江江豚康复保护项目”，负责专项调查、保护宣传和提供技术支持。

2011 年春夏之间，长江流域连续干旱，引发了水域生态危机，水生哺乳动物面临的威胁更加严重。北京海洋馆和多家水族馆一道，协助全国水生野生动物保护分会召开了“长江流域大旱 水生生物危机——我们在行动”新闻发布会，就水生野生动物的保护向社会发出呼吁。

北京海洋馆是国家科普教育基地，传播海洋知识、水生生物知识、保护水生动物的文化理念，已成为日常经营行为。开展大众科普宣传活动丰富了场馆展示，也延伸着水生动物保护的文化与声音。

4 水族馆是水生濒危动物保护系统的重要组成部分

水生动物保护有就地保护和易地保护两种形式，易地保护是保护工作的重要组成部分，是恢复珍稀濒危物种种群的重要手段，是对就地保护形式的一种补充。

水族馆在水生动物保护中具有独特的潜力。水族馆模拟、创造了类似于自然的水生生物生存环境与条件，各种水生动物由专业人员呵护，在这一环境中得到繁育、成长。这一优势，使水族馆在保护水生生物方面具有独特的潜力，并逐步成为水生生物饲养、繁育、展示、科普教育、资源保护和科学研究的场所。

北京海洋馆开展的海豚、海狮繁育研究，已成功实现人工环境下的动物繁育。这为解决水生野生动物的保护和利用关系，提供了一条有效途径。

人们完全可以将水族馆作为水生动物易地保护场所，利用其独特的潜力优势，不断提高驯养繁殖水平，在做日常经营展示同时，发挥水族馆水生濒危动物保护基地的作用，做更多的水生动物保护工作，为逐步减缓水生野生动物濒危程度加剧的趋势，起一定的推动作用。

5 开展水生动物保护可促进水族馆发展

随着需要保护的水生动物数量增加，人们更需要对珍稀濒危物种有了解与认识，这为海洋馆的持续发展提供了空间和关注度。

水生动物保护工作已成为北京海洋馆的一种社会责任，成为企业运营的一个组成部分，其主要内容包括日常经营展示、参加水生野生动物保护活动、科普宣传、开展保护物种的技术合作。保护工作的开展，体现了企业的社会价值和品质，同时还为企业带来良好收益。2005 年以来，北京海洋馆经济效益持续提升。

北京海洋馆的发展历程和水生动物保护工作实践，给了我们这样的启示：

（1）海洋馆开展水生动物保护工作可以成为国家水生动物保护系统中的一个环节、一个组成部分；

（2）开展水生动物保护既吻合国家对水族馆的关注，也实现了水族馆的社会价值、社会责任，还拓宽了发展途径，带来经济价值，有利于水族馆的可持续发展。

水族馆在水生动物保护方面可以有更大作为。

The Progress of Aquarium Business and the Attention from the Government in China: the Progress and Conservation in Beijing Aquarium

Hu Weiyong[1] Fan Xiangguo[2] Guo Zhijie[2]
Wang Yuanqun[1] Zhang Junying[1]

(1. Beijing Aquarium, Beijing 100081, China;
2. Administration Center of China Fishery Law Enforcement, Ministry of Agriculture, Beijing 100125, China)

Abstract Can aquarium business play the extant role in conversation aquatic animals after its eighty years advance in China? What will be done and how to start? An example of Beijing Aquarium in conservation works from 2004 to now will be introduced in this paper. The efficiency of different dimension, such as preservation and exhibition of one Chinese sturgeon population in aquarium, promotion the standardization level through engaging in development business standard and conservation technique standard, enhancement the opportunities of influence visitor to aquarium and so on, were further studied. Supporting by the government, the endangered species can be transported to aquarium, and the native academy supporting can be achieved, the development of business standard is agreed by government management. Altogether, the conservation efforts in Beijing Aquarium make the social and profitable responsibility of the business enterprise realized, which is in favor of its progress and animal protection. Aquarium has been as the part of the constitution of animal conservation. As a result, Attention from the government impact the most positively on the progress of aquarium business in China. It is close to relate the progress aquarium has achieved to its effectiveness of conservation works. It outlines the more important role of aquarium business in animal conservation filed in future in China.

Key words Aquarium Conservation Progress

科企联合：实现中华鲟保护新突破*

张显良**

（中国水产科学研究院长江水产研究所　湖北武汉　430223）

摘　要　中华鲟是我国一级水生野生保护动物，野生数量急剧下降，亟须加强保护工作。我国政府高度重视中华鲟的保护进程。企业由于设备先进、经营灵活、面向大众的优势，可以与研究院所开展科企合作，优势互补，推进中华鲟保护进程。本文经过一年北京海洋馆与中国水产科学研究院长江水产研究所合作，野生中华鲟突破主动摄食瓶颈，产后康复，子一代中华鲟蓄养群体建立。实践证明，通过科企合作开展珍稀濒危水生野生动物保护和科研，是完全可行的，是实施珍稀濒危动物保护的一个有效途径。

关键词　中华鲟　水族馆　科学机构　合作　保护

1　引言

地球上的生物物种正在以惊人的速度减少。根据世界自然保护联盟（IUCN）公布的濒危物种红色名录，现在物种灭绝的速度和恐龙大量灭绝时代的速度已经相近，有将近1.6万种动植物面临灭绝，其中约3 000种动物是“极度濒危”，“可能已灭绝”的有200多种。斯坦福大学的科学家用计算机模拟的方法对9 787种现存鸟类和129种已灭绝鸟类作了分析，得到的结论是100年内10%的鸟类将消失。

我国疆域辽阔，具有多样化的气候条件和复杂的自然环境，孕育了丰富的生物物种。以水生生物为例，由于我国海岸线长达18 000 km，内陆河流、湖泊众多，水生生物资源极为丰富。据不完全统计，我国有各种水生动物约2万种，其中海洋水生动物约16 200种，淡水水生动物约3 300余种。同样的原因，我国也有相当数量的水生动物已经灭绝或濒临灭绝边缘。为保护这些珍贵的水生生物，我国政府颁布了相应的保护物种名录。其中已经列入国家一级保护动物名录的水生动物有中华鲟、白鲟、达氏鲟等13种，列入国家二级保护动物名录的水生动物有胭脂鱼、江豚、大鲵、鯨鱼等67种。而据估计，我国目前处于濒危状态、需要国家重点保护的水生野生动物已达近400种。

中华鲟（*Acipenser sinensis* Gray）是属于鲟形目（Acipenseriformes）、鲟科（Acipenseridae）、鲟属（*Acipenser*）的一种大型鱼类，近代主要分布在黄海、东海、南海北部近海，以及长江、珠江、闽江、钱塘江、黄河等江河，朝鲜西南部和日本九州西部也有过记录。目前，中华鲟主要见于我国黄海、东海大陆架水域和长江，闽江、钱塘江、黄河已基本绝迹，珠江数量也极少。为保护中华鲟自然资源，1988年，我国已将其列为国家一级重点保护的野生动物。1996年，世界自然保护联盟濒危动物红色名录将其列为濒危（EN）等级；1997年，濒危野生动植物种国际贸易公约（CITES）将其列为附录II物种。

中华鲟在海洋中生长发育，在江河中产卵繁殖，江河产卵孵化的幼鲟和产后亲鲟又降河洄

---

*本文在2006年由北京海洋馆主办的中华鲟保护交流会上宣读。

**作者简介：中国水产科学研究院院长，研究员。

游至海洋栖息和肥育,是一种典型的过河口溯河产卵洄游的鱼类。葛洲坝截流前,中华鲟产卵场分布在长江上游雷波冒水至重庆木洞之间约 600 km 江段。1981 年葛洲坝截流后,中华鲟被滞留在坝下,由于坝下江段多属于平原河道,其地形结构与长江上游有较大差异,适宜中华鲟产卵的江段有限。20 多年的监测表明,葛洲坝截流以后,在葛洲坝下游江段仅发现一处中华鲟产卵场,而且产卵场的江段长度仅 7 km,面积相当狭小。再加上近年来经济建设高速发展,人类活动加剧,水域水质污染日趋严重,导致中华鲟的资源状况逐渐衰退。根据长江水产研究所的调查,每年洄游至长江的中华鲟产卵群体数量,已从葛洲坝截流初期的约 2 176 尾,下降到目前的每年约 500 尾左右。加强对中华鲟物种的保护和研究工作,已成为我国水产科技工作者和物种保护工作者的紧迫任务。

2　科企联合是开展中华鲟保护和研究的有效途径

2.1　我国中华鲟保护工作的进展以及存在的问题

近 30 年来,我国政府对中华鲟的保护高度重视,各有关部门投入了大量的人力物力,研究和保护这一珍贵物种,对缓解该物种资源衰退起到了积极作用。到目前,已经初步建立了实施中华鲟物种保护的政策法规、技术、管理和保护措施等方面的基本体系结构,其中最主要的进展包括四点。

(1) 明确禁止对中华鲟的商业捕捞,并实施重点保护。中华鲟曾是长江重要的渔业对象,针对其自然资源不断下降的现实情况,我国政府于 1982 年就明确规定,禁止对中华鲟的各种商业捕捞活动,并于 1988 年将中华鲟列为国家一级保护动物。

(2) 积极开展中华鲟人工繁殖放流,人工增殖中华鲟自然种群。为弥补葛洲坝工程建设对中华鲟自然资源的影响,我国政府在加强对中华鲟自然种群保护的同时,还积极开展中华鲟人工繁殖和放流。从 1983 年以来,基本上每年都向长江放流一定数量的中华鲟幼鱼,以对自然种群进行补充。截至 2006 年,已向长江放流各种规格中华鲟苗种 700 多万尾。

(3) 加强中华鲟产卵场及关键栖息地的保护。为保护对中华鲟种群延续至关重要的唯一产卵场和重要栖息地,湖北省人民政府于 1996 年设立了长江湖北宜昌中华鲟省级自然保护区;2002 年,上海市人民政府也设立了长江口中华鲟幼鱼自然保护区。

(4) 加强了中华鲟物种保护技术研究工作,取得了一批实用的中华鲟保护技术成果。如长江水产研究所建立的中华鲟人工繁殖技术、大规格苗种规模化培育和成鱼养殖技术、标志放流技术以及中华鲟自然繁殖生态研究等成果,对中华鲟物种保护工作提供了坚实的技术支撑和理论指导。

虽然如此,每年的监测结果表明,中华鲟自然种群数量逐年下降的趋势并没有得到根本缓解,中华鲟的保护还面临着许多问题。这其中除了因中华鲟分布较广,管理难度大,以及水利工程建设、水质污染、长江及近海密集的渔捞活动等问题近期无法完全消除等客观因素外,还存在每年在中华鲟保护上投入的资金有限,使得一些对中华鲟物种保护极为重要的关键问题,如人工繁殖中华鲟产后亲鲟的恢复,野生亲鲟的蓄养成熟等困扰人们多年的难题,虽然有了较好的设想,却由于设施条件的限制而无法进行。一些有效的保护措施,如中华鲟人工备份种群的建立、异地种群保护和繁殖等工作,也因资金限制而不能正常开展。因此,探索更多的中华鲟物种保护技术和方法,寻找更多的中华鲟物种保护途径,是加强中华鲟物种保护的必由之路。

2.2 科企联合保护中华鲟的尝试

将人工繁殖后的中华鲟亲鲟进行护理，使其在人工环境中摄食，恢复生理机能，并进而实现人工环境下的蓄养成熟，不仅可以提高产后亲鲟成活率，实现亲鲟的重复利用，保护中华鲟自然种群，同时也是建立全人工中华鲟种群的必要基础。中华鲟属秋季繁殖的鱼类，产后亲鲟的护理期处于冬季低温期。长江水产研究所经多次试验发现，低温环境对产后亲鲟的恢复非常不利，同时野生中华鲟个体硕大，在自然环境条件下，产过卵的亲鲟均需要迅速降河洄游至海洋环境肥育，因此，尽量创造类似于自然的驯养环境，对亲鲟的产后护理以及驯养成功非常重要。

北京海洋馆是目前世界内陆地区最大的海洋生物科普馆，集观赏、科普教育和娱乐为一体。整体建筑面积42 000 m^2，总水量近18 000 t。该馆有适用于大型水生动物驯养的养殖设施，并配备有完善的维生系统，可以对水温、水质、盐度等进行全天候监控和精确调节，具有非常先进的硬件条件，饲养淡水及海水鱼类千余种。通过多年运作，在水生生物驯养、管理等方面，也积累了较多的成熟技术和人员。为此，2005年初，长江水产研究所与北京海洋馆达成协议，决定实施科企联合，利用长江水产研究所较强的技术优势和北京海洋馆良好的设施条件，合作开展中华鲟驯养展示以及保护相关的科学研究。

合作协议签订后，北京海洋馆将鲨鱼馆改造成了鲟鱼馆。该馆占地920 m^2，驯养池长28 m，宽11 m，深4.5 m，容积1 400 m^3，按照中华鲟的生活习性对池底进行了调整，铺设了洁净的沙石。该驯养池配有4组密闭式砂过滤罐水处理系统，水处理量达28 m^3/min，每50 min可使全池水体循环一次；有4组冷热交换器，水温可在17～28 ℃间任意调控，调控后的温差小于±1 ℃；水体盐度可通过淡水或人造海水在0～30 ‰之间进行调节。该驯养池还有一个规格18 m×3 m的雅克力展窗可供游人观赏。

驯养池改造完成后，对驯养池水体环境进行了调整，将原驯养鲨鱼的海水逐渐调制成2.5 ‰的淡水，重建硝化系统等微生态环境，根据中华鲟饲养要求调节水质，将水温控制在20.0～23.0 ℃，酸碱度稳定在7.7～8.2，溶氧量控制在8.0～9.0 mg/L；同时监测水体总细菌、真菌及大肠杆菌状况，并根据实际情况及时进行调整，营造了目前国内最大且较为理想的大型中华鲟驯养池。

2005年3月26日和4月2日，经过20多小时、1 200 km的长途运输，顺利将26尾人工繁殖的中华鲟和1尾野生中华鲟运至海洋馆。2006年年初，又将2尾产后中华鲟长途运输至海洋馆进行驯养试验和展示。

2.3 科企联合实施中华鲟保护工作的效果

到目前为止，虽然双方合作的时间不长，实际开展工作的时间仅有一年半，但经过双方的共同努力，已在中华鲟运输、驯养、科研以及展示等方面取得一系列的成效，尤其在野生中华鲟驯养方面，取得了突破性的进展。总结这一年半的合作成绩，主要体现在以下几个方面。

（1）实现野生中华鲟全人工环境的开口摄食和产后亲鲟的良好恢复。通过一系列技术措施，3尾野生中华鲟已在水族馆驯养成功，均已开口摄食。其中2尾产后中华鲟亲本经过护理和培育，身体机能状况已恢复正常状态，目前均已生长良好。这为继续开展野生中华鲟全人工蓄养成熟等科学试验奠定了基础。

（2）成功建立了子一代中华鲟的水族馆蓄养群体。长江水产研究所人工繁殖获得的

26尾中华鲟子一代，在运输至海洋馆的第二天即开始摄食，表现出对蓄养环境较高的适应性，至目前成活率达100％，且长势良好。这个结果表明，利用海洋馆现有条件，保存一定规模的中华鲟后备人工群体，是完全可行的，为中华鲟物种的易地保护以及人工备份提供了一个新途径。

（3）实现了大型野生中华鲟的安全长途运输。野生中华鲟个体大，产后亲鲟体质虚弱，活体运输尤其长途运输难度一直较大，运输成活率低。为解决这个问题，长江水产研究所专门设计制作了一套配备了完善维生系统的活鱼运输箱，然后通过合作双方的良好组织和协作，成功将3尾野生中华鲟成体和26尾大规格人工群体从长江水产研究所基地运至北京海洋馆，运输成活率达到100％。该方法的建立，为今后实施中华鲟易地保护提供了技术支撑。

（4）北京海洋馆经营效益和社会影响有了提高。因为中华鲟这一珍稀物种的引入，增加了北京海洋馆对社会公众的影响力，吸引了更多的游客前来参观，增加了企业经营收入。更重要的是，中华鲟在海洋馆的安家、展示，使更多的社会大众知道了保护中华鲟、保护珍稀濒危水生动物的重要性，这将极大地促进全社会来保护中华鲟，使中华鲟保护事业呈现一种良性互动的局面。

3 经验和体会

此次长江水产研究所与北京海洋馆在中华鲟保护工作上的成功实践，给了我们很多有益的启示。

（1）科研单位与相关企业合作进行珍稀濒危水生动物保护和科研，使科研单位技术力量强，企业设施条件好、资金充裕的优势都得到充分发挥，实现了优势互补，从而使得合作成效明显。

（2）通过与企业的合作，不仅达到了保护中华鲟的效果，科研单位在合作中也可以发现更多的问题，利用更多的资源开展研究，促进科研水平的提高。

（3）科研单位与企业的合作必须以良好的社会责任感为基础。物种保护是一项公益性事业，参与合作的科研单位和企业双方必须对此达成共识，合作才可能成功，这一点对有盈利需求的企业方尤其重要。

（4）合作前，合作双方应充分了解自己及合作方的优势与不足，这样才能真正实现强强联合，优势互补。

实践证明，通过科企合作开展珍稀濒危水生野生动物保护和科研，是完全可行的，是实施珍稀濒危动物保护的一个有效途径。

Union of Scientific Research Institutions and Enterprises: Realizing New Breakthrough in Protecting Chinese Sturgeon

Zhang Xianliang

(Yangtze River Fisheries Research Institute Chinese Academy of Fishery Sciences, Wuhan 430223, China)

Abstract Chinese sturgeon has been listed as the first level protected wild aquatic animals, conversation of which should be strengthened, for its population size in nature is decreasing significantly ever since. The efficiency of conversation this endangered species has been paid close attention by Chinese government. The union directed by the government between academic institutions and business enterprises, can be a new approach to advance conservation effectiveness through their advantages complemented for each other, with high-quality equipments, flexible function, and directly popular connection in business, and professional research on sturgeon in academy. An example of this union between Beijing Aquarium and Yangtze River fisheries research institute, Chinese academy of fisheries sciences, the effectiveness of one-year cooperation will be discussed in this paper. The result showed, the handicap of feeding of wild adult Chinese sturgeon under the cultured condition was broken through in Beijing Aquarium. Further, the recovery of wild post-spawned parents sturgeons was fulfilled. The first-filial generation group was developed successfully the same time. Above all, the transportation technique by truck for a long distance was developed, which make the base technique for this cooperation. As a conclusion, this allied approach is innovative and efficient in conservation of rare and endangered wild aquatic animals. The further co-research of conservation on the wild species would be conducted through this union.

Key words *Acipenser sinensis* Endangered Aquarium Academy Union Conservation

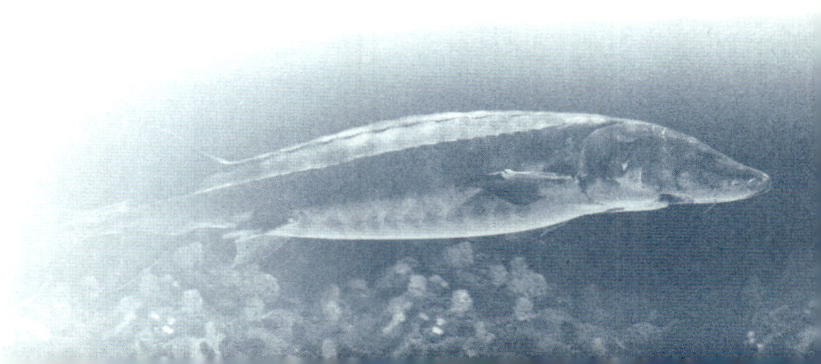

中华鲟保护性展示:北京海洋馆经营模式的新探索*

王元群**

(北京海洋馆　北京　100081)

摘　要　北京海洋馆是旅游单位,有责任为我国特有的濒危物种——中华鲟的保护做出贡献。北京海洋馆与长江水产研究所开展科企合作,充分发挥设施条件好、养殖能力强、社会知名度高的优势,结合长江所技术、科研能力,突破野生及子一代中华鲟在大型驯养池的驯养瓶颈,实现了公众展示、教育,开拓了中华鲟新的保护途径,扩大了中华鲟保护的公众影响力。科企联盟保护中华鲟,促进了企业市场化运营优势的发挥,拓宽了企业经营思路,提升了企业的形象,不仅使企业实现盈利的社会功能,更促进企业实现社会责任。以保护为中心、科企结合、市场化运作的中华鲟项目,是北京海洋馆经营模式的一次成功探索,可以为水族馆行业的经营模式创新提供借鉴。

关键词　中华鲟　水族馆　研究所　联盟　保护　经营模式

中华鲟,古老物种,已有1.4亿年历史。其祖先与恐龙同时代,恐龙灭绝了,中华鲟却顽强地延续下来。中华鲟在长江里繁育,在大海中成长,是中国的独有物种,具有很高的经济价值和研究价值;但由于种种原因,如今它已处境濒危。

北京海洋馆是大型的内陆水族馆,是首都靓丽的旅游景点。企业宗旨是"陶怡大众,教益学生,维系生态",因此有责任为保护中华鲟做点贡献。2005年开始,北京海洋开展中华鲟易地保护项目,进行养殖、展示、经营模式的新探索。

1　科企合作:养殖、展示的新模式、新成果

在农业部和有关政府部门的支持下,北京海洋馆与中国水产科研院长江水产研究所合作,以自己设施良好、驯养能力强、公众知名度高的条件,结合长江水产所技术力量、研究能力,创造性地做成了中华鲟保护性展示项目,取得了成效。实现了野生中华鲟在人工环境的开口摄食和产后亲鲟的康复。五年来,北京海洋馆和长江所通过一系列措施,9尾野生中华鲟已在海洋馆驯养成功,均已开口摄食;8尾产后中华鲟亲体,经过护理和培育,身体机能得以恢复。这为今后开展野生中华鲟人工蓄养等科学试验奠定了基础。

(1) 成功建立了子一代中华鲟的水族馆蓄养群体。现在北京海洋馆的子一代中华鲟存有56尾,驯养、生长良好。

(2) 协助开展了野生和子一代中华鲟在海洋馆条件下的繁育研究。

(3) 根据五年的经验,正在形成野生中华鲟救护、驯养和子一代中华鲟驯养的技术规范。

(4) 实现了康复的野生中华鲟的放流。2尾在北京海洋馆康复的野生中华鲟在2007年4月22日放流长江、回归自然。这为野生中华鲟资源重复利用提供了新范式。

(5) 宣传了中华鲟及其生态环境。硕大的野生中华鲟与几十尾子一代中华鲟的展示,有着巨大的震撼力。与此同时,馆内外宣传中华鲟的生态环境,传递生态信息。

---

* 本文在2009年第七届世界鲟鱼大会上宣读。

** 作者简介:王元群,硕士,北京海洋馆副总经理。

2　以企业经营方式进行项目运作

对中华鲟易地保护项目，我们没有采取封闭的方式而是以企业经营、市场化的方式进行运作。中华鲟从长江引入北京，我们利用京城和北京海洋馆的区位优势、企业优势，有效地扩大了保护中华鲟的影响力。

(1) 项目运作与公众相结合。围绕“关爱国宝中华鲟，保护海洋水生动物”的中心，开展丰富多彩的活动，与社会公众互动。通过互动、参与，使公众更多地了解中华鲟的历史，了解中华鲟的价值，了解中华鲟生存的生态环境。公众在陶怡中受到教益，激发了保护中华鲟的热情。

(2) 项目运作与媒体相结合。国内大量媒体和网站对中华鲟项目给予了积极的报道，使中华鲟保护之声唱响大江南北。五年连续不断。

(3) 项目运作与境外推广活动相结合。将中华鲟送往香港特别行政区展出，引起轰动。海外水族界对中华鲟很感兴趣，愿意交流。

公众参与、多方参加的运作方式，放大了项目的社会效应，也为中华鲟保护探索了新的途径。

3　中华鲟保护项目促进企业的发展

中华鲟易地保护项目，对北京海洋馆的发展起到了积极的作用。

3.1　拓宽了展示和经营思路

中华鲟项目是以北京海洋馆为平台，融合多方资源，开展保护性展示，而不仅仅是买鱼展示。

项目强调生态保护，引入文化创意的理念，将中华鲟保护项目作成文化产品，提升工作层次，丰富了经营内涵。

项目带动了系列衍生品：

中华鲟旅游，考察中华鲟生态环境；

人鲟共舞表演，体现人鲟和谐；

中华鲟夏令营产品，对儿童进行科普；

以及中华鲟卡通形象、纪念品、音像制品、观鲟宴和尝试中华鲟动漫策划等。

系列化的中华鲟展示产品，有利于中华鲟保护文化的形成，对中华鲟保护的持续与传播起着重要作用。

3.2　提升了企业形象

通过中华鲟在北京海洋馆的展示，进一步唤起了关爱中华鲟、保护水生野生动物、保护生态环境的公众意识与行动。这一项目提升了北京海洋馆的社会功能和企业形象，同时也吸引了游客前来参观，为企业带来收益。

北京海洋馆连续两届荣获首都旅游紫禁杯最佳先进集体称号。

项目得到农业部和有关部门的大力支持和帮助。多位领导到北京海洋馆考察，对中华鲟保护项目给予肯定。全国人大常委会副委员长、中国科学院院长路甬祥两度题词，“关爱海洋动物、保护地球家园”。

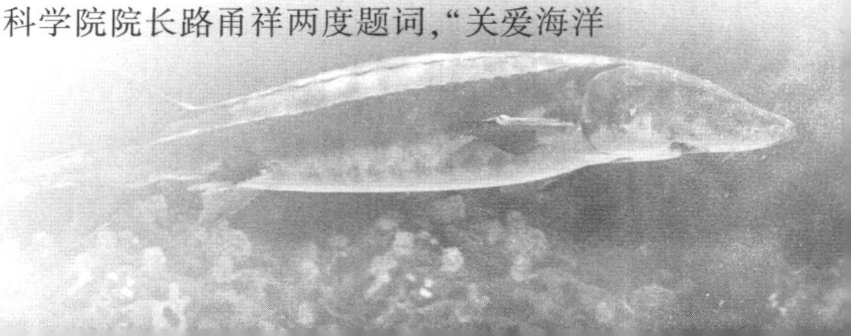

因此可以判断，中华鲟保护性展示项目，即以保护为中心、科企结合、市场化运作，是北京海洋馆经营模式的一次成功探索。这条路可以继续走下去。

The Exhibition of Chinese Sturgeon in Beijing Aquarium: the Explore of the New business Model

Wang Yuanqun

(Beijing Aquarium, Beijing 100081, China)

Abstract Beijing Aquarium (BA) as a scene enterprise should be responsible for the contribution to conservation of Chinese sturgeon, the endangered species special in China. The alliance between Beijing Aquarium and Yangtze River fisheries academic institution (YRFAI) held successfully the exhibition of Chinese sturgeon in BA, with the breakthrough of the breeding wild adult and younger filial sturgeons in captivity, more important role in ecology education, due to the integral advantages both in high-quality equipment, professional culturing technique, wonderful esteem in BA, and in pioneering research on Chinese sturgeon in YRFAI. The result showed the approach of Chinese sturgeon conservation in public aquarium was innovative, which can offer much more opportunities to influence the visitors than ever. The social responsibility of education and preservation of endangered species was fulfilled the way the alliance cooperated in conservation, the responsibility of enterprise of going up profit ability was achieved the same time. Further, believe of the management of business was developed, mission of the business impressed on the visitor. It concluded this Chinese sturgeon exhibition, the way the conservation is as the nucleus, with ally supporting and market running, is a successful explores of behavior model in BA, which can therefore be the baseline for aquarium business behavior in China.

Key words *Acipenser sinensis* Aquarium Academy Alliance Conservation Function

香港海洋公园中华鲟饲养及展示概要

杨月媚

（香港海洋公园）

摘　要　香港海洋公园中华鲟馆——长江足迹展览于2010年开馆，以促进淡水生态环境及生物资源的保护为主旨，是香港海洋公园中重要的展馆之一。每年有超过11万的游客参观此展馆。此外，该展馆也成为中华鲟人工繁育的子一代和子二代的养殖试验场所。2011年9月开始对养殖的中华鲟进行常规体检，主要对鱼体的生长指标进行测量。根据与全国水生野生动物保护协会达成的协议，在2015年3月9日，将6尾生长到合适规格的中华鲟子一代中华鲟转运至长江水产研究所荆州基地用于中华鲟的保护与长期的研究工作。目前水族馆中养殖有13尾中华鲟，均由中国水产科学研究院长江水产研究所提供，它们在水族馆中接受专业人员的饲养和照料。

关键词　中华鲟　香港海洋公园　养殖

1　展馆筹备

中华鲟馆的前身是鲨鱼馆，于1990年落成。2010年海洋公园完成原鲨鱼馆的改建工程，耗资逾千万港元打造全港首个以长江为主题的新景点“中华鲟馆——长江足迹展览”。这个全港最大的淡水鱼类展馆设有一个容量达42万升的主鱼池，设计以长江自然生态环境及其急速水流为蓝本，可因不同情况而调校水流的强弱，为中华鲟营造了优良的生活环境。展馆于2010年6月17日正式开幕。

2　科普教育展示

中华鲟馆——长江足迹展览以淡水河流生境为主题，由一开始主池以长江流域为主题，展出中华鲟及其他长江淡水鱼类，到展出本地淡水鱼类，及最后部分的河口生境动物，加上有趣的教育展板、模型和影片，向游人提供保育讯息，体现公园“寓教育于娱乐”的理念。

部分主题展板及影片可向游客介绍长江水产研究所及北京海洋馆等单位科研人员的工作，包括野外拯救、人工繁育及鱼苗流放等，极具创意的装潢更可根据研究情况而作定期更新。

除了加强游人对野生水生动物保育的重视，展馆的另一重要讯息是保育水资源。近出口处设有宣扬善用水资源的展板，让游人当场对淡水资源保育工作许下承诺，积极从生活上的小改变实践珍惜用水。

此外，展馆的“动物趣谈”教育节目，由公园的教育大使向游人作深度讲解，亦会按需要举办相关专题教育团供学校团体参加。自开馆以来超过 94 万人次参加有关活动。根据统计数字，每年到中华鲟馆——长江足迹展览参观的人数高达 110 万人次。展览提供了让中华鲟充分扮演其动物保育大使角色的有效平台，加强了科普教育及推广保育的效果。

3 中华鲟子一代展示及护理成果

2010 年 6 月中华鲟馆正式开幕，展出 10 尾由全国水生野生动物保护分会赠送的中华鲟。于长江水产研究所专家的指导及协助下，海洋公园于 2011 年 9 月进行了首次中华鲟体检工作，包括体长和体重纪录作对比，抽血样本和超声波等，以便更准确地设定合适喂饲量，并确保中华鲟在香港海洋公园健康成长。此外持续累积体检结果，可有助于海洋公园对中华鲟饲养及展示计划的开展。

几年间，展馆内的中华鲟已长大了很多。根据海洋公园与长江水产研究所及北京海洋馆签订的合作协议，当中华鲟子一代生长至某一指标后会运回荆州继续饲养。经过三次的体检结果，海洋公园及长江水产研究所人员决定将 6 尾中华鲟子一代运回内地。海洋公园对此次中华鲟运输工作十分重视，特安排了两个大型运鱼箱进行运送。运鱼箱内置两用供氧系统。运送途中，长江水产研究专家与海洋公园水族员全程监察水质情况和鱼类状态，并准备后备水源进行换水，以保持水质稳定。最终 6 尾中华鲟子一代于 2015 年 3 月 9 日顺利到达荆州基地，其中一尾雄性中华鲟还参与了 2015 年 10 月的人工繁殖计划。

4　中华鲟子二代的护理及其中挑战

为了向大众展示中华鲟子二代的繁殖成果，海洋公园于2014年2月18日获得长江水产研究所赠送的15尾中华鲟子二代。该批子二代出生于2012年，由荆州运至香港海洋公园。在公园检疫期间，兽医团队与饲养人员悉心照料，为子二代进行身体检查，包括抽血及刮腮检查，紧密与专家合作，汇报子二代情况和交流有关饲养技术。完成检疫的中华鲟子二代现时已于中华鲟馆与其余三尾子一代一同展出。

为了全面了解中华鲟的生长及健康情况，海洋公园持续定期为中华鲟进行体检工作。累积至2015年6月的体检结果中，子一代平均增长百分比为15％及22％[注]，而子二代平均增长百分比为19％。这些数据反映中华鲟于园内稳定成长，同时对饲养中华鲟的研究具有很重要的参考价值。

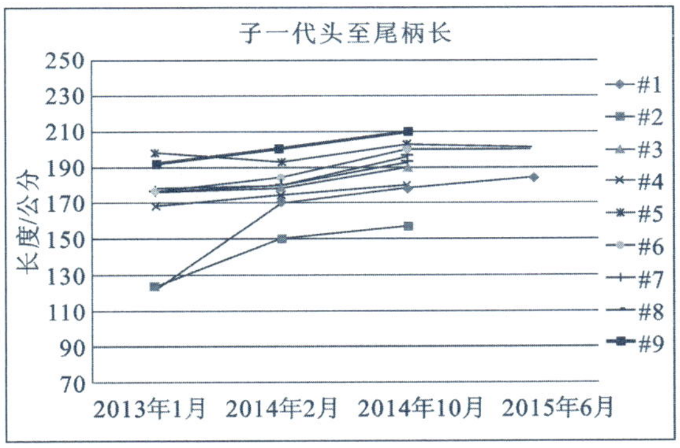

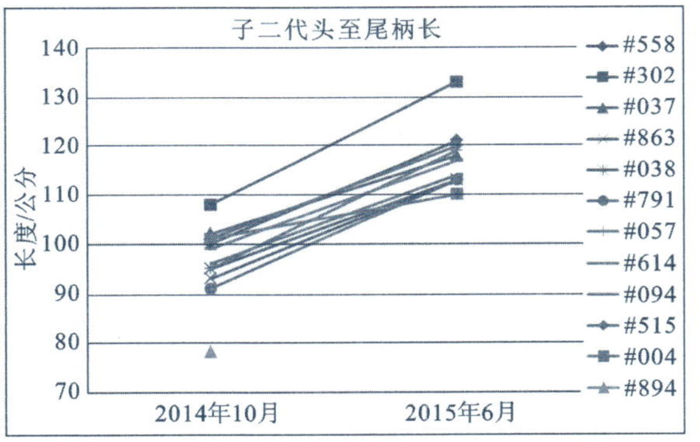

2015年4月中旬，水族员发现有一尾中华鲟子二代（#004）腹部肿胀。经过潜水员检查、拍照核实等一系列跟进工作，隔天将该尾鲟鱼搬离展区并独立饲养。兽医团队进行多项检查，并与专家讨论病征，最后#004被确诊为分枝杆菌感染。在三个月的治疗过程中，兽医团队与水族员定期为#004灌喂鱼浆和口服药，并注射抗生素。直到6月下旬，#004开始持续表现虚弱，海洋公园护理及兽医团队与长江水产研究所专家共同决定将#004人道毁灭，以减低其疾病痛楚及感染其他中华鲟的风险。此外兽医团队亦对#004进行核磁共振与计算机扫描，希望可进一步了解内脏器官状态和病理研究。

注：因为6尾中华鲟子一代于2015年3月运回内地而没有参与2015年6月的体检，所以分别计算两组数值，2013年1月至2014年10月9尾（#1-#9）中华鲟的平均增长为15％，而2013年1月至2015年6月3尾（#1，#5及#6）的平均增长为22％。

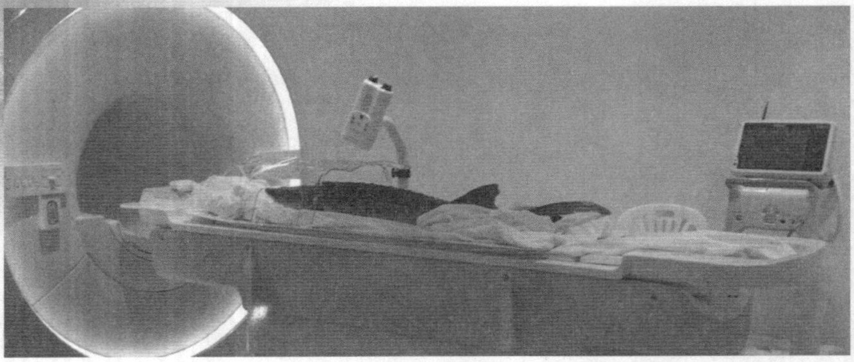

5 结语及展望

目前中华鲟馆内每一尾中华鲟都健康良好，海洋公园会持续进行定期的中华鲟体检，并根据结果调整每一尾中华鲟的喂饲量，使其能获得充足的营养，继续健康成长。通过分析累积得来的生长数据，海洋公园希望可以进一步了解中华鲟于水族馆的人工环境下的生长情况，并与各中华鲟专家、科研及饲养单位分享，为中华鲟保育出一分力。与此同时，海洋公园亦会继续于馆内进行各项科普教育工作，加强公众对此珍稀物种的珍惜和爱护。

The Introduction of Chinese Sturgeon Breeding and Exhibition in Hong Kong Ocean Park

Yang Yuemei

(Ocean Park Hong Kong)

Abstract Opened in 2010, the Chinese Sturgeon Aquarium — Yangtze Exploration has become an important exhibit of Ocean Park Hong Kong to encourage conservation of freshwater habitat and freshwater resources, which has over 110 000 guests visit per year. Meanwhile Ocean Park has also created this facility as a valuable platform to understand the knowledge and challenges of keeping F1 and F2 Chinese sturgeon under human care. Starting from September of 2011, regular body checks was conducted, which obtained useful growth data of this rare species. Under a previous agreement with the National Aquatic Wildlife Conservation Association (NAWCA), six sturgeons, which had grown to the requisite size, returned to the Yangtze River Fisheries Research Institute on 9 March 2015 to participate in the research and conservation programme as a continuation of the long term research in understanding the species' growth rate and life history. Currently, there are 13 captive bred Chinese sturgeons at the Park, all donated by the YRFRI, residing at the Chinese Sturgeon Aquarium, which are under the ever advancing husbandry and veterinarian care provided by Ocean Park Hong Kong.

Key words Chinese Sturgeon Breeding Hong Kong Ocean Park

水族馆中华鲟(*Acipenser sinensis*)行为观察与技术应用*

张艳珍[1,2,**]　张晓雁[1]　危起伟[2,***]
王彦鹏[1]　杜　浩[2]　蔡经江[1]

(1.北京信沃达海洋科技有限公司(北京海洋馆)　北京　100081
2.农业部淡水生物多样性保护重点实验室 中国水产科学研究院长江水产研究所　湖北武汉　430223)

摘　要　采用现场观察方法,结合数字监控设备,对北京海洋馆饲养的野生、子一代及子二代共40尾中华鲟(*Acipenser sinensis*)开展行为特征研究,分析常规行为与异常行为(性腺发育后)差异,筛选中华鲟的主要行为特征指标,建立水族馆行为观察的技术方法与应用。结果显示,活动区域、游泳方式、呼吸频率和摄食行为积极性是养殖中华鲟主要行为特征指标;野生成熟、W-2009,F1-1998,F1-2001,F1-2005 和 F2-2012 中华鲟常规游速分别为(8.60±2.06)s/TL,(5.52±1.62)s/TL,(5.96±0.43)s/TL,(5.45±0.82)s/TL,(4.22±0.77)s/TL 和(3.03±0.45)s/TL,尾摆频次与游速成正比;呼吸频率为年龄越小频率越高,F2-2012 中华鲟为 35 次/min;摄食积极性强。性腺发育后中华鲟行为:泳层上移、游速变快、呼吸频次增加、出现摄食积极性弱或不进食。行为技术可用于水族馆中华鲟健康保育,数据结果可为中华鲟的科研提供基础参考。

关键词　中华鲟　水族馆　行为特征　观察技术　应用

鱼类行为学是水产养殖的重要理论基础,鱼类的行为特征可以快速、准确地反映出自身行为应激变化,以及其所在水体环境的相关变化。近年来应用现代实验生态学的原理和方法,很多学者研究了生态因子、生理发育等因素与鱼类的行为的关系[1,2]。国内外鲟鱼行为相关研究较多,包括生态习性和摄食行为特点,对底质、水流、水温、光照等生态因子的选择与适应行为,以及环境胁迫下的应激行为等[3-11]。

中华鲟(*Acipenser sinensis*)是国家一级保护动物,由于多种原因,目前自然种群数量急剧减少[12-15]。北京海洋馆是中华鲟重要的易地保护和保育科教基地之一。开展野生及产后亲本的康复护理驯养以及子一代和子二代中华鲟的蓄养保育工作,及时掌握鱼况动态对保证全人工条件下中华鲟养殖安全至关重要。关于中华鲟行为研究已有繁殖行为生态学、感觉器官发育与机能、生态因子的影响、环境偏好性、水族馆适应性以及产后摄食与恢复护理等研究[8,15-18],然而对于中华鲟的行为特征指标及其监测技术报道较少。养殖条件下生物与非生物环境均不同于自然,水族馆展窗、水质优势及监控设备有利于进行中华鲟的行为研究。本次研究在保持环境稳定的前提下,摸索中华鲟常规行为反应,与性腺发育后的异常行为比较,筛选重要行为指标,建立行为观察技术方法。

1　材料与方法

1.1　实验对象

实验对象为北京海洋馆养殖的野生中华鲟、子一代及子二代中华鲟,共40尾,来自于中国

---

* 基金项目:水利部公益性行业科研专项经费项目(201501001);973 计划课题(2015CB150702)。

** 第一作者,女,1985 年出生,硕士,研究方向为濒危物种保护生理生态学,E-mail:zhyanzh1985@163.com

*** 通讯作者(Corresponding author),男,1960 年出生,研究员,博导,主要研究方向为珍稀鱼类保护,E-mail:weiqw@yfi.ac.cn

水产科学研究院长江水产研究所中华鲟养殖基地，在北京海洋馆养殖时间1年以上。生物学指标结果如表1所示。

表1　实验样本情况

Table 1　The conditions of the samples

组别 Group	数量及年龄 Number and age	全长/cm Totle length	体重/kg Body weight
W-成体	4(>30龄)	335±8	220±15
W-2009	9(7龄)	170±9	28±3
F1-1998	10(17龄)	265±23	121±29
F1-2001	6(14龄)	247±19	93±24
F1-2005	3(10龄)	239±13	200±13
F2-2012	8(3龄)	122±12	8±2

注：W代表野生中华鲟，F1为人工繁殖子一代中华鲟，F2为人工繁殖子二代中华鲟，数字为繁殖年份。

Note: W means the wild sturgeon; F1 means the first ilial generation sturgeons; F2 means the second generation sturgeons; The number mean the breeding year.

1.2　养殖条件

中华鲟驯养池体规格29.0 m×11.0 m×4.4 m，展示观察窗规格20.0 m×3.0 m，循环水过滤系统的循环率为50 min/次，养殖密度2.4 kg/m^3，水质指标控制如表2所示。池体水流方向：两侧为水平水流，展窗底部约30°角向后上方。饵料种类包括生鲜饵料和特制混合料，投喂方式采用水下固定食点每尾单独投喂，配合部分陆上撒喂方式。

表2　水质指标控制范围

Table 2　The range of the water quality index

指标 Index	水温/℃ Temp.	溶解氧/(mg/L) DO	酸碱度 pH	浊度(NTU) Turbidity	总菌/(cfu/100ml) Total bacteria
控制范围 Index range	21~22	7.0~7.8	7.5~8.0	<0.07	<8 000

1.3　观察记录方式

每日不同时段(9:00，13:00，17:00)进行三次观察，每次时间30~40 min，观察点包括展窗前和缸体上方，主要用具包括：记录本、秒表、白光手电、Canon相机、Panasonic监控设备。实验数据获取始于2011年，截至2015年。

1.3.1　观察内容

(1) 形态特征。体型特征、眼部特征、体表形态，包括表观特征变化、是否受伤、体色变化、眼睛、腹部、泄殖孔。

(2) 行为特征。日常行为和特殊行为，包括鱼群分布、活动范围、泳层分布、游速、游姿、呼吸、摄食行为以及繁殖行为等。

1.3.2　记录方法

(1) 体态特征。是否有变化或异常，如有则描述。

(2) 游泳区域。中华鲟在观察时段内经常游动的区域和泳层。区域划分方法为,驯养池按水平长度分为7个区域,从左至右依次记为1~7区,各区长约4 m,各区以中部造景为界分为前区、后区;池体按高度垂直分为三个泳层,从水体底部向表面顺序依次为下层、中层和上层,各层高度约1.5 m,贴水面记录为表层,贴砂层记录为底层。

(3) 游速。以游过其身体全长所用的时间为其游速,单位 s/TL。

(4) 游姿。通过观察中华鲟游动时躯体左右摇摆程度、尾部摆动幅度及上浮和下潜的平稳性确定。平稳性好:游动时身体平稳,尾摆幅度中等,可自如下潜或上浮。平稳性较好:游动时身体左右摆动,但不明显,可自如下潜或上浮。平稳性较差:游动时身体左右摆动较明显,下潜或上浮时不仅尾部摆动幅度明显加大,身体左右摆动较明显。平稳性差:游动时身体左右摆动明显,下潜或上浮时尾部摆动幅度大,且下潜或上浮困难。

(5) 呼吸频率。在有规律呼吸时,记录20秒呼吸次数,以口部张合一次为一次呼吸,不能看到口部时以鳃动频率代替,单位为次/min;无规律呼吸时描述呼吸方式。呼吸幅度:口部和鳃部张合的幅度描述,大幅为≥4 cm,中等为1~3 cm,小幅为≤1 cm

(6) 摄食行为。潜水员潜入水底喂食前后的游泳区域、游泳水层、游泳平稳性,摄食积极性,吞咽饵料行为过程,摄食量等。摄食积极性:通过观察潜水员投递饵料时口的伸展程度以及进食结果表示。积极性强:食点进食,口膜完全伸展,进食用时短。积极性中等:食点进食,口膜伸展长度1/2左右,进食用时长。积极性弱:不到食点,跟游给食或时有不吸食,口膜伸展长度<1/2,进食时间长。不进食:不进食,口膜不主动伸展。

1.4 数据统计

采用SPSS分析软件及Excel对数据进行分析处理,采用Paired-Samples *T* Test对成对结果组间差异性检验,$P<0.05$为显著性差异。

2 结果与分析

2.1 中华鲟常规行为

2.1.1 活动区域

中华鲟常规活动区域及泳层,按年龄组划分,规律性不明显,组内个体之间差异较大,大龄组尤其明显。结合各水层鱼群组分布结果,野生成熟中华鲟及部分F1-1998中华鲟分布泳层更广,不同泳层均有出现,个体活动区域相对固定,而其他4个低龄组泳层多趋向于中下层,活动区域不稳定(表3和表4)。

表3 不同年龄组分布情况

Table 3 The distribution of different age groups

组别 Group	活动区域 Swimming-area	活动泳层 Swimming-layer
W-成体	不同区域范围直径:5~20 m	下层、中层、上层
W-2009	1~7区	下层
F1-1998	不同区域范围直径:7~10 m	中层、下层,个别上层
F1-2001	1~7区	中层、下层
F1-2005	1~7区	中层、下层
F2-2012	1~7区	下层

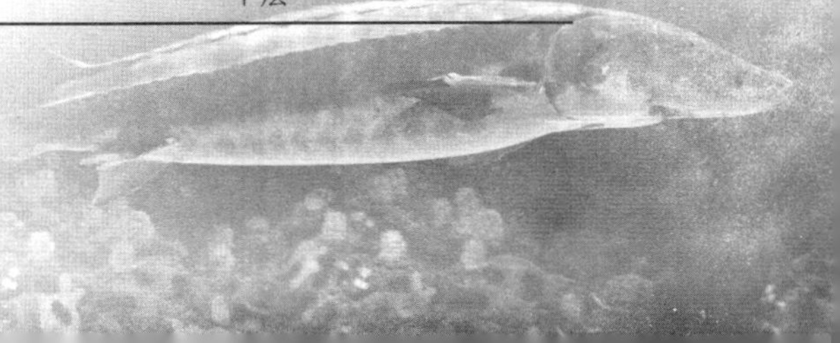

表 4　各泳层鱼组分布情况

Table 4　The distribution of different swimming layer

泳层 Swimming-layer	鱼群组 Group
上层	野生成熟,F1-1998 部分,偶尔其他龄组
中层	野生成熟,F1-1998,F1-2001,F1-2005
下层	野生成熟,F1-1998,F1-2001,F1-2005,F2-2012,W-2009

按照样本个体大小(全长)分组分析,结果显示个体较大的野生成熟中华鲟和 F1-1998 中华鲟分布泳层范围较大,低龄组个体相对小,泳层比较单一,且多分布在下层。

2.1.2　游泳方式

水族馆中华鲟多为绕圈游,绕游直径范围变化大,不同个体之间差别较大,常规条件下中华鲟游动平稳性好。

中华鲟游速各龄组比较稳定,野生成熟中华鲟游速相对较缓慢((8.60 ±2.06)s/TL),与其他龄组存在显著差异($P<0.05$),游速最快的为 F2-2012 中华鲟((3.03 ±0.45)s/TL);中华鲟游动尾摆频次一般介于 25 ~45 次/min,详见表 5。尾摆频率与游速成正比,越低龄组游速越快,尾摆频次越高,尾摆幅度不均一。

表 5　各龄中华鲟游速及尾摆频次

Table 5　The swimming velocity and wiggle frequency of tail of different age groups

组别 Group	游速/(s/TL) Swimming velocity	尾摆频次/(次/min) Wiggle frequency of tail
W-成体	8.60 ± 2.06^{a}	25 ±6
W-2009	5.52 ± 1.62^{b}	32 ±4
F1-1998	5.96 ± 0.43^{b}	29 ±6
F1-2001	5.45 ± 0.82^{b}	34 ±4
F1-2005	4.22 ± 0.77^{b}	32 ±5
F2-2012	3.03 ± 0.45^{b}	45 ±3

注:角标不同字母间有显著差异。

Note:the different letters on the angle means have significant differences.

2.1.3　呼吸频率

野生中华鲟呼吸频率不同,介于 7 ~16 次/min,4 尾存在显著性差异,其他不同年龄组中华鲟呼吸频率差别较大,结果显示年龄越小呼吸频次越高,F2-2012 中华鲟为 35 次/min,对不同年龄组结果进行差异显著性分析,结果见图 1 和表 6。

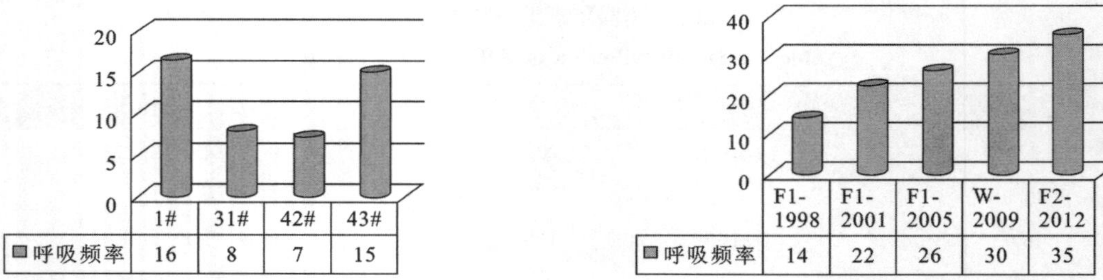

(a) 野生成熟组　Wild sturgeon　　(b) 其他各龄组　Other ages

图 1　中华鲟呼吸频率

Fig. 1　The Respiratory Frequency of the samples

表 6　各龄组呼吸频率结果的差异性分析

Table 6　Difference analysis of respiratory rate in different age groups

	W-2009	F1-1998	F1-2001	F1-2005	F2-2012
W-成体	b	a	b	b	b
W-2009		b	b	a	a
F1-1998			b	b	b
F1-2001				a	b
F1-2005					b

注：a 为无差异（$P>0.05$）；b 为有显著差异（$P<0.05$）。

Note：a represents no significant difference；b represents significant differences.

2.1.4　摄食行为

中华鲟水下喂食积极性强，快速聚集至食点，喂食时间内多在食点绕游，泳层降至下层，吸力大，口部伸长幅度大。中华鲟咀嚼吞咽过程中口部动作以及身体幅度较大，时出现竖游、转圈、折返等行为，一般咀嚼时间介于 10～30 s，无吐食行为。

2.2　性腺发育行为变化

2.2.1　体态变化

性腺发育后中华鲟体态发生明显变化，主要为腹部明显隆起，腹部血丝分布密集；生殖孔形状发生变化。

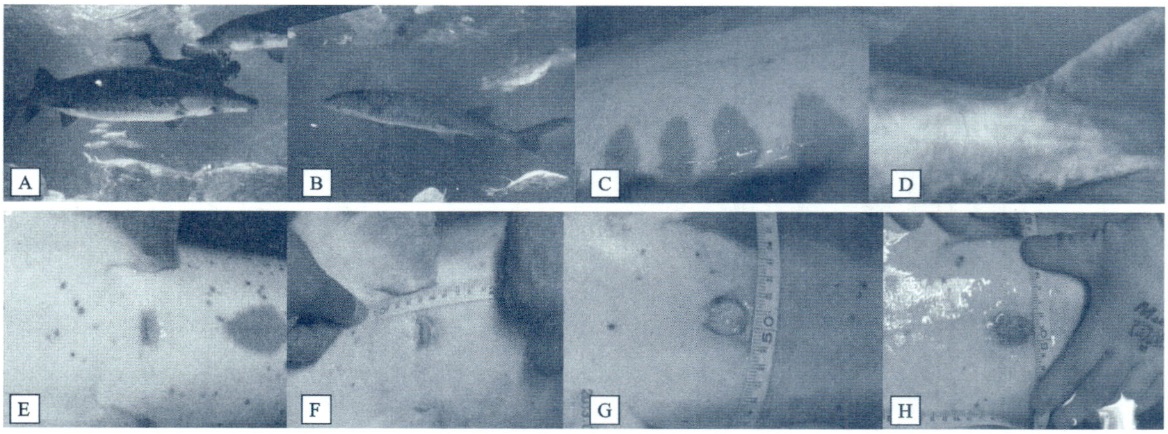

图 2　发育后中华鲟

（A：雌性；B：雄性；C：腹部血丝；D：尾柄血丝；E：雄性发育前泄殖孔；F：雄性发育后泄殖孔；G：雌性发育前泄殖孔；H：雌性发育后泄殖孔）

Fig. 2　The changes of developed Chinese sturgeon

（A：female；B：male；C：blood streak on abdomen；D：blood streak on caudal peduncle；E：genital pore of pro-developed male；F：genital pore of post-developed male；G：genital pore of pro-developed female；H：genital pore of post-developed female）

2.2.2　活动区域

性腺发育后，中华鲟活动区域发生明显变化。区域范围变化大且不稳定，多出现贴壁绕游或折返现象，且泳层明显上移。表 7 中所示为 9 尾发育后中华鲟的活动区域和泳层的变化结果。

表 7 发育前后活动区域及泳层变化

Table 7 The swimming-area and swimming-layer of developed sturgeon

编号 Num.	年份 Age	活动区域(Swimming-area)		泳层(Swimming-layer)	
		发育前 Pro-developing	发育后 Post-developing	发育前 Pro-developing	发育后 Post-developing
1#	W-N	0 ~ 3/4 ~ 6 顺	范围不定,贴壁折返	中层	不定
42#	W-N	不定范围	贴壁绕游	不定	上层
43#	W-N	0 ~ 3 顺	贴壁绕游	中上层	上层
38#	F1-1998	1 ~ 4 顺	贴壁绕游或折返	中层	中上层
25#	F1-2001	不定范围	1 ~ 5 大圈	中下层	中上层
40#	F1-1998	2 ~ 5 逆	4 ~ 6 逆	中下层	中上层
46#	F1-1998	1 ~ 4 逆	3 ~ 6 逆	下层	中上层
45#	F1-1998	不定范围	3 ~ 6 顺或贴壁	中层	中上层
49#	F1-1998	砂区	2 ~ 5 顺	底	中层

2.2.3 游动方式

性腺发育后游速有明显变化,除 2 尾野生中华鲟游速有所缓慢,其他发育中华鲟游速均变快,个别游速加快 1 倍或以上,结果见图 3。

游姿有明显变化,个别出现平衡性较差或差的水平。

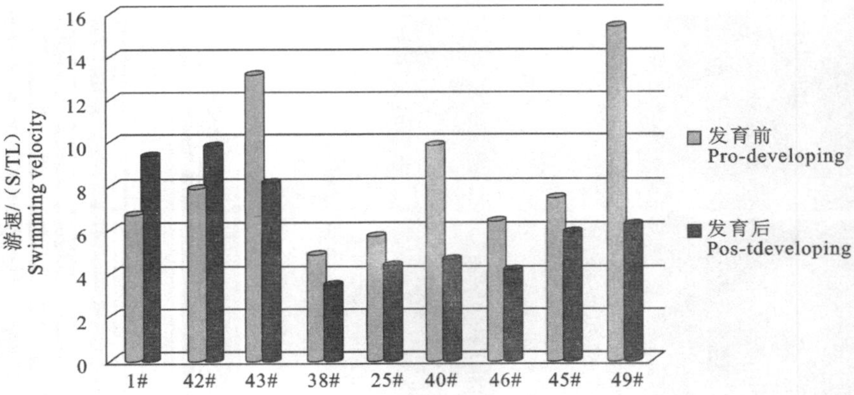

图 3 发育中华鲟游速变化

Fig. 3 The swimming velocity of the developed sturgeon

2.2.4 呼吸频次

性腺发育中华鲟呼吸频率均有增加,部分个体增加幅度超过 1 倍,结果见图 4,发育前后有显著性变化($P<0.05$)。

2.2.5 摄食行为

性腺发育后一定时间内出现进食积极性下降,通常表现为积极性弱,持续几周到几个月不等,最长达约 10 个月,之后出现不进食,持续时间最长的为 1 年 4 个月;而野生中华鲟在水族馆性腺多次发育,摄食情况不同,结果见表 8。

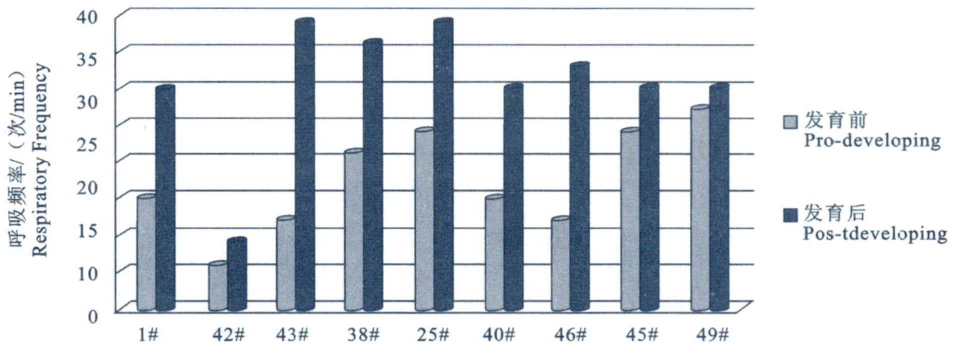

图 4 发育中华鲟呼吸频率变化

Fig. 4 The respiratory frequency of the developed sturgeon

表 8 发育后摄食行为变化

Table 8 The feeding results after development

编号 Num.	年份 Age	持续时间 Duration 不进食 Don't eat	持续时间 Duration 积极性弱 Inactivity
1#	W-N	10 个月/3 周	4 个月
42#	W-N	1 年 4 个月	5 个月
43#	W-N	无/3 个月	3 个月
38#	F1-1998	个别周	不定期间断性不积极现象
25#	F1-2001	1 ~2 周	持续不积极超过 10 个月
40#	F1-1998	3 周	2 个月
46#	F1-1998	无	3 个月
45#	F1-1998	个别次	2 周
49#	F1-1998	近 2 个月	3 个月

注:/ 前后文本表示发育 2 次的各自结果。

Note:"/" the results of twice developments.

3 讨论

鱼类的行为特征一般包括活动区域的分布情况、游动方式、游动的速度与加速度、呼吸频率、尾部摆动频率、捕食行为、对危害的规避行为、死亡现象等,这些行为特征指标可以通过人员观察、电子传感器、视觉传感器等方式获取。行为指标可以快速、准确地反映出自身行为应激变化,以及其所在水体环境的相关变化,中华鲟在水族馆饲养条件下常规行为有一定的行为轨迹,活动区域、游泳方式、呼吸频率和进食行为积极性是养殖中华鲟主要行为特征指标,这与多数其他鱼类一致[1]。

本次结果显示中华鲟活动分布区域为个体大小决定型,个体越大分布泳层范围越广,个体小中华鲟多集中在中下层活动且易受干扰。中华鲟游动缓慢,常规游速(即巡航速度)介于 3 ~10 s/TL(表示游过一个全长用时),而报道中牙鲆平均游速约为 1.0 BL/s(表示每秒游过体长数),虹鳟约为 3.2 BL/s[1],性腺发育后中华鲟尽管游速明显增加,但相比较其他鱼类仍缓慢。与多数鱼类一致,中华鲟尾摆频次与游速成正相关[1,2]。对于鱼类的呼吸频率相关报道较少,多数研究集中于呼吸与耗氧率的相互关系[19],而本次研究中水体溶氧量为系统调控,其呼吸频率只与个体状况有关,中华鲟个体间呼吸频率差异性以及如何反应器官机能可做进一步研究。在自然条件下,中华鲟繁殖前一年开始溯河上游并停食至产卵后回到大海再次进食[13-15],这与本次研究性腺发育后中华鲟不同程度停食一致;然而养殖条件下,中华鲟活动范围小,生态环境优越,其能量消耗情况完全不同于自然环境,因此也决定了其发育过程中摄食行为的差异性。性腺发育后中华鲟内分泌指标变化显著[20,21],这也是导致其行为反应异于常规行为的根本原因,其活动区域、泳层变化、游速、呼吸频率等变化均与此相关,历史报道中多提到中华鲟繁殖期间产卵场可见中华鲟频繁跃出水面等现象。

行为技术应用于水族馆,在未经捕捞生理检查情况下,通过每日行为观察建立行为特征的基础参考值,可启示中华鲟健康状况;利用行为技术判断个体性格,通过亲和训练,捕捞训练等减少应激,降低健康风险。行为技术可用于水族馆中华鲟健康保育,对于各行为表现同时本次结果为中华鲟的基础科研提供基础参考。

参考文献

[1] 何大仁,蔡厚才. 鱼类行为学[M]. 厦门:厦门大学出版社,1998.
[2] 周应祺. 应用鱼类行为学[M]. 北京:科学出版社,2011.
[3] Carmona R, Omezain A, Arcia-Gallego M. Bology, conservation and sustainable development of sturgeons[M]. Springer publishers, 2009.
[4] LeBreton G T O, Beamish F W H, Mckinley R S. Sturgeons and paddlefish of North America[M]. Kluwer Academic publishers, 2004.
[5] Peake S. Substrate preference of juvenile hatchery-reared lake sturgeon[J]. Envbiol fish, 1999, 56: 367-374.
[6] 石小涛. 史氏鲟在拥挤胁迫中的应激反应[D]. 武汉:华中农业大学水产学院,2006.
[7] 李大鹏,庄平,王明学,等. 史氏鲟稚鱼的趋光性及不同光照周期对其生长的影响[J]. 华中农业大学学报,2001,20(6):564-567.
[8] 柴毅. 中华鲟感觉器官的早期发育及其行为机能研究[D]. 武汉:华中农业大学水产学院,2006.
[9] 吴常文,江成国,陈国业. 海水养殖杂交鲟的行为习性[J]. 应用生态学报,2006,17(2):320-324.
[10] 李丹,林小涛,李想,等. 水流对杂交鲟幼鱼游泳行为的影响[J]. 淡水渔业,2008,38(6):46-51.
[11] 庄平,章龙珍,罗刚,等. 长江口中华鲟幼鱼感觉器官在摄食行为中的作用[J]. 水生生物学报,2008,32(4):475-481.
[12] Wang J H, Wei Q W, Zou Y C. Conservation strategies for the Chinese sturgeon, *Acipenser sinensis*, an overview on 30 years of practices and future needs[J]. Journal of Applied Ichthyology, 2011, 27(2): 176-180.
[13] 危起伟,李罗新,杜浩,等. 中华鲟全人工繁殖技术研究[J]. 中国水产科学,2013,20(1):1-11.
[14] 陈细华. 鲟形目鱼类生物学与资源现状[M]. 北京:海洋出版社. 2007.
[15] 危起伟. 中华鲟繁殖行为生态学与资源评估[D]. 武汉:中国科学院水生生物研究所,2003.
[16] Wei Qiwei, Zhang Xiaoyan, Zhang Xianfeng, *et al*. Acclimating and maintaining Chinese sturgeon *Acipenser sinensis* in a large public aquarium environment[J]. Journal of Applied Ichthyology, 2011, 27(2): 533-540.
[17] 张晓雁,刘鉴毅,危起伟,等. 不同喂食方法对野生产后中华鲟的摄食促进[J]. 动物学杂志,2007,42(2):142-146.
[18] 王恒. 中华鲟子二代环境偏好性研究[D]. 武汉:华中农业大学水产学院,2014.
[19] Kind P K, Grigg G C, Booth D T. Responses to prolonged hypoxia in the Australian lungfish, *Neoceratodus forsteri*[J]. Resp Physiol & Neurobiol, 2002, 132: 179-190.
[20] 张艳珍. 养殖中华鲟性腺发育及血液相关生理指标变化的观察[D]. 武汉:华中农业大学水产学院,2009.
[21] 张艳珍,陈细华,危起伟,等. 中华鲟血清卵黄蛋白原水平初步观察[J]. 淡水渔业,2008,38(5):10-14.

Observation of Behavior of Chinese Sturgeon (*Acipenser sinensis*) and Its Application in the Aquarium

Zhang Yanzhen[1,2]　Zhang Xiaoyan[1]　Wei Qiwei[2]
Wang Yanpeng[1]　Du Hao[2]　Cai Jingjiang[1]

(1. Beijing Xinwoda Marine Technology Co. Ltd (Beijing Aquarium), Beijing 100081, China;
2. Key Laboratory of Freshwater Biodiversity Conservation, Ministry of Agriculture of China, Yangtze River Fisheries Research Institute Chinese Academy of Fishery Sciences, Wuhan 430223, China)

Abstract The study was on the 40 Chinese sturgeons in Beijing Aquarium, including the wild, the first filial generation and the second generation sturgeons, using observation method and the digital monitoring equipments. By comparison between conventional and abnormal behavior, to find the main behavior indexes of Chinese sturgeon, and establish the observation methods of behavior of Chinese sturgeon and its application in the aquarium. The results show that the activities region, swimming, respiratory frequency and eating behavior were the main behavior indexes. The swimming velocity of the wild matured sturgeons, the W-2009, F1-1998, F1-2001, F1-2005 and F2-2012 were (8.60 ± 2.06) s/TL, (5.52 ± 1.62) s/TL, (5.96 ± 0.43) s/TL, (5.45 ± 0.82) s/TL, (4.22 ± 0.77) s/TL and (3.03 ± 0.45) s/TL respectively. The wiggle frequency of tail was in proportion to the velocity. The respiratory frequency was reduced with age, the F2-2012 Chinese sturgeon is 35 times /min. The behavior of development Chinese sturgeon were changed, such as the swimming layer was upward, swimming velocity and respiratory frequency were increased, and eating enthusiasm was weak or refused. In a word, behavior of technology can be used for the Chinese Sturgeon in Aquarium.

Key words *Acipenser sinensis*　Aquarium　Observation　Behavior index　Application

野生中华鲟水族馆驯养观察*

刘鉴毅[1,3,**]　张晓雁[2]　危起伟[1,3,4,***]　张　辉[4]
杨道明[2]　朱永久[1,3]　陈细华[1,3]　杨德国[1,3]

(1. 农业部淡水鱼类种质资源与生物技术重点开放实验室 中国水产科学研究院长江水产研究所　湖北荆州　434000
2. 北京海洋馆　北京　100081　3. 中国水产科学研究院淡水渔业研究中心　江苏无锡　214081
4. 华中农业大学水产学院　湖北武汉　430070)

摘　要　2005 年 4 月 3 日 ~ 7 月 31 日期间以捕捞于长江宜昌江段的野生中华鲟(*Acipenser sinensis*)为研究材料,对其在水族馆中的驯养方法进行了研究。结果表明,野生中华鲟摄食的饲料品种由 1 种逐渐增加至 6 种,日摄食量由 120 g 逐渐增加至 2 000 g 并逐渐趋于稳定,说明水族馆驯养池的环境适合野生中华鲟的生存。

关键词　中华鲟　驯养　水族馆

世界上现存鲟形目鱼类共 27 种,其中有 17 种处于濒危或极危状态,9 种处于易危状态,如何才能有效地保护这些珍贵鱼类在全世界范围内引起了广泛关注[1]。中华鲟(*Acipenser sinensis*)是一种典型的溯河洄游性鱼类,主要分布于我国的东海、黄海、珠江和长江干流[2]。20 世纪后期,由于过度捕捞、水利工程建设、航运、水体污染等因素,其野生种群资源量不断下降,虽然在 1988 年被列为国家一级保护动物,但 1981 ~ 1999 年的 19 年间,中华鲟的幼鲟补充群体和亲鲟补充群体仍然分别减少了 80 %和 90 %左右[3]。为了更好地保护中华鲟,需要开展一些易地保护研究。北京海洋馆具有先进的技术和饲养条件,在珍稀濒危水生动物的保护方面可进行一些有益的探索。将野生成熟中华鲟在水族馆中驯养成功,进而实现其性腺的再次成熟对于中华鲟的易地保护将是一个重大的突破。

根据野生中华鲟的生活习性,在产卵前会停止进食,在产卵后一般回到大海时才会进食[4]。此次进行驯养的这尾野生中华鲟在人工环境下如何主动进食成为一项研究课题。野生鲟鱼驯养和摄食行为的研究,国外如美国和俄罗斯有所突破,但是中国尚未成功,仅有一些其他鲟鱼驯养研究的报道[5-7]。根据研究方案,本文对一尾野生雌性中华鲟在水族馆的驯养和摄食行为进行了观察记录。

1　材料与方法

1.1　试验鱼

进行驯养的野生中华鲟是中国水产科学研究院长江水产研究所于 2004 年 10 月 12 日在长江宜昌江段捕捞的亲本。经检查,该尾中华鲟为雌性,全长 320 cm,体重 222 kg,年龄估计为

* 基金项目:国家公益研究专项(2004DIB3J099);国家基础性工作重点专项(2002DEA1004);国家自然科学基金重大项目(30490231)。

** 第一作者,男,副研究员,研究方向为主要从事淡水珍稀水生动物保护生态和繁殖生理研究,E-mail:liu jy999@ vip. 163. com

*** 通讯作者(Corresponding author),E-amil:weiqw@ yfi. ac. cn

20龄。捕获后暂养于湖北省荆州市中华鲟增殖放流基地中华鲟亲鱼暂养池,没进行人工繁殖。暂养池为圆柱形,直径12 m、高2.5 m。对鲟鱼体表的损伤,涂抹云南白药和凡士林混合物加以治疗;对于内伤,则采用肌肉注射硫酸庆大霉素2 ml(8万单位×1支)、维生素C 2 ml(0.5 g×2支)、氯霉素2 ml(0.25 g×1支)、鱼蚌康复剂2 ml(4万单位×2支)、ATP 2 ml(20 mg×3支)以及氨苄西林钠2 ml(0.5 g×1支)加以保护治疗。在暂养期间鲟鱼不摄食,因此不投喂任何饲料,只做一般行为观察。暂养池内保持微流水环境,暂养期间池内水体pH为7.2,DO 6.5~9.3 mg/L,水温8~21 ℃,每天排污一次,每周清洗一次;越冬期间水温过低时,抽取地下水升温,防止冻伤。在起运前10 d打一针保健针,前2 d再打一针。2005年4月2日上午,在专用运输箱内采用麻醉剂进行轻度麻醉,并在其背部注射PIT标记(420B4F570F)后,起吊至特制的专用活鱼运输箱运往北京。4月3日上午11:45,在经历23 h,1 400 km的路程之后,该中华鲟安全抵达北京海洋馆。共同进京的还有另外25尾人工养殖的不同年份出生的子一代中华鲟,规格为全长100~200 cm,体重8~100 kg,其中年龄最大者为8龄,均与野生中华鲟安放在同一驯养池内。

1.2 水族馆驯养池

由鲨鱼馆改建而成的鲟鱼馆具备良好的基础设施条件,具有目前世界先进的生命支撑系统(life support system),可以对水温、水质进行人工监控和调节,还具有淡水、海水转换调控能力。驯养水池规格为29.0 m(长)×11.0 m(宽)×4.4 m(高),总水量达1 400 t,展窗20.0 m(长)×3.0 m(高),经过改造,将原先池内摆放不合理的岩石取出,按照中华鲟的生活习性将池底重新进行调整,并对水池采取清池消毒、清除杂藻、堵塞漏洞、池底铺设洁净的沙石等措施。由此营造了目前国内水族馆中最大的中华鲟驯养池。

1.3 水质监测方法

中华鲟对水质的要求比较高,例如,DO要在5 mg/L以上,透明度在30 cm以上,最佳水温应保持在21 ℃左右等[9]。为了满足这一要求,必须每天对驯养池内的水质进行监测,以满足中华鲟的要求。pH和温度采用HANNA便携pH计(复合pH电极,带自动温度补偿ATC,产自意大利)测量;溶氧量采用HACH溶氧仪(产自美国),渗透膜法测量;NH_3/NH_4^+,NO_2^-,NO_3^-含量采用HACH分光光度计(试剂为HACH套装,产自美国)测量,细菌含量则采用HACH试剂,Millipore膜过滤法测量。

1.4 饲料及处理方法

根据中华鲟的食性,准备了两大类饲料,天然饲料和人工配合饲料,种类多达20余种。这些饲料都经过严格的检查和消毒,以确保中华鲟的饮食安全。天然饲料要求新鲜,大小要求适口,为整鱼或经适当加工处理后的鱼块;人工配合饲料为直径8 mm的圆柱状颗粒。日投饵两次,并记录每天的摄食量,新鲜鱼块或整鱼按摄食的实际量计算,人工配合饲料按每次摄食平均量乘以摄食次数计算。天然饲料来源为市售,并按饲料性质的不同主要采用两种处理方法:(1)死饵(鲢鱼肉、鱿鱼、白斑虾、小河鱼等),先置于−25 ℃冷库中冷冻杀菌保存,用自来水解冻后加工,紫外线杀菌处理后使用;(2)活饵(鲫鱼等),30‰人工海水杀菌20 min后检疫池中饲养2~3 d,使用前再用30‰人工海水杀菌20 min,然后用清洁淡水洗净,紫外杀菌处理后投喂。

1.5 喂食方法

中华鲟进入鲟鱼池后次日开始喂食,喂食方法有两种:固定食场喂食人工颗粒饲料;水下喂食天然饲料。在池底中间位置建立食场,每天上午 9:00 ~ 10:00 投喂人工配合饲料,下午 3:00 ~ 4:00 由潜水员水下喂食天然饲料。人工颗粒饲料的投喂方式为,通过长 4 m,直径 50 mm的 PVC 管直接将饲料导入池底食场。这样可以最大限度上的保证投喂位置的固定,同时减少饲料在水中的飘浮时间,最大限度降低饲料污染水质的几率。天然饲料的投喂方式为,潜水员水下喂食。

1.6 行为观察方法

野生中华鲟进入驯养池后最初 72 h 连续观察其行为,包括游动方向(顺时针或逆时针)、游动水层(底层、中层或表层)、游动速度 v(绕行 1 周的路程/时间)、呼吸频率 f(指鳃盖在 1 min 内张合次数,一张一合记为 1 次)、尾部摆动频率 n(1 min 内尾部左右来回摆动的次数)、游姿是否正常、吃食是否主动、有无应激反应(游速突然变化、皮肤肌肉紧张等),并进行记录,待其恢复正常后每 4 h 观察一次并进行记录。对其摄食前、摄食中和摄食后的行为进行重点观察和记录,并不定期测试其对潜水员和不同饲料的反应。

2 结果与分析

2.1 驯养期间水质变化

由于驯养池具有先进的生命支撑系统,4 月 11 日 ~ 7 月 31 日,水温、DO,ORP,pH 均未出现较大程度的波动;NH_4^+/NH_3 含量一直为 0.01 mg/L,未出现波动;NO_2^- 含量也一直为 0.01 mg/L,中间出现 2 次轻微的波动,波动值分别为 0.04 和 0.5;链肠杆菌共测量 22 次,其中有 13 次测量结果为 0;但总细菌含量变化较大。其具体情况见表 1 和图 1。

表 1 4 月 11 日 ~ 7 月 31 日水质监测结果

Table 1 Results of water quality monitored between Apr. 11 and Jul. 31

项目 Item	测量次数 Times of measuring	变化范围 Range	平均值 Average
水温/℃ Temperature of water	100	19.4 ~ 23.4	20.816
DO/(mg/L)	104	7.6 ~ 9.5	9.000
氧化还原电位值 ORP	100	246 ~ 355	312.33
pH	103	7.86 ~ 8.80	8.084
Ca^{2+}/(mg/L)	2	40 ~ 60	50
KH	2	8.5 ~ 9.0	8.75
NH_4^+/NH_3/(mg/L)	82	0.01 ~ 0.01	0.01
NO_2^--N/(mg/L)	82	0.01 ~ 0.50	0.016
NO_3^--N/(mg/L)	3	0.12 ~ 26.40	10.893
大肠杆菌/(个/100 ml) Coliform	22	0 ~ 570	68.318
总细菌/(个/100 ml) Bacilli	32	346 ~ 7 000	3 794.41

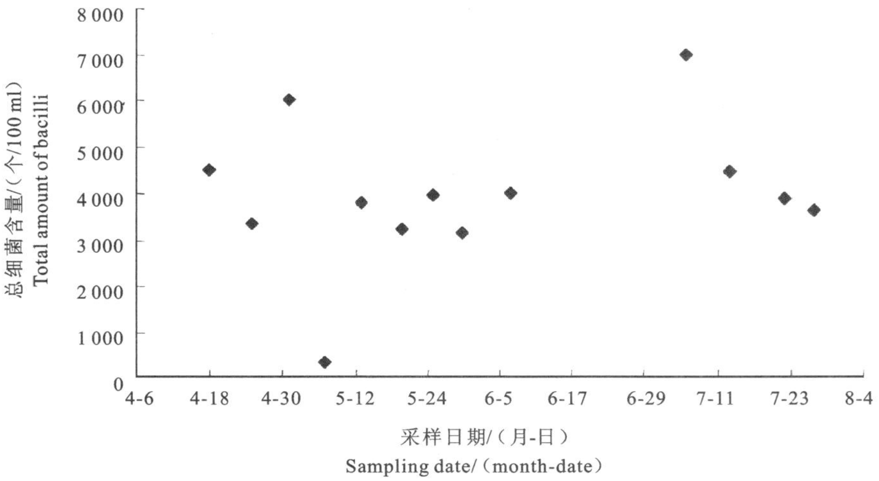

图 1　驯养池内总细菌含量随采样日期的变化

Fig. 1　Change of total amount of bacilli with sampling date in the aquarium

2.2　喂食过程

野生中华鲟对潜水员及饲料有一个逐渐适应的过程。4 月 4 日下午 3: 30 潜水员第一次喂食野生中华鲟一尾小鲫鱼,但不久后便吐出;约 1 h 后再喂,其口部蠕动,有鲫鱼鳞片溢出,未吐出鲫鱼,证实已经进食。进食后游动路线发生改变,由上层改为中、下层游动,并有觅食的表现;10 min 后又改为上层游动。4 月 5 日上午投喂人工配合饲料,野生中华鲟未摄食人工配合饲料,虽然 10 min 后在食场上方游动,但没有下潜摄食的迹象;下午 2: 30 喂食生鲜饲料,野生中华鲟对潜水员无受惊反应,但是改变游动路线;2: 52 开始进食第一条小鲫鱼,吃后绕展窗中央部分小范围游动;2: 58 进食第二条小鲫鱼,共 400 g。此后,随着时间的推移,野生中华鲟的摄食行为逐渐发生了变化,主要表现在以下几个方面。

(1) 摄食量逐渐增大,平均每 8 d 摄食的饲料重量随驯养时间变化的关系如图 2 所示。

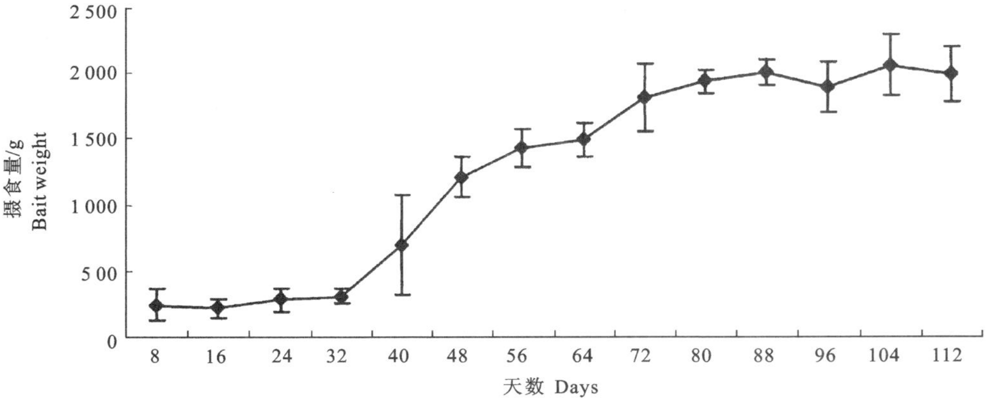

图 2　野生中华鲟平均每 8 d 摄食的饵料质量与驯养天数之间的关系

Fig. 2　The relationship between bait amount taken each 8 days and domestication duration

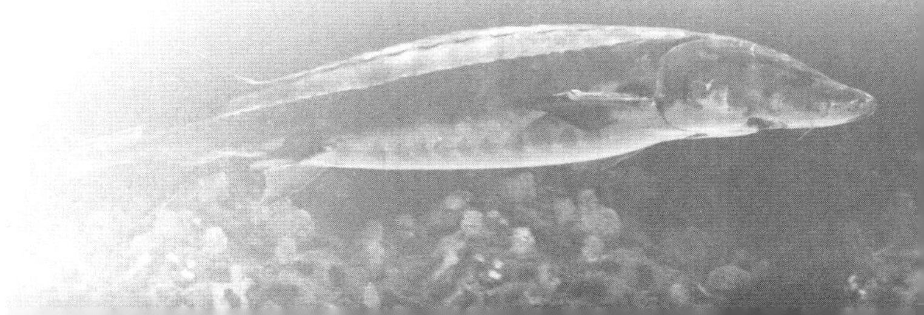

(2) 摄食的饲料品种越来越多,摄食的饲料品种随时间变化的关系见表2。

表2 野生中华鲟摄食的饲料品种的变化

Table 2 The relationship between bait items taken with the duration of domestication of Chinese Sturgeon and date

时间(月-日) Time (month-date)	进入驯养池的天数 Days in the aquarium	第一次摄食的饲料 Bait first eaten	已摄食的饲料品种 Prey species
4-4	2	鲫鱼 *Carassius auratus*	鲫鱼
4-26	24	鲢鱼肉 Muscle of Silver carp	鲫鱼、鲢鱼肉
6-22	81	南美白对虾肉 Muscle of *Penaeus vannamei*	鲫鱼、鲢鱼肉、南美白对虾肉
6-23	82	牙鲆 *Paralichthys olivaceus*	鲫鱼、鲢鱼肉、南美白对虾肉、牙鲆
7-11	100	小黄鱼 *Pseudosciaena ployactis*	鲫鱼、鲢鱼肉、南美白对虾肉、牙鲆、小黄鱼
7-21	110	大虾 *Cherax quadricarinatus*	鲫鱼、鲢鱼肉、南美白对虾肉、牙鲆、小黄鱼、大虾

(3) 进食行为越来越主动,从潜水员下水到第一次开始摄食的时间间隔越来越短,并能够主动接近潜水员,后期还出现了投饵时间未到而吸食潜水员脚蹼或用头部撞人索食的现象。每次喂食时摄食饲料的次数也越来越多,摄食一次即吃一次饲料,从最初每次喂食时摄食2次发展到后期每次摄食达10余次,并且每2次摄食时间间隔也越来越短,后期甚至出现了连续摄食的情况。

(4) 摄食过程越来越呈现规律性,潜水员在水下喂食的整个过程中,进食过程一般分为2个时间段,2个时间段之间游离潜水员。每个时间段连续进食数次,每2次进食的时间间隔较短,并停留在潜水员附近。进食时间共为30 min左右,进食时吸食的力量非常大,游动有力,在摄食过程中喜欢贴沙层游动。

(5) 对潜水员的适应性逐渐增强。最开始是躲避潜水员,后来逐渐允许潜水员接近,继而主动接近潜水员,再到摄食过程中允许潜水员抚摩,再到能够在不摄食的时候配合潜水员做一些动作。这期间野生中华鲟对潜水员越来越适应。

虽然在整个过程中,野生中华鲟摄食生鲜饲料的情况越来越好,但从驯养开始至今,野生中华鲟始终未能摄食人工配合饲料。最开始在投喂人工配合饲料时,虽然没有摄食,但还能靠近食场,并在食场上方游动,其行为发生了一定的变化。到后期投喂人工配合饲料时,其基本上已没有任何反应,既不摄食,也不靠近。

2.3 行为观察

4月3日下午13:50野生中华鲟进入驯养池,刚下水时表现出强烈的应激反应:快速向前冲游撞墙后而立即下沉。3名潜水员入池进行助游过程中,鱼体不用力,各鳍几乎不动,全靠人力帮助游动。14:00中华鲟停留在池底背光区(无沙层),但身体无倾斜。14:00~14:03统计呼吸频率(f)分别为31次/min,39次/min,36次/min,38次/min;后又在14:07和14:11统计呼吸频率分别为 36 次/min,36 次/min;14:15 呼吸频率减慢为 34 次/min;14:21 呼吸频率又减慢为 31 次/min。在此过程中,潜水员靠近鱼体时,鱼体紧张,肌肉收缩。14:21 潜水员靠近鱼体时,鱼受惊开始游动,比较急速,并跃出水面一次,但遇到障碍物时可以避开,卡住时可以退出来。14:31 停止游动,并停留于池体沙层处,呼吸频率为 37 次/min。14:35 开始游动,窜出水面一次,后游动状态稳定,游动缓慢,呼吸频率为 30 次/min,腮部张开程度较大。后又停留在池底不动,14:56 重新开始游动,状态正常。17:00 观察,中华鲟状态稳定,游动正常。夜晚观察时游动正常,在中上水层沿池壁进行游动,游动缓慢,无碰撞展窗及造景的现象。

此后,该中华鲟的行为逐渐恢复正常。开始 10 d 内一般在展窗右侧或左侧的中下层水域做小范围环行运动。4 月 14 ~ 15 日,运动范围逐渐向中上水层转移,环行运动的范围也越来越大。但从 4 月 16 日开始,又喜在底层角落处做小范围环行运动,偶尔也会在中上层水域运动,没有什么规律。5 月 20 ~ 7 月 19 日,野生中华鲟一般在水体上层沿整个池壁逆时针游动,并且游速较快,无其他明显特征。

3 讨论

和野生中华鲟处于同一驯养池,人工养殖的、不同年份出生的子一代中华鲟,除了摄食人工配合饲料外,绝大多数已开口摄食生鲜饲料。这说明人工养殖的中华鲟能较快的恢复摄食生鲜饲料的本能,而野生中华鲟不但至今未摄食过人工配合饲料,而且其对人工配合饲料的反应也越来越冷淡。野生中华鲟究竟能不能摄食人工配合饲料的问题还需要进一步研究。

在整个驯养期间,野生中华鲟意外受伤的事件时有发生,虽然采取了包裹管道等一些措施,后期中华鲟受伤的事件确实也大有减少,但仍然不能完全杜绝。这是否因为造景设置不合理,还是养殖密度过大(偶尔会出现鱼体相互碰撞的现象)等问题,也需要进一步的实验和研究。

另外,在整个驯养过程中,野生中华鲟与子一代中华鲟一直处于同一驯养池中,野生中华鲟的行为是否受到子一代中华鲟的影响也是一个值得探讨的问题。

参考文献

[1] Billard R, Lecointre G. Biology and conservation of sturgeon and paddlefish. Reviews in Fish Biology and Fisheries, 2001, 10(4): 355-392.

[2] Wei Qiwei, Ke Fu'en, Zhang Jueming, et al. Biology, fisheries, and conservation of sturgeons and paddlefish in China. Environmental Biology of Fishes, 1997, 48(1/4): 241-255.

[3] 柯福恩. 论中华鲟的保护与开发. 淡水渔业, 1999, 29(9): 4-7.

[4] 四川省长江水产资源调查组. 长江鲟鱼类生物学及人工繁殖研究. 成都: 四川科学技术出版社, 1988: 88.

[5] 陈凌云, 章秋虎, 项卫民, 等. 匙吻鲟孵卵驯养与无公害网箱养殖新技术. 渔业致富指南, 2005(7): 33-34.

[6] 张彤晴, 葛家春. 小体鲟幼鱼水库坝下流水池塘驯养技术. 江苏农业科学, 2002(5): 59-61.

[7] 罗晓松, 曾令兵. 工厂化驯养鲟鱼病害防治技术. 科学养鱼, 2000(11): 32-33.

[8] 颜远义. 中华鲟生物学特性及养殖方法. 水产科技, 2003(5): 14-16.

[9] 刘勇. 中华鲟人工繁殖技术研究的现状和展望. 水利渔业, 1988(4): 20-25.

Observation on Domestication of Wild Chinese Sturgeon in an Aquarium

Liu Jianyi[1,3] Zhang Xiaoyan[2] Wei Qiwei[1,3,4] Zhang Hui[4]
Yang Daoming[2] Zhu Yongjiu[1,3] Chen Xihua[1,3] Yang Deguo[1,3]

(1. Key Lab of Freshwater Fish Germplasm Resources and Biotechnology, Ministry of Agriculture; Yangtze River Fisheries Research Institute, Chinese Academy of Fisheries Sciences, Jingzhou 434000;
2. Beijing Aquarium, Beijing 100081
3. Freshwater Fisheries Research Center, Chinese Academy of Fisheries Science, Wuxi 214081
4. Fishery College, Huazhong Agricultural University, Wuhan 430070, China)

Abstract A wild mature Chinese Sturgeon (*Acipenser sinensis*), caught from the Yichang section of Yangtze River was kept in an aquarium in Beijing in 2005. It was the first practice to domesticate Chinese Sturgeon in China. The food items taken by the Chinese Sturgeon was recorded from April 3 to July 31. The results showed that the mature Chinese Sturgeon could take food from the trainers'hands. The food items had been added to 6 kinds. The quantity of fish ate by the sturgeon a day was gradually increased from 120 g to 2 000 g, and became stable soon after. It indicated that the environment of aquarium was suitable to domesticate the wild Chinese Sturgeon.

Key words Chinese Sturgeon (*Acipenser sinensis*) Domestication Aquarium

中华鲟水族馆驯养研究*

刘鉴毅[1,3,**]　张晓雁[2]　危起伟[1,3,4]　张　辉[4]　杨道明[2]
朱永久[1,3]　陈细华[1,3]　杨德国[1,3]　郑跃平[4]

(1. 农业部淡水鱼类种质资源与生物技术重点开放实验室 中国水产科学研究院长江水产研究所　湖北荆州　434000
2. 北京海洋馆　北京　100081
3. 中国水产科学研究院淡水渔业研究中心　江苏无锡　214081
4. 华中农业大学水产学院　湖北武汉　430070)

摘　要　以25尾人工繁殖中华鲟(*Acipenser sinensis*)和1尾捕捞于长江宜昌江段的野生中华鲟为研究材料,对其在水族馆中的驯养方法进行了研究。结果表明,26尾中华鲟已全部开口进食,其中有15尾人工繁殖原进食人工配合饲料的中华鲟开口进食生鲜饲料,且摄食生鲜饲料的品种越来越多;2005年4~8月,25尾人工繁殖中华鲟摄食的饲料数量稳步增加,并呈现出良好的生长态势;野生中华鲟摄食的饲料品种由1种逐渐增加至6种,日摄食量由120 g逐渐增加至2 000 g并逐渐趋于稳定。说明水族馆驯养池的环境适合中华鲟生长。

关键词:中华鲟　驯养　水族馆

世界上现存鲟形目鱼类共27种,其中17种处于濒危或极危状态,9种处于易危状态,鲟鱼类保护是全世界关注的问题[1]。中华鲟(*Acipenser sinensis* Gray)是一种典型的溯河洄游性鱼类,主要分布于我国的东海、黄海、珠江和长江干流[2]。20世纪后期,由于过度捕捞、水利工程建设、航运、水体污染等因素,其野生种群资源量不断下降。虽然在1988年中华鲟被列为国家一级保护动物,但在1981~1999年的19年间,其幼鲟补充群体和亲鲟补充群体仍然分别减少了80%和90%左右[3]。为更好地保护中华鲟,需要尝试开展易地保护研究。北京海洋馆具有先进的饲养技术和条件,是驯养中华鲟和研究中华鲟生长状况的理想场地。有关鲟鱼的驯养研究,国内已有一些简要报道[4-7],本课题组对26尾中华鲟在水族馆中的驯养方法和生长情况进行了较为深入的研究。

1　材料与方法

1.1　试验鱼

进行驯养试验的中华鲟共26尾。其中,25尾为人工养殖的不同年份出生的子一代中华鲟,年龄最大者为8龄;另1尾为雌性野生中华鲟,是长江水产研究所于2004年10月12日在长江宜昌江段捕捞的亲本,年龄估计为20龄。这些鱼分3批分别于2005年3月26日、4月3日、4月16从湖北省荆州市中华鲟增殖放流试验基地经20多小时1 400 km的路程之后到达北京海洋馆。经过4~5 h的兑水及稳定,分别用自制帆布鱼担架或直径为1.2 m的PVC圆筒经吊机吊入池内。为便于后期研究,对每一尾中华鲟进行了编号,中华鲟编号及全长、体重的

*基金项目:国家公益研究专项(2004DIB3J099);国家基础性工作重点专项(2002DEA1004);国家自然科学基金重大项目(30490231)。

**第一作者,男,1966年出生,副研究员,主要从事淡水珍稀水生动物保护生态和繁殖生理研究,E-mail:liu jy999@vip.163.com

详细情况见表1。

表1　进行驯养试验中华鲟的全长、体重及其编号

编号	全长/cm	体重/kg	编号	全长/cm	体重/kg
1	320	222.0	14	125	5.2
2	185	35.0	15	119	5.2
3	182	35.0	16	150	27.3
4	122	11.0	17	125	17.3
5	121	10.5	18	150	18.2
6	117	7.0	19	101	15.5
7	116	8.5	20	126	17.0
8	104	11.5	21	124	16.5
9	125	9.0	22	112	13.5
10	117	6.5	23	110	15.1
11	116	8.9	24	121	14.0
12	115	12.5	25	121	13.8
13	101	4.0	26	117	13.5

注：编号1为野生中华鲟，其余为人工养殖中华鲟。

1.2　水族馆驯养池

鲟鱼馆由鲨鱼馆改建而成的，基础设施完备，具有目前世界上先进的生命支撑系统（life support system），可以对水温、水质进行人工监控和调节，还具有淡水、海水转换调控能力。驯养水池规格为29.0 m×11.0m×4.4 m，总水量达1 400 t，展窗20.0 m×3.0 m。工作人员按照中华鲟的生活习性对池子进行了改造，将原先池内摆放不合理的岩石取出，将体积大、重量沉、无法移出的造景尽可能地向池边聚拢；池底也重新进行了调整，并对水池采取了清池消毒，清除杂藻，堵塞漏洞，池底铺设洁净的沙石等措施，由此营造了目前国内水族馆中最大且较为理想的中华鲟驯养池。

1.3　水质调整与监测方法

由于驯养池原先驯养的是鲨鱼，驯养中华鲟时需要将原池中的海水调制成淡水，经过周密计算，连续7 d换水，每次换水400 t，7次换水后将原池中30 ‰海水调制成了2.5 ‰的淡水。驯养池从海水转化成淡水后，原生物砂床中培养的硝化细菌等有益菌基本消亡，维持池中水质平衡的生化系统遭到彻底破坏。为了重建硝化系统，减缓中华鲟入池后的水质波动，从3月1日开始对驯养池进行植铵工作，共持续20 d，植铵浓度控制在0.5～0.8 mg/L。4月20日左右硝化系统趋于平稳。硝化系统建立全程历时约50 d。依据中华鲟饲养标准水质要求，建立驯养池饲养水质标准（*T*，pH，ORP，DO，生物指数等）[8]。将水温（*T*）控制在20.0～23.0 ℃；酸碱度（pH）稳定在7.7～8.2；ORP控制在280～350 mV的耐受度之内；溶氧（DO）在8.0～9.0 mg/L；同时监测水体总细菌、真菌以及大肠杆菌的波动范围，并根据实际情况及时进行调整。水质监测的项目及其仪器与方法简列于表2。

表 2　驯养池内水质监测的项目及其仪器方法

测定项目	仪器或方法
pH; T	HANNA 便携式 pH 计(复合 pH 电极,带自动温度补偿 ATC)
DO	HACH DO 仪(渗透膜法)
NH_3/NH_4^+; NO_2^-; NO_3^-	HACH 分光光度计,试剂为 HACH 套装
细菌数	HACH 试剂(Millipore 膜过滤法)

1.4　饲料及处理方法

根据中华鲟的食性,准备了两大类饲料,天然饲料和人工配合饲料,种类多达 20 余种。这些饲料都经过严格的检查和消毒,以确保中华鲟的饮食安全。天然饲料要求新鲜,大小适口,通常为整鱼或经适当加工处理后的鱼块;人工配合饲料为直径 8 mm 圆柱状颗粒。日投饵两次,并记录每天的摄食量。新鲜鱼块或整鱼按摄食的实际量计算,人工配合饲料按每次摄食平均量乘以摄食次数计算。天然饲料参考中华鲟自然生态下的饵料组成,市场购买,并按饲料性质的不同主要采用如下两种处理方法:

(1) 死饵(鲢肉、鱿鱼、白斑虾、小河鱼等),先置于 −25 ℃冷库中冷冻杀菌保存,用自来水解冻后加工,紫外线杀菌处理后使用;

(2) 活饵(鲫等),30 ‰人工海水杀菌 20 min 后检疫池中饲养 2 ~ 3 d,使用前再用 30 ‰人工海水杀菌 20 min,然后用清洁淡水洗净,紫外杀菌处理后投喂。

1.5　驯饵方法

人工繁殖的 25 尾中华鲟在原试验基地饲养期间,进食人工颗粒饲料(山东升索饲料厂生产)比较稳定。在中华鲟进入驯养池后次日开始喂食,喂食方法有两种:固定食场喂食人工颗粒饲料;水下喂食天然饲料。在池底中间位置建立食场,每天 9:00 ~ 10:00 投喂人工配合饲料,15:00 ~ 16:00 由潜水员水下喂食天然饲料。人工颗粒饲料的投喂方式为,通过长 4 m,直径 5 cm PVC 管直接将饲料导入池底食场,这样可以最大限度地保证投喂位置的固定,同时减少饲料在水中的飘浮时间,最大限度降低饲料污染水质的几率。天然饲料的投喂方式为,潜水员水下喂食。潜水员在前期水下驯饵过程中,每天水下工作 3 ~ 4 h,针对每一尾中华鲟,逐渐接近,消除中华鲟的恐惧感,通过亲密接触,不断摸索每一尾中华鲟的行为和习性,在确切掌握中华鲟的行为后,开始水下直接喂食生鲜饲料。

1.6　行为观察方法

中华鲟进入驯养池后最初 72 h 连续观察其行为,包括游动方向(顺时针或逆时针)、游动水层(底层、中层或表层)、游动速度(绕行 1 周的路程/时间)、呼吸频率(指鳃盖在 1 分钟内张合次数,1 张 1 合记为 1 次)、尾部摆动频率 n(1 分钟内尾部左右来回摆动的次数)、游姿是否正常、吃食是否主动、有无应激反应(游速突然变化、皮肤肌肉紧张等),并进行记录,待其恢复正常后每 4 h 观察一次并进行记录。对其摄食前、摄食中和摄食后的行为进行重点观察和记录,并不定期测试其对潜水员和不同饲料的反应。

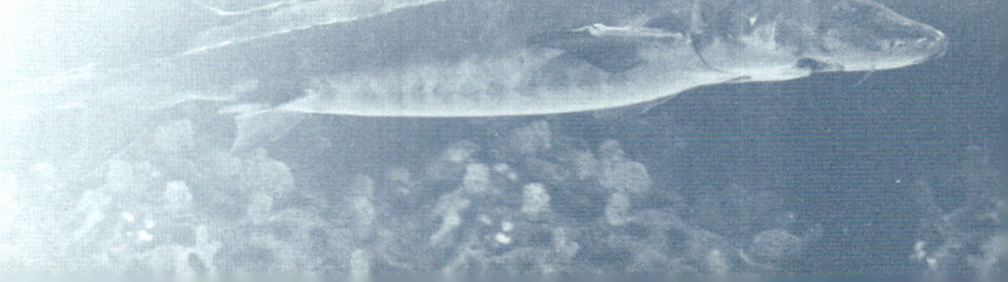

2 结果与分析

2.1 驯养期间水质变化

由于驯养池具有先进的生命支撑系统，从4月11日到7月31日，水温、DO，ORP，pH均未出现较大程度的波动；NH_4^+/NH_3 含量一直为0.01 mg/L，未出现波动；NO_2^- 含量也一直为0.01 mg/L，中间出现2次轻微的波动，波动值分别为0.04 mg/L和0.5 mg/L；大肠杆菌共测量22次，其中有13次测量结果为0；但总细菌含量变化较大，变幅346～7 000 cfu /ml，主要在3 000～5 000 cfu/ml范围内。其具体情况见表3。

表3 4月11日～7月31日水质监测结果

项目	测量次数	变化范围	平均值	项目	测量次数	变化范围	平均值
水温/℃	100	19.4～23.4	20.816	NH_4^+/NH_3/(mg/L)	82	0.01～0.01	0.01
DO/(mg/L)	104	7.6～9.5	9.000	NO_2^--N/(mg/L)	82	0.01～0.50	0.016
ORP	100	246～355	312.33	NO_3^--N/(mg/L)	3	0.12～26.40	10.893
pH	103	7.86～8.80	8.084	大肠杆菌/(cfu/ml)	22	0～570	68.318
Ca^{2+}/(mg/L)	2	40～60	50	总细菌/(cfu/ml)	32	346～7 000	3 794.41
KH	2	8.5～9.0	8.75				

2.2 驯饵结果

按照上述饲料处理及驯饵方法，综合考虑水质波动和营养调整等因素。8月底，26尾中华鲟全部开口进食，其中有15尾人工繁殖的中华鲟开口进食生鲜饲料，其余10尾仍按上述方法持续进行训练。25尾人工繁殖中华鲟4～8月进食饲料品种及数量见表4。野生中华鲟进入驯养池次日即摄食小鲫鱼1尾，以后随着时间的推移，其摄食量逐渐增大。在前1个月内平均每8 d的摄食量增长缓慢，均在500 g以下；从第2～3个月间平均每8 d的摄食量增长迅速，由最初500 g增加至2 000 g，之后摄食量增加趋缓；第4个月平均每8 d的摄食量维持在2 000 g左右。随着摄食量增加，摄食的饲料品种也越来越多(表5)。在此过程中，多数中华鲟进食的行为越来越主动，从潜水员下水到开始摄食的时间间隔越来越短，并能够主动接近潜水员。后期一些中华鲟还出现了投饵时间未到而吸食潜水员脚蹼或用头部撞人索食的现象。野生中华鲟的摄食过程逐渐呈现出一定的规律性，其进食过程一般分为2个时间段；2个时间段之间游离潜水员，每个时间段连续进食数次；每2次进食的时间间隔较短，并停留在潜水员附近；进食时间共为30 min左右，进食时吸食的力量非常大，游动有力，在吃食过程中喜欢贴沙层游动。

表4 25尾人工繁殖中华鲟各月摄食饲料数量

月份	鲫鱼/g	鲢鱼肉/g	鱿鱼/g	南美白对虾/g	小野鱼/g	饲料/g	总计/g
4	100	800	100	320	0	1 040	2 360
5	200	400	100	540	0	1 540	2 780
6	200	700	50	500	0	2 040	3 490
7	200	600	0	500	300	2 040	3 640
8	300	300	0	1 000	300	2 040	3 940

表 5 野生中华鲟摄食饲料品种随时间的变化

时间(月-日)	进入驯养池的时间/d	首次摄食的饲料	已摄食的饲料品种
04 -04	2	鲫鱼	鲫
04 -26	24	鲢鱼肉	鲫、鲢鱼肉
06 -22	81	南美白对虾	肉鲫、鲢鱼肉、南美白对虾肉
06 -23	82	牙鲆	鲫、鲢鱼肉、南美白对虾肉、牙鲆
07 -11	100	小黄鱼	鲫、鲢鱼肉、南美白对虾肉、牙鲆、小黄鱼
07 -21	110	大虾	鲫、鲢鱼肉、南美白对虾肉、牙鲆、小黄鱼、大虾

2.3 生长情况

驯养期间,不同规格的中华鲟均呈现出良好的生长态势。根据 2005 年 7 月 20 日、10 月 27 日和 2006 年 2 月 9 日对 25 尾人工繁殖中华鲟全长的测量表明,全长增长比率处于 0.86 %～22.77 %(图 1)。良好的生长表现说明水族馆驯养池环境适合中华鲟的生存。

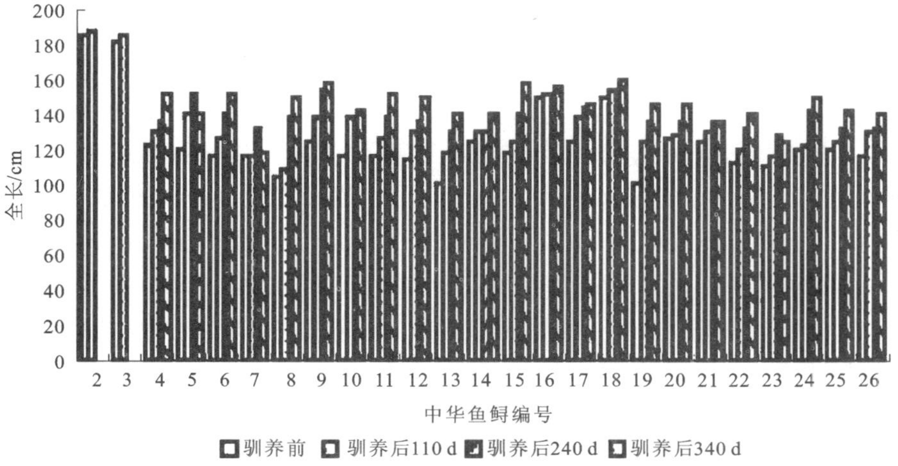

图 1 25 尾原人工繁殖中华鲟驯养前后生长情况对比

注:2 号中华鲟驯养后全长测量时间为 2005 年 6 月 24 日。

2.4 疾病诊治

主要包括外伤医治、水霉感染和消化道(鳔)炎症三类。

(1) 外伤医治。中华鲟进入驯养池初期频繁发生剐蹭造景的现象,据此采取了包裹管道等一些措施,并对池体局部进行了改造,有效地避免了剐蹭进一步发生,同时使用红霉素软膏对伤口处进行涂抹。目前基本上不再出现剐蹭造景造成外伤并感染的现象。

(2) 水霉感染。对第 3 批进入海洋馆后感染水霉的中华鲟 5 尾进行治疗后,除其中 1 尾因应激反应过于强烈死亡外,其余全部治愈。

(3) 消化道(鳔)炎症。其中一尾人工繁殖的全长 1.8 m 的中华鲟发生痉挛后,腹部胀气并向上翻倒,脊柱从腹鳍基部开始呈 S 形弯曲。采取人工排气等一些抢救措施后可正向游动,但游速缓慢,平衡能力较差。后期偶能进食,但不稳定。

2.5 行为观察

从整个驯养过程来看，中华鲟对潜水员的适应性逐渐增强。开始躲避潜水员，逐渐允许潜水员接近，到后来主动接近潜水员，再到吃食过程中允许潜水员抚摩，再到能够在不吃食的时候配合潜水员做一些动作，野生中华鲟对潜水员越来越亲近。从不同大小规格中华鲟对潜水员的反应来看，一般规格越大，适应性越强。

开始10 d内，野生中华鲟一般在展窗右侧或左侧的中下层水域做小范围环行运动。4月14～15日，运动范围逐渐向中上水层转移，环行运动的范围也越来越大。但从4月16日开始，又喜在底层角落处做小范围环行运动，偶尔也会在中上层水域运动，没有什么规律。从5月20日到7月19日，野生中华鲟一般在水体上层绕整个池体逆时针游动，并且游速较快，无其他明显特征。人工繁殖中华鲟的行为比较杂乱，未呈现出明显的规律性。

3 问题与分析

3.1 驯食

和野生中华鲟处于同一驯养池，人工养殖的不同年份出生的子一代中华鲟，除了摄食人工配合饲料外，绝大多数已开口摄食生鲜饲料。这说明人工养殖的中华鲟能较快地恢复摄食生鲜饲料的本能。野生中华鲟不但至今未摄食过人工配合饲料，而且其对人工配合饲料的反应也越来越冷淡。野生中华鲟究竟能不能摄食人工配合饲料的问题还有待进一步研究。

用活鲫鱼使中华鲟开口进食。进食状况趋于稳定后，使用冷冻鲫鱼喂食，中华鲟偶尔进食，大多时间吐出。后单独使用其余品种饵料进行投喂试验，进食不稳定，对其余品种饵料进食欲望差，多次试验多次吐出。野生中华鲟对饵料的种类变化反应极为敏感，其对活鲫鱼偏爱是个体原因，还是由于使用活鲫鱼首次投喂存储了记忆，对其余品种饵料的适应能力变差，或是野生中华鲟原本偏爱鲫鱼，原因待进一步试验求证。

3.2 水质波动

饲养早期池中硝化系统未完全建立，NO_2^- 不能被完全转化，产生小范围波动。待硝化系统建立完全后，NO_2^- 稳定于0.01 mg/L。池体硝化系统完全建立历时50 d。

中华鲟饲养池水质采用密闭式循环过滤，通过臭氧进行杀菌处理；同时，系统定期进行清洁，换水。然而随着中华鲟投喂量逐渐加大，系统负载逐渐上升，再加上池体生化系统尚未成熟，系统运作尚未达到最佳值，导致细菌含量出现波动。

通过降低日饵料投喂量，加强水下清洁，加强臭氧工作效率等措施，细菌含量下降，趋于稳定。

3.3 中华鲟伤病

池中造景为人工珊瑚礁，主要材质为水泥，礁石边缘粗糙，由于材质颗粒粗，较为锋利，中华鲟游动时不小心碰到或刮擦，即可造成表皮受伤。鉴于中华鲟刚进入时绕小圈游动频繁（圈体大小类似于原长江所暂养池），后逐渐加大游动空间，可满池游动，同时满池游动率日渐提高，受伤几率下降。分析其原因为，中华鲟习惯了在无障碍暂养池游动，对于海洋馆中鲟鱼池环境不熟悉。随着时间的推移，加上同时对造景锋利部分进行包覆，最终撞击及受伤现象基本消失了。

为了增强鱼体抗病能力,在所有中华鲟的生鲜饲料中周期性添加了药物。目前主要添加维生素等,但由于周期较短目前还无明显效果。抗病药物的选择和如何合理使用的问题还需要进一步探讨。

参考文献

[1] Billard R, Lecointre G. Biology and conservation of sturgeon and padd lefish [J]. Revi Fish Biol Fish, 2000, 10(4): 355-392.

[2] Wei Qiwei, Ke F, Zhang J, et al. Biology, fisheries, and conservation of sturgeons and paddlefish in China [J]. Environ Biol Fish, 1997, 48(1/4): 241-255.

[3] 柯福恩. 论中华鲟的保护与开发[J]. 淡水渔业, 1999, 29(9): 4-7.

[4] 陈凌云, 章秋虎, 项卫民, 等. 匙吻鲟孵化驯养与无公害网箱养殖新技术[J]. 渔业致富指南, 2005(7): 33-34.

[5] 罗晓松, 曾令兵. 工厂化驯养鲟鱼病害防治技术[J]. 科学养鱼, 2000(11): 32-33.

[6] 危起伟. 长江中华鲟繁殖行为生态学与资源评估[D]. 武汉: 中国科学院水生生物研究所, 2003: 23-112.

[7] 张彤晴, 葛家春. 小体鲟幼鱼水库坝下流水池塘驯养技术[J]. 江苏农业科学, 2002(5): 59-61.

[8] 颜远义. 中华鲟生物学特性及养殖方法[J]. 水产科技, 2003(5): 14-16.

[9] 刘勇. 中华鲟人工繁殖技术研究的现状和展望[J]. 水利渔业, 1988(4): 20-25.

Research on Domestication of Chinese Sturgeon (*Acipenser sinensis*) in an Aquarium

Lin Jianyi[1,3] Zhang Xiaoyan[2] Wei Qiwei[1,3,4] Zhang Hui[4] Yang Daoming[2] Zhu Yongjiu[1,3] Chen Xihua[1,3] Yang Deguo[1,3] Zheng Yueping[4]

(1. Key Lab of Freshwater Fish Germplasm Resources and Biotechnology, Ministry of Agriculture; Yangtze River Fisheries Research In stitute, Chinese Academy of Fisheries Sciences, Jingzhou, Hubei, 434000;
2. Beijing Aquarium, Beijing, 100081;
3. Freshwater Fisheries Research Center, Chinese Academy of Fisheries Science, Wuxi, Jiangsu, 214081;
4. Fishery College of Huazhong Agricultural University, Wuhan, 430070)

Abstract 25 artificially propagated Chinese sturgeons (*Acipenser sinensis*) of and 1 wild Chinese sturgeon caught in Yichang section of Yangtze River were used to carried on are search about domesticating method in the aquarium. The results showed: all 26 Chinese sturgeons had begun to take food, 15 artificially propagated Chinese sturgeons that originally eat artificialy mixed feed began to eat the fresh fodder, and the variety of the fresh fodder was getting more and more; from April to August, the quantity of feed taken by 25 artificially propagated Chinese sturgeons increased steadily, and these 25 sturgeons presented a good growth appearance; the wild Chinese sturgeon could eat as more as 6 kinds of baits by the time, and the quantity of the bait eaten every day increased from 120 g to 2 000 g gradually, and became steady. It indicated that the aquarium was suitable for the Chinese sturgeon to live.

Key words *Acipenser sinensis* Domestication Aquarium

中华鲟全人工繁殖技术研究*

危起伟[1,5,6,**]　李罗新[1,5]　杜　浩[1,5,6]　张晓雁[2]　熊　伟[1]
张　辉[1,5]　沈　丽[1,5]　吴金明[1]　张书环[1]　王成友[1,5]　李创举[1,5]
柴　毅[3]　李奕慰[4]　乔新美[1]　刘志刚[1]　高宇鹏[1]　甘　芳[1]

(1. 农业部淡水生物多样性保护重点实验室 中国水产科学研究院长江水产研究所　湖北武汉　430223
2. 北京海洋馆　北京　100081　　3. 长江大学动物科学学院　湖北荆州　434023
4. 厦门松浩实业有限公司　福建厦门　361100
5. 中国水产科学研究院淡水渔业研究中心　江苏无锡　214081
6. 华中农业大学水产学院　湖北武汉　430070)

摘　要　中华鲟(*Acipenser sinensis*)在人工驯养条件下性腺发育并最终成熟是突破中华鲟全人工繁殖技术的关键环节,有效的养殖模式对促进中华鲟性腺发育成熟或提前成熟,以及实现雌雄同步成熟具有非常重要的作用。本研究在不同养殖模式(水温和营养调控)下对中华鲟性腺发育状况进行长期观测。结果显示,中华鲟在仿自然变温或恒温的养殖环境中性腺均可发育至III期。采用配合饲料、饲料中添加鲜活鱼或冰鲜鱼以及仅用鲜活鱼或冰鲜鱼的喂养方式,均能够使部分中华鲟性腺发育启动;但添加鲜活鱼或冰鲜鱼组性腺发育启动的比例较高,其中有3尾中华鲟性腺达IV期。成功对性腺发育至IV期的1尾雌鲟(12龄,体质量57 kg)和1尾雄鲟(14龄,体质量64 kg)实施了人工催产和授精,采获卵6.25万粒,精液2 850 ml,受精率为60.1%(36.3%~80.4%)。在(19.3±0.2)℃下经过约125 h仔鱼大量孵出,获初孵仔鱼2.3万余尾。此次中华鲟全人工繁殖技术研究的有效尝试可为今后养殖中华鲟的规模化繁育提供借鉴。

关键词　中华鲟　水温　饵料　生态调控　人工繁殖　性腺　发育

中华鲟(*Acipenser sinensis*)为大型溯河产卵洄游性鱼类,是国家一级重点保护动物。由于涉水工程建设、捕捞、污染及航运等人类活动的干扰,其赖以生存的栖息生境退化或功能丧失,导致其自然种群的严重衰退[1-2]。目前中华鲟自然产卵场仅发现有葛洲坝下一处,且适宜其繁殖和栖息的有效面积非常有限[3],自然繁殖效果和物种自我维持的境况不佳,自然种群有灭绝的危险。建立中华鲟人工养殖群体,实现全人工繁殖是保存中华鲟物种、实施易地保护和自然资源增殖不可或缺的途径。

我国早在1976年就通过在江边栓养的方式实现了野生中华鲟人工繁殖的成功[4],葛洲坝截流后,1983年又实现了坝下中华鲟人工繁殖[5]。之后国家采取了捕捞野生亲鲟实施人工繁殖及增殖放流的方式,开展了中华鲟的资源养护工作。20世纪90年代初仅进行了人工种群的零星蓄养,直到1997年中华鲟苗种规模化培育技术取得突破以后,人工种群蓄养才形成一定的规模。由于中华鲟寿命长,性成熟晚,体型大,且具有在海水中生长,淡水中繁殖的生活习

*基金项目:公益性(农业)行业科技专项(201203086);国家自然科学基金(31172413);中央科研院所科研基本业务费(2008jbfb01);农业财政项目(1251216200501)等。

**通讯作者(Correspording author),研究员,博士,主要从事濒危水生动物保护研究工作,E-mail:weiqw@ yfi. ac. cn

性,人工养殖中华鲟达到性成熟所需的时间、人力、物力投入巨大,导致中华鲟全人工繁殖进展缓慢。其中,养殖中华鲟性腺发育往往仅停留在II期,很难向III或IV过渡。对于雌性而言,初级卵母细胞难以进入大生长期,养殖中华鲟的性成熟问题成为其全人工繁殖的主要障碍[6]。这也是欧洲鲟(*A. sturio*)[7]、高首鲟(*A. transmontanus*)[8]和湖鲟(*A. fulvescens*)[9]等养殖鲟类繁殖过程中遇到的普遍问题。养殖环境不利于刺激内分泌系统发挥功能,从而不能促进性腺启动发育或发育成熟[10]。因此,探索适合的养殖模式以促进养殖中华鲟的性腺发育成熟或提前成熟,以及实现雌雄同步成熟是突破中华鲟全人工繁殖技术的关键。郭柏福等[11]初步报道了中华鲟全人工繁殖的结果,但未涉及中华鲟性腺发育过程及养殖模式的详细报道,且所获仔鱼成活率较低,产后亲鲟死亡。因此,中华鲟全人工繁殖技术尚需要进一步研究。

本研究通过对不同养殖模式下中华鲟性腺发育状况的连续监测,观测到2尾雌鲟和5尾雄鲟性腺由II期发育至III期或IV期。并成功对2尾性腺发育至IV期的中华鲟实施产前生态调控,实现了雌雄成熟同步,取得全人工繁殖成功。本文对人工驯养中华鲟养殖模式、性腺发育进程及全人工繁殖技术进行系统总结,以期为今后养殖中华鲟的规模化繁育提供借鉴。

1 材料与方法

1.1 实验养殖池及实验鱼

实验鱼饲养在4个不同养殖基地,即荆州中华鲟保育和增殖放流中心(以下简称"荆州基地",养殖池规格:直径16 m,水深2.5 m圆形流水水泥池4口)、福建厦门中华鲟繁育保护基地(以下简称"厦门基地",养殖池规格:面积600 m^2,水深1.5 m矩形流水水泥池4口)、湖北宜昌秭归三峡库区网箱(以下简称"三峡网箱",网箱规格为10 m×10 m×8 m深网箱4口)和北京海洋馆(养殖池规格长29.0 m,宽11.0 m,水深4.4 m,全循环水恒温系统)[12]。

实验用鱼共250尾,来自中国水产科学研究院长江水产研究所通过野生中华鲟人工繁殖的后代(子一代),繁殖分别在1997年、1998年、1999年和2000年秋季进行。至2012年,年龄为12~15龄。2012年期间对这250尾实验鱼的测量数据见表1,即平均体长212 cm(范围175~286 cm),平均体重58.8 kg(范围34.6~125.3 kg)。其中,厦门基地和荆州基地养殖中华鲟子一代群体是自仔鱼阶段开始饲养。北京海洋馆和三峡网箱的中华鲟是在荆州基地养殖至6~8龄后转入。实验鱼采用PIT进行标记识别,部分采用锚标标志识别。

表1 实验用养殖中华鲟的生物学特征

Tab.1 Biological characteristics of cultured *Acipenser sinensis* in the study

养殖场所 Farm	数量 Number	平均全长(范围)/cm Average total length(range)	平均体长(范围)/cm Average body length(range)	平均体重(范围)/kg Average body weight(range)
荆州基地 Jingzhou Farm	44	212(182~249)	174(145~215)	55.1(34.6~93.2)
三峡网箱 Three Gorges Reservoir	16	196(175~213)	163(145~173)	43.9(34.8~60.5)
北京海洋馆 Beijing Aquarium	10	230(199~286)	196(171~240)	78.7(42.4~125.3)
厦门基地 Xiamen Farm	180	211(182~230)	174(145~195)	57.4(40.8~73.6)

1.2 不同养殖模式下性腺发育监测

根据中华鲟饲养场所的设施条件,在不同养殖基地开展了不同的养殖模式下中华鲟性腺发育对比观测试验,其中养殖模式主要设置水温节律调控及营养调控。水温节律控制试验主要是根据养殖场所的水源、控温设施及控温能力设置。共设置4种养殖水温节律:荆州基地采用地下水和湖水混合方式进行适度控温养殖,水温7月份最高,月均温度(28±1.5)℃;1月份最低,月均温度(7±2.0)℃(图1)。厦门基地采用水库水养殖,水温8月份最高,月均温度(27±3.0)℃;1月份最低,月均温度(14±2.0)℃。三峡网箱水温同三峡库区,水温8月份最高,月均温度(24±0.5)℃;2月份最低,月均温度(10.5±0.5)℃。北京海洋馆采用循环水控温养殖,水温恒定在(21.0±0.5)℃。另外,三峡网箱水体交换与水库水位及水库内的流速有关。其他各养殖基地水体交换量约24 h交换一次。各养殖基地实验池中由进排水系统形成的微流水养殖,流速介于0~0.5 m/s。

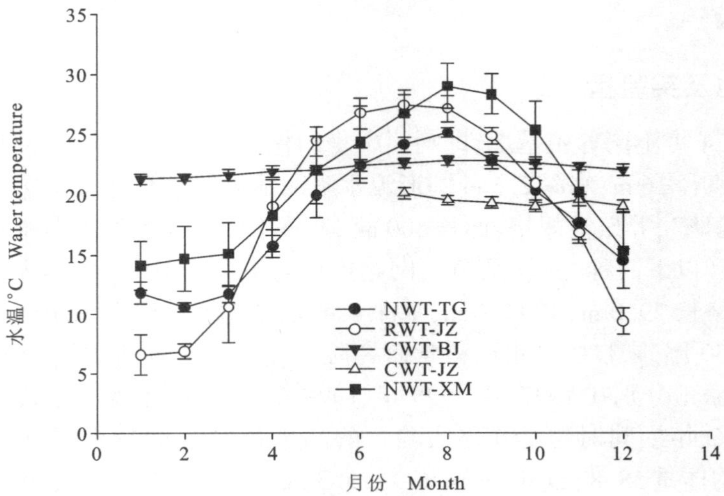

图1 各养殖基地水温周年变化节律图(月平均值±标准差)

NWT-XM:天然水温(厦门基地);NWT-TG:天然水温(三峡网箱);CWT-BJ:恒温(21±0.5)℃(北京海洋馆);RWT-JZ:调控水温(荆州基地);CWT-JZ:恒温(荆州基地)

Fig.1 Water temperature profile at different sturgeon farms(the monthly mean value ± standard deviatio)

NWT-XM:Natural water temperature(Xiamen Farm);NWT-TG:Natural water temperature(Three Gorges Reservoir);CWT-BJ:Constant water temperature(21 ± 0.5)℃(Beijing Aquarium);RWT-JZ:Regulated water temperature(Jingzhou Farm);CWT-JZ:Constant water temperature(Jingzhou Farm)

各养殖基地饵料来源及投喂方式:

(1)荆州基地中华鲟10龄前采用人工配合饲料投喂,日投喂0.5%~0.7%。人工配合饲料各成分如下:粗蛋白质43%,粗脂肪3%,粗纤维2%,粗灰分18%,钙5%,磷3%,赖氨酸2%,NaCl 3%,H_2O 10%。10龄后采用冰鲜蓝点马鲛(*Scomberomorus niphonius*)与多春鱼(*Mallotus villosus*),日投喂0.5%~1.0%。

(2)厦门基地采用两种饵料喂养方式,一种仅采用人工配合饲料投喂,配方同前,90尾;另一种在2009年以前采用人工配合饲料投喂,配方同前,日投喂量0.4%~0.7%;2009年以后采用冰鲜多春鱼,日投喂量0.5%-0.7%,90尾。

(3)三峡网箱采用两种养殖模式,一种仅投喂人工配合饲料,配方同前,日投喂量0.8%~1.0%,共10尾;另一种除投喂人工配合饲料,外还投喂鲜活饵料银鱼(*Hemisalanx brachyrostralis*)、鲫鱼(*Carassius auratus*),虾(*Macrobranchium nipponense*),及野杂鱼等;人工饲料日投喂量0.4%~0.8%,鲜活饵料日投喂量0.1%~0.5%,共6尾。

(4)北京海洋馆采用鲜活鲫鱼、乌贼(*Sepia officinalis*)、海蟹(*Portunus trituberculatus*)、大陇六线鱼(*Hexagrammos otakii*)及南美白对虾(*Penaeus vannamei*)和冰鲜蓝点马鲛与多春鱼,日投喂量0.4%~0.6%。

采用外科手术、腹腔穿刺获取中华鲟性腺[13],参考陈细华对雌雄鲟鱼性腺发育进行鉴别和分期[6]。同时还采用B超、内窥镜、血液生化及激素指标测量等方法进行了辅助判别[14]。

1.3 产前中华鲟雌雄同步发育监控实验

产前中华鲟雌雄同步发育调控实验在长江水产研究所荆州基地完成。该基地建有大型中华鲟养殖生态调控系统(32 m×12 m×3.8 m循环水池2口),采用YSI6820水质监测仪24小时连续监测水温、pH、溶氧和氨氮等水质指标。雌雄同步发育监控方法:将性腺发育至IV的雌鱼(♀2-IV)与雄鱼(♂3-IV)分别于2012年7月和9月转入生态调控池,水温控制在19.0~19.5 ℃,溶氧6.0~8.4 mg/L,流速0.2~0.5 m/s。对雌雄鱼的性腺发育程度进行连续监测,主要采用外科手术、腹腔穿刺获取性腺,计算核偏位值(polarisation index, PI)确定卵的成熟度[15]。还采用B超、内窥镜、血液生化及激素指标测量等方法进行了辅助判别。

1.4 中华鲟全人工繁殖及苗种培育

以LHRH-A_2为催产剂,雄鲟采用1针注射,注射剂量4 μg/kg体重;雌鲟采用2针注射进行催产[16],注射剂量10 μg/kg体重,14 h后注射第二针,注射剂量为第一针剂量的10%。效应期后,精卵均采用腹部节律按摩法活体采集。精液分批次采集后充氧并于2~4 ℃储存,定期检测精液质量,统计精子剧烈运动时间、快速运动时间和30%死亡时间。卵子分批次采集后静置约2 h进行人工授精,采用黄泥浆脱粘,后置于孵化框中孵化,孵化水温控制在19.0~19.6 ℃。原肠中期(约24 h)后统计受精率,对胚胎发育进行观察,仔鱼孵出后转入玻璃钢桶(直径1 m,高60 cm)中培育,8~11日龄逐渐投喂切碎的水蚯蚓进行开口驯化实验。

1.5 数据统计分析

依据PIT标记或锚标标记编号详细跟踪记录性腺发育启动(性腺发育由II期向III或IV过渡)的中华鲟养殖背景,包括养殖场地,喂养饲料种类,性腺发育状况,监测指标等,以进行分析对比。

参考刘鉴毅等对中华鲟精液和卵子质量进行测量[17]。采用体积法估算产卵量。胚胎发育至原肠中期后分批次统计受精率,每批次抽取200颗以上胚胎,正常发育胚胎占抽样胚胎比例计为受精率。孵化期内在解剖镜下观察中华鲟子二代的胚胎发育全过程,发育前期每0.5 h观察一次,到达囊胚期后1.5~3 h观察一次。为了达到较好的观察和拍摄效果,需要剥离卵膜,现场在解剖镜下拍摄图片。用Excel和SigmaPlot 12.0软件统计分析和作图。

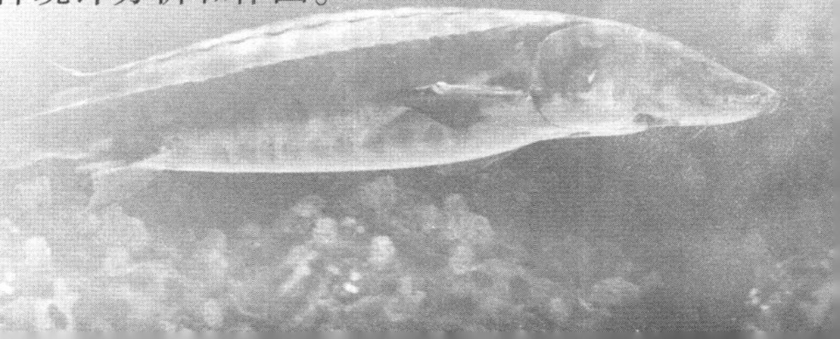

2 结果

2.1 不同养殖模式下中华鲟性腺发育情况

自2010年至2012年,在250尾12~15龄中华鲟中共监测到性腺发育至III期及以上的雄鱼5尾,雌鱼2尾,其余性腺发育均处于II期。7尾性腺发育启动中华鲟的体重范围是52.0~111.2 kg,全长范围185~247 cm,体长范围152~200 cm(表2)。对这7尾性腺发育启动的中华鲟进行饵料来源分析,可知不同的饵料组性腺发育情况有明显不同(表2~表4)。仅采用全人工配合饲料投喂的中华鲟,仅有1尾雄鱼(厦门基地)达到III成熟,其他2尾性腺发育至III和3尾性腺发育至IV的鱼后期都投喂了鲜活鱼和冰鲜鱼(表3)。而且,后期全部投喂鲜活饵料(北京海洋馆)的中华鲟性腺发育启动的比例明显高于同时投喂人工饲料和鲜活鱼组(荆州基地和三峡网箱)(表3,表4)。这表明后期采用鲜活鱼和冰鲜鱼的投喂方式确实促进了中华鲟性腺发育的启动。

性腺发育进入III期的鲟鱼摄食状况未有明显改变,而进入IV期后的3尾鱼中,有2尾雄鱼摄食量明显减少,1尾雌鱼停止摄食(表3)。

这7尾性腺从III期发育到IV期的中华鲟有2尾来自北京海洋馆,饲养环境水温恒定在(21±0.5)℃;5尾处于季节性变动水温中,其中有1尾雌鲟(♀2-IV)来自三峡网箱,养殖期间水温变幅10~24 ℃;有2尾来自厦门基地,水温变幅14~28 ℃;有2尾来自荆州基地,水温变幅6~26 ℃(图1,表3)。

表2 性腺发育至III和IV期的中华鲟生物学指标

Tab. 2 Biological characteristics of cultured *Acipenser sinensis* with gonad maturity at stage III or IV

编号 Fish ID	全长/cm Total length	体长/cm Body length	体重/kg Body weight	性腺发育期 Gonad maturity stage	监测年龄/a Age at monitoring	监测月份 Monitoring month	养殖地点 Farm
♂1-III	195	165	55.8	III	12	2012.04	厦门基地 Xiamen Farm
♂2-III	187	152	52.0	III	12	2012.03	北京海洋馆 Beijing Aquarium
♂3-IV*	215	181	64.0	IV	14	2012.03	北京海洋馆 Beijing Aquarium
♂4-IV*	185	159	52.0	IV	11	2012.08	荆州基地 Jingzhou Farm
♂5-IV	192	168	50.7	IV	10	2010.08	厦门基地 Xiamen Farm
♀1-III	247	200	111.2	III	14	2012.08	荆州基地 Jingzhou Farm
♀2-IV*	215	180	57.0	IV	12	2012.08	三峡网箱 Three Gorges Reservoir

*表示该鱼参与了全人工繁殖。

*The sturgeon being used to the controlled reproduction.

表 3 中华鲟不同性腺发育阶段的食物和养殖水温

Tab. 3 Diets and water temperature during the gonad development of cultured *Acipenser sinensis*

编号 Fish ID	性腺发育期 Gonad maturity stage				摄食状况 Ingestion
	II	II ~ III	III	IV	
♂1-III	FF/NWT-XM	FF/NWT-XM	FF/NWT-XM	–/–	摄食正常 Normally feed
♂2-III	FF/RWT-JZ	LF + IF/CWT-BJ	IF/CWT-BJ	–/–	摄食正常 Normally feed
♂3-IV	FF/RWT-JZ	LF + IF/CWT-BJ	IF/CWT-BJ	IF/CWT-BJ	摄食明显减少 Obviously reduce food intake
♂4-IV	FF/RWT-JZ	IF/RWT-JZ	IF/RWT-JZ	IF/CWT-JZ	摄食明显减少 Obviously reduce food intake
♂5-IV	FF/NWT-XM	FF/NWT-XM	FF/NWT-XM	FF/NWT-XM	未观察 Not observed
♀1-III	FF/RWT-JZ	IF/RWT-JZ	IF/RWT-JZ	-/-	摄食正常 Normally feed
♀2-IV	FF/RWT-JZ	LF + FF/NWT-TG	IF/RWT-JZ	-/CWT-JZ	不摄食 Take no food

注：FF 为配合饲料（含 60 %鱼粉，23 %预糊化淀粉）；LF 为鲜活鱼；IF 为冰鲜鱼；NWT-XM 为天然水温（厦门基地）；NWT-TG 为天然水温（三峡网箱）；CWT-BJ 为恒温（21 ±0.5）℃（北京海洋馆）；RWT-JZ 为调控水温（荆州基地）；CWT-JZ 为恒温（荆州基地）。

Note: FF represents formula feed with 60 % fishmeal and 23 % pregelatinizedstrach; LF represents live fish; IF represents iced fish; NWT-XM represents natural water temperature (Xiamen Farm); NWT-TG represents natural water temperature (Three Gorges Reservoir); CWT-BJ represents constantwater temperature (21 ± 0.5) ℃ (Beijing Aquarium); RWT-JZ represents regulated water temperature (Jingzhou Farm); CWT-JZ represents constantwater temperature (Jingzhou Farm).

表 4 不同饵料喂养下的性腺发育至 III 或 IV 期的中华鲟比例

Tab. 4 Percentage of *Acipenser sinensis* with gonad developed to III or IV maturity stage in different types of diets

养殖基地 Farm	实验个体数量/ind Number of treated sturgeon	饲料 Diet	性腺发育至 III 或 IV 期中华鲟数量/ind. Number of sturgeon with III or IV gonad	比例/% Percentage
荆州基地 Jingzhou Farm	44	FF + IF	2	4.5
三峡网箱 Three Gorges Reservoir	6	FF + LF	1	16.6
	10	FF	0	0
厦门基地 Xiamen Farm	90	FF + LF	1	1.1
	90	FF	1	1.1
北京海洋馆 Beijing Aquarium	10	LF + IF	2	20

注：FF 为配合饲料（含 60 %鱼粉，23 %预糊化淀粉）；LF 为鲜活鱼；IF 为冰鲜鱼。

Note: FF represents formula feed with 60 % fishmeal and 23 % pregelatinizedstrach; LF represents live fish; IF represents iced fish.

2.2 中华鲟雌雄同步发育监测结果

催产前 4 个月，雌鲟（♀2-IV）被转入生态调控池中，通过穿刺采卵后计算核偏位值。在入池的 4 个月中核偏位值由 0.20 逐渐下降至 0.05，达到可催产的极化程度 0.05（图 2）。雄鱼在北京海洋馆七月份检查时性腺乳白色，质地较密，判定精巢发育至 IV 初（IV-a）。于产前一个月（9 月 26 日）由北京海洋馆转运至荆州基地生态调控池中，至 10 月产前检查时，轻压腹部有大量精液溢出，判断精巢发育至 IV 末（IV-c）（图 2）。

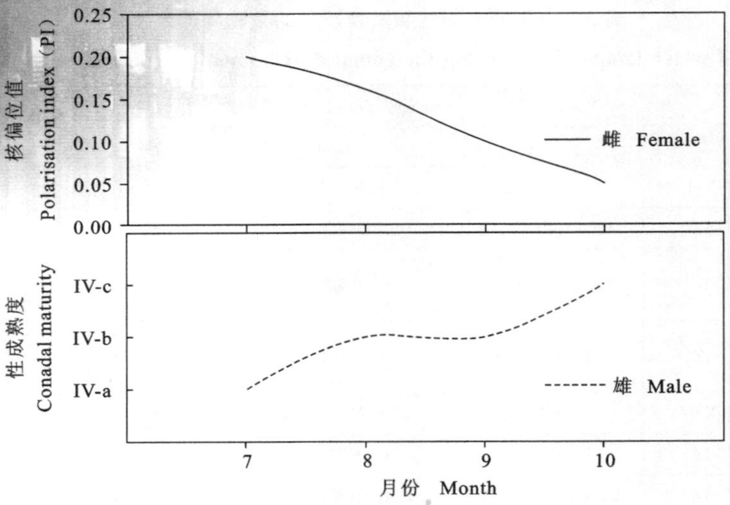

图2 参与全人工繁殖的雌、雄中华鲟在产前的成熟度变化情况

Fig. 2 Gonad maturity staging in a male and a female *Acipenser sinensis* used to the controlled reproduction

2.3 中华鲟全人工繁殖效果

在性腺发育监测和室内生态调控基础上,2012 年 10 月 28 ~ 30 日成功使 1 尾雌鲟(12 龄)和 1 尾雄鱼(14 龄)达到同步成熟,并实施人工催产授精。雌鲟在 LHRH-A_2注射后的 15 h 开始排卵,采用腹部节律按摩法一次性收集卵子。雌鲟产前体重 57 kg,产卵 1 天后体重下降至 51 kg,产空率约 95 %。产卵量 62 500 粒,平均卵径 3.71 mm(范围 3.30 ~ 4.44 mm),平均单卵重 0.033 g,估算产卵总重量 2.06 kg,约占体重的 3.6 %。雄鱼在 LHRH-A2 注射 12 h 后分5 批次采集精液,合计采精量 2 850 ml。雄鲟产前体重 64 kg,产后体重 58 kg,下降约 6 kg。所采集的精液剧烈运动时间平均达 36 s,寿命超过 4 min 以上(表 5)。雌雄亲鱼健康存活,产后第 10 天开始摄食。

表 5 中华鲟人工催产及繁殖效果

Tab. 4 Results of controlled reproduction of cultured *Acipenser sinensis*

项目 Items		平均值 Mean	范围 Range
卵径/mm	Egg diameter	3.71	3.30 ~ 4.44
卵粒重量/g	Egg weight	0.033	0.039 ~ 0.0341
产卵量/ind	Egg quantity	62 500	—
精子剧烈运动时间/s	Swirling movement time of sperm	36	20 ~ 55
精子快速运动时间/s	Fast movement time of sperm	78	50 ~ 89
精子寿命/min	Sperm life	183	165 ~ 260
采精量/ml	Milt volume	2 850	—
催产水温/℃	Inducing water temperature	19.3	19.0 ~ 19.5
效应时间(雌)/h	Effective time for female	15	—
效应时间(雄)/h	Effective time for male	12	—

续表

项目 Items		平均值 mean	范围 range
受精率/%	Fertilized rate	60.1	36.3 ~ 80.4
孵化率/%	Hatching rate	36.8	—
出苗量/ $\times 10^4$ ind	Quantity of hatching-out larvae	2.3	—
开口摄食仔鱼量/ $\times 10^4$ ind	Quantity of initial-feeding larvae	1.8	—

2012 年 10 月 30 日选择精子活力最好批次的精液(保存时间 30 ~ 36 h)进行人工授精,受精率统计结果显示,不同批次采集的卵子受精率不同,最先采集的卵子受精率达 80.4 %,部分批次采集卵子的受精率仅有 36.3 %,平均受精率为 60.1 %。受精后 2.5 h 发育至 4 细胞期,11 h到达囊胚期,18 h 到达原肠中期,46 h 神经管闭合,64 h 长心管期,80 h 尾部达心脏,125 h 左右大量出苗(表 6 和图 3)。获初孵仔鱼约 2.3 万尾,初孵仔鱼全长 9.6 ~ 10.2 mm。11 月 8 ~ 14 日,约有 1.8万尾仔鱼顺利开口摄食。

表 6　中华鲟子代胚胎发育典型特征

Tab. 6　Typical characteristics of embryonic development in the offspring of cultured *Acipenser sinensis*

胚胎发育时期 Embryonicdevelopment stage	主要特征 Typical characteristics	受精后时间/h Hours after fertilization
未受精卵 Unfertilized egg	球形、黑色、不透明,卵膜紧贴于卵的表面	*
刚受精的卵 Fertilized eggs	动物极中央出现由暗色色素环包围的明亮极性斑	0
4 细胞期 4-cell stage	第二次分裂,为经裂,与第一次分裂相垂直,将动物极分为 4 个几乎相同的分裂球	2.5
多细胞期 Morula stage	细胞变多、变小,形成多细胞胚体,植物极被完全分裂	8
囊胚中期 Midblastula stage	动物极细胞变多变小,且分裂不同步,细胞间有明显间隙	11
原肠早期 Early gastrula stage	赤道附近有一深色的色素带出现	18
大卵黄栓期 Big yolk plug stage	大卵黄栓的形成	28
隙状胚孔期 Lyriform blastopore formation	卵黄栓消失,胚孔两侧唇靠近,呈隙状	37
宽神经板期 Wide neural plate formation	神经板明显增厚,且分为内外两部分,未来中脑处的神经板最厚	40
神经管闭合期 Neural tube closing stage	神经褶闭合成神经管,排泄系统原基显著加长	46
眼囊形成期 Eye sac formation	眼原基形成,在中脑两侧可见上突呈弧形的第一对咽弧的原基	48
长管心脏期 Long tubular heart formation	心脏呈长管状,视泡明显	64
尾达心脏期 The tail touches the heart	头抬起,尾的末端到达心脏,并可做大幅度左右摇摆	80
尾达头部期 The tail touches the head	尾的末端接触头部,尾部鳍褶变宽	90
出膜期 Hatching stage	尾的末端略过头部,卵黄囊上血管明显,仔鱼大量出膜	125

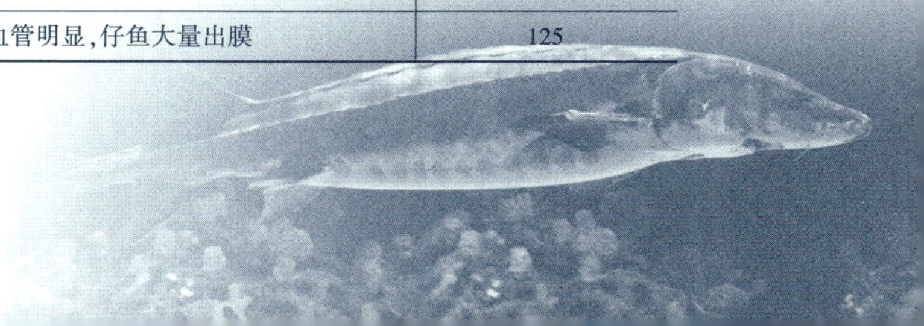

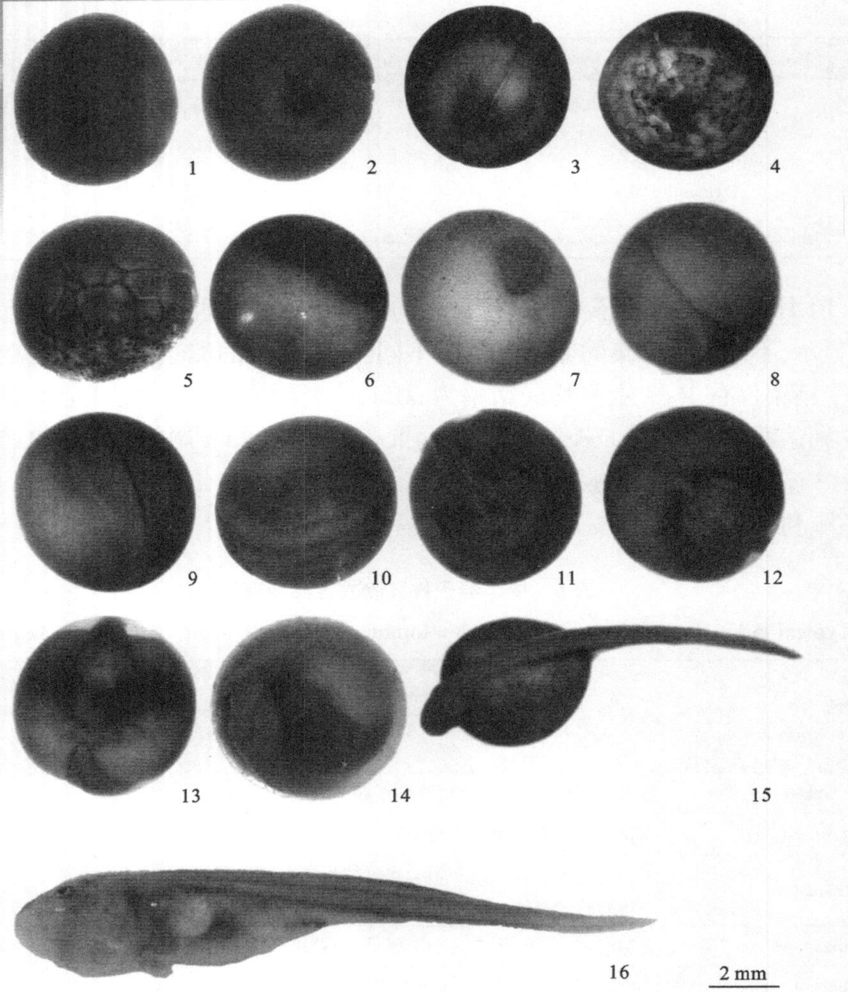

图 3　中华鲟子二代胚胎发育照片

1. 未受精卵;2. 刚受精的卵;3. 4 细胞期;4. 多细胞期;5. 囊胚中期;6. 原肠早期;7. 大卵黄栓期;8. 隙状胚孔期;9. 宽神经板期;10. 神经管闭合期;11. 眼囊形成期;12. 长管心脏期;13. 尾达心脏期;14. 尾达头部期;15. 出膜期;16. 初次摄食仔鱼

Fig. 3　Thephotos of embryonic development stages of the F2-generation Chinese sturgeon

1. Unfertilized egg;2. Fertilized eggs;3. 4-cell stage;4. Morula stage;5. Midblastula stage;6. Early gastrula stage;7. Big yolk plug stage;8. Lyriform blastopore formation;9. Wide neural plate formation;10. Neural tube closing stage;11. Eye sac formation;12. Long tubular heart formation;13. The tail touches the heart;14. The tail touches the head;15. Hatching stage;16. The initial feeding larva.

3　讨论

3.1　中华鲟性腺发育成熟适宜养殖条件和营养需求

鱼类在长期进化过程中形成的对生态环境因子的选择和适应是种群延续的重要基础。在鱼类繁殖阶段,生态环境因子在诱导和启动鱼类内分泌系统,进而通过下丘脑-垂体-性腺作用轴促进性腺发育、成熟、排卵过程中发挥着重要的功能[10]。常规的养殖环境由于空间的局限性和环境条件的单一性往往不利于启动与性腺发育相关的生殖内分泌调节途径[18],因而,对繁殖相关的生态环境需求研究不仅在自然资源保护中发挥着重要作用,而且也是实现养殖群体成功繁殖的有效途径。基于对自然繁殖生态需求的研究,短吻鲟(*A. brevirostrum*)[19]、中吻鲟(*A. medirostris*)[20]等已实现了人工模拟环境下的自然产卵,对于珍稀濒危物种的保护有重要意义。

现有研究表明,水温、光周期及食物等是鲟类性腺发育及成熟的主要影响因素[10],也是后备亲鲟养殖管理的重要内容。本研究发现养殖中华鲟在仿自然变温或恒温环境中性腺均可以启动发育或发育至成熟,表明在适合中华鲟生长范围内,水温高低及其变化与否可能并不是中华鲟性腺发育启动的决定性因素。但在不同的水温节律下,中华鲟的生长状况不同,也可能最终导致性腺发育速度及成熟度存在差异。进入 III 期后,鲟鱼需要特殊的水温变化以改善最终成熟度及性产物质量,这对于实施繁殖和鱼子酱生产都至关重要[20]。

Doroshov 等[21]研究表明,高首鲟性成熟为体重依赖型,其他鲟类也有类似的情况。目前,发现人工养殖中华鲟子一代性成熟雌性个体比自然种群明显偏小,约为长江野生雌性最小成熟个体体重的 40 %~50 %,这不利于中华鲟的保护及规模化繁育。其原因,可能与养殖空间尺度有关:养殖空间不足,包括养殖池塘过浅,转弯半径过小,影响其正常行动,导致生长迟缓和鱼体多擦伤,鱼体长期处于空间胁迫状态,因"长不大"而影响其性成熟。中华鲟雌性平均或最小性成熟个体在 27 种鲟类中为最大[22,23],国内高龄中华鲟子一代养殖池水体普遍偏小,这可能是导致人工养殖中华鲟性成熟晚的原因之一,值得深入研究。

在食物供给方面,本实验采用投喂配合饲料和添加鲜活鱼或冰鲜鱼的方式,均能够使部分中华鲟性腺发育达到最终成熟,而添加鲜活鱼或冰鲜鱼的养殖组性腺发育启动的比例明显要高于仅投喂配合饲料组。研究表明,鲜活或冰鲜鱼类或虾类体中的营养物质在体内具有更高的吸收转化效率,因此,它们更有利于卵黄的积累,从而有利于性腺发育,同时也会改善性产物的质量,最终对繁殖有利[24]。一些小型海水鱼类的繁殖试验表明,投喂鱼肉确实能够提高性腺发育成熟度、提高卵质及子代养殖成活率[24]。结合本实验结果,中华鲟养殖群体在后备亲鲟培育过程中适时添加鲜活或冰鲜饵料是促进其性腺成熟或提前成熟的有效途径。

除水温和营养外,养殖环境中的水质、光照、流速以及养殖密度等都可能对鲟鱼的生长和最终成熟产生影响,开展相关研究对于完善中华鲟亲鲟培育途径及最终实现规模化养殖中华鲟繁育至关重要。

3.2 养殖中华鲟繁殖生物学特性

在多种鲟类养殖过程中发现,适宜的养殖环境条件,充足的营养供应等往往使得养殖鲟鱼达到性成熟的时间较自然繁殖群体明显提前[21,25]。中华鲟性成熟时间长,自然种群雄性 8 年以上性成熟(平均初次性成熟 12.5 龄),雌性 14 年以上(平均初次性成熟 16.8 龄)[17,26]。与郭柏福等人的研究结果不同[11],本研究参与全人工繁殖的中华鲟雌鲟为 2000 年出生世代,繁殖时仅有 12 龄,低于中华鲟自然繁殖群体的平均年龄和记录到的最小年龄(14 龄)。本次参与全人工繁殖的雄鲟达 14 龄,但早在 2010 年在厦门基地监测到一尾 10 龄雄鲟的性腺发育达到 IV 期,轻挤压腹部有精液溢出,已经成熟,也明显低于自然种群平均初次性成熟年龄。本研究结果表明,养殖中华鲟性腺可提前成熟,意味着中华鲟养殖成熟年限可以缩短,相应的人力、物力投出成本将大大减少,这对于中华鲟人工群体的建立及维持,以及易地保护有重要意义。

与本次全人工繁殖中华鲟雌鲟性成熟年龄小相一致的是,雌鲟体重较轻,仅有 57 kg,怀卵量也相对较小,仅有 6.2 万粒左右,明显低于野生鱼的产卵量[17]。而且,本次全人工繁殖采获卵的平均卵径只有 3.7 mm,明显小于自然繁殖群体的平均卵径以及郭柏福等人的研究结果,与施氏鲟初次全人工繁殖结果相一致[25]。

参与全人工繁殖的雄鲟体重 64 kg，精液未采集完全，其采精量达 2 850 ml，已和野生成熟个体接近，估计全产采精量可达 4 000 ml，表明人工养殖雄鲟的繁殖能力要明显高于野生雄鲟个体[17]。近年来，已观测到由于污染的加重，中华鲟野生繁殖亲鲟体内的环境污染物，如有机锡、麝香类环境雌激素等的含量明显升高[27]，对中华鲟繁殖群体的性腺发育产生明显影响，进而影响中华鲟自然繁殖效果。就这些环境污染物而言，人工养殖雄鲟所处的环境可能优于野生亲鲟生存环境，这可能是导致养殖雄鲟产精量和精液活力明显高于野生雄鱼的原因。

3.3 中华鲟全人工繁殖技术研究探讨

中华鲟寿命长，性成熟年龄大，开展其全人工繁殖研究所耗费的时间和周期漫长。在长达十多年的人工梯队饲养管理中，饲料供应、水温、水质、养殖空间等某一环节的改变都可能导致不同的养殖效果；同时，由于养殖群体资源的有限性，中华鲟全人工繁殖技术研究也很难实现像其他小型鱼类的多次重复对比，这也是中华鲟全人工繁殖研究进展缓慢的原因。

采用低温刺激被认为是调控鲟鱼类性腺发育成熟的重要措施[10,25,28]，本研究通过对产前中华鲟亲鲟的生态环境调控和对中华鲟性腺发育状况进行连续监控，实现了雌雄中华鲟的同步发育成熟，使得雌雄鲟顺利产卵，而且产后能顺利康复成活，这一成功经验可为以后开展相关研究提供重要借鉴，也为开展养殖中华鲟亲鲟资源的重复利用提供了支撑。

本研究过程中所采用的 B 超、内窥镜、血液生化指标、外科手术及腹腔穿刺等中华鲟性腺发育监测技术手段，为这次全人工繁殖的成功提供了重要的参考，在今后的研究中仍需要进一步熟化和整合。

养殖中华鲟性成熟或提前成熟是中华鲟全人工繁殖的关键，本研究所采用的多样化养殖模式尝试对未来实现中华鲟规模化全人工繁殖有重要借鉴意义。外源激素诱导等其他有效促进中华鲟性腺发育的相关研究值得尝试。随着中华鲟全人工繁殖技术的完善，通过中华鲟人工养殖种群实现中华鲟的资源储备、增殖和资源养护将成为可能。

致谢

农业部渔业局、财务司、水生野生动植物保护办公室、长江渔业资源管理委员会、湖北省水产局、中国水产科学研究院等部门和单位的领导和专家长期对本研究给予支持和帮助；长江水产研究所的领导、同事，特别是濒危鱼类保护研究组杨德国、陈细华、刘鉴毅（调入东海水产研究所）、朱永久、王凯、王科兵、谭从刚，以及未署名的同事、已毕业和在读研究生、荆州太湖试验场的全体工作人员参与了本研究工作；合作单位北京海洋馆和厦门基地等许多未提名的人员，在中华鲟全人工繁殖技术研究中做出了贡献。在此一并致以诚挚的谢意！特别感谢长期关心支持本工作的老师和同行；感谢两位匿名审稿人对论文提出的宝贵意见和建议！

参考文献

[1] Wang J H, Wei Q W, Zou Y C. Conservation strategies for the Chinese sturgeon, *Acipenser sinensis*: an overview on 30 years of practices and future needs[J]. J Appl Ichthyol, 2011, 27(2): 176-180.

[2] Qiao Y, Tang X, Brosse S, et al. Chinese sturgeon (*Acipenser sinensis*) in the Yangtze River: a hydroacoustic assessment of fish location and abundance on the last spawning ground[J]. J Appl Ichthyol, 2006, 22: 140-144.

[3] Wang C, Kynard B, Wei Q, et al. Spatial distribution and habitat suitability indices for non-spawning and spawningadult Chinese sturgeons below Gezhouba Dam, Yangtze River: Effects of riveralterations[J]. J Appl Ichthyol, 2013, 28

[4] 长江水产资源调查组. 长江鲟鱼类生物学及人工繁殖技术研究[M]. 成都:四川科学技术出版社, 1988:284.

[5] 傅朝君, 刘宪亭, 鲁大椿, 等. 葛洲坝下中华鲟的人工繁殖[J]. 淡水渔业, 1985(1):1-5.

[6] 陈细华. 中华鲟胚胎发育和性腺早期发育的研究[D]. 广州:中山大学, 2004:100.

[7] Williot P, Rouault T, Pelard M, et al. Artificial reproduction and larval rearing of captive endangered Atlantic sturgeon *Acipenser sturio*[J]. Endang Spec Res, 2009, 6:251-257.

[8] Doroshov S I, Clark Jr W H, Lutes P B, et al. Artificial propagation of the white sturgeon, *Acipenser transmontanus Richardson*[J]. Aquacult, 1983, 32(1/2):93-104.

[9] Anderson, E. R. Artificial propgation of lake sturgeon *Acipenser fulvescens*(Rafinesque) under hatchery conditions in Michigan[J]. Fisheries research report N01898, 1984, 43.

[10] Webb M A, Doroshov S I. Importance of environmental endocrinology in fisheries management and aquaculture of sturgeons[J]. Gen Comp Endocrinol, 2011, 170(2):313-321.

[11] 郭柏福, 常剑波, 肖慧, 等. 中华鲟初次全人工繁殖的特性研究[J]. 水生生物学报, 2011, 35(6):940-945.

[12] 杨德国, 朱永久, 危起伟, 等. 淡水人工培育中华鲟亚成体的海水驯化试验[J]. 海洋水产研究, 2007, 28(3):120-124.

[13] 陈细华, 危起伟, 朱永久, 等. 低龄中华鲟外科手术性别鉴定技术[J]. 中国水产科学, 2004(4):371-374.

[14] Wei Q W, Zhang X Y, Zhang, X F, et al. Acclimating and maintaining Chinese sturgeon *Acipenser sinensis* in a large public aquarium environment[J]. J Appl Ichthyol, 2011, 27(2):533-540.

[15] Williot P, Brun R, Rouault T, et al. Artificial spawning in cultured sterlet sturgeon, *Acipenser ruthenus* L., with special emphasis on hermaphrodites[J]. Aquacult, 2005, 246(1/4):263-273.

[16] 刘鉴毅, 危起伟, 杜浩, 等. 中华鲟人工繁殖关键技术的改进效果研究[J]. 经济动物学报, 2006, 10(02):96-100.

[17] 刘鉴毅, 危起伟, 陈细华, 等. 葛洲坝下中华鲟繁殖生物学特性及其人工繁殖效果[J]. 应用生态学报, 2007, 18(6):1397-1402.

[18] Nazari R M, Modanloo M, Ghomi M R, et al. Application of synthetic hormone LHRH-A(2) on the artificial propagation of Persian sturgeon *Acipenser persicus*[J]. Aquacult Int, 2010, 18(5):837-841.

[19] Kynard B, Pugh D, Parker T, et al. Using a seminatural stream to produce young sturgeons for conservation stocking: maintaining natural selection during spawning and rearing[J]. J Appl Ichthyol, 2011, 27:420-424.

[20] van Eenennaam J P, Linares-Casenave J, Doroshov S I. Tank spawning of first generation domestic green sturgeon[J]. J Appl Ichthyol, 2012, 28:505-511.

[21] Doroshov S I, Moberg G P, van Eenennaam J P. Observations on the reproductive cycle of cultured white sturgeon, *Acipenser transmontanus*[J]. Environ Biol Fish, 1997, 48:265-278.

[22] Wei Q, Ke F, Zhang J, et al. Biology, fisheries, and conservation of sturgeons and paddlefish in china[J]. Environ Biol Fish. 1997, 48:241-255.

[23] Holcik, J., Freshwater Fishes of Europe(Vol I part II): General Introduction to Fishes and Acipenseriformes[M]. Wiesbaden: Aula Verlag, 1989:1-469.

[24] Zeng C. Development of captive breeding techniques for marine ornamental fish: a review[J]. Rev Fish Sci, 2010, 18(4):315-343.

[25] 曲秋芝, 孙大江, 马国军, 等. 施氏鲟全人工繁殖研究初报[J]. 中国水产科学, 2002, 9(3):277-279.

[26] 危起伟, 陈细化, 杨德国, 等. 葛洲坝截流 24 年来中华鲟产卵群体结构的变化[J]. 中国水产科学, 2005, 12(4):452-457.

[27] Hu J, Zhang Z, Wei Q, et al. Malformations of the endangered Chinese sturgeon, *Acipenser sinensis*, and its causal agent[J]. P Natl Acad Sci USA, 2009, 106:9339-9344.

Research on Technology for Controlled Propagation of Cultured Chinese Sturgeon(*Acipenser sinensis*)

Wei Qiwei[1,5,6] Li Luoxin[1,5] Du Hao[1,5,6] Zhang Xiaoyan[2] Xiong Wei[1]
Zhang Hui[1,5] Shen Li[1,5] Wu Jinming[1] Zhang Shuhuan[1] Wang Chengyou[1,5]
Li Chuangju[1,5] Chai Yi[3] Li Yiwei[4] Qiao Xinmei[1]
Liu Zhigang[1] Gao Yupeng[1] Gan Fang[1]

(1. Key Laboratory of Freshwater Biodiversity Conservation, Ministry of Agriculture of China, Yangtze River Fisheries Research Institute, Chinese Academy of Fishery Sciences, Wuhan 430223, China;
2. Beijing Aquarium, Beijing 100081, China;
3. School of Animal Science, Yangtze University, Jingzhou 434023, China;
4. Xiamen Songhao Industrial Co. , LTD, Xiamen 361100, China;
5. Freshwater Fisheries Research Center, Chinese Academy of Fishery Sciences, Wuxi 214081, China;
6. Fisheries College, Huazhong Agricultural University, Wuhan 430070, China)

Abstract Sexual maturity is a critical step to breakthrough of controlled reproduction of cultured Chinese sturgeon(*Acipenser sinensis*). The suitable breeding management strategy will be beneficial to promoting the maturation progress of cultured Chinese sturgeon and to spawning synchronously of female to male. In this study, we observed the gonadal development of 250 cultured Chinese sturgeon(ages: 12-15 years old, total length: 175-286 cm, body weight: 34.6-125.3 kg) under different water temperature profiles or different feeds. The result showed that gonadal development could reach stage III under natural water temperature or a near-constant temperature. The gonadal development could reach stage III or IV in different types of diets, however the feed added live or iced fish is better for the development from stage II to IV than only single feed source. Controlled propagation was then successfully conducted with one female(12 years old with 57 kg body weight) and one male(14 years old with 64 kg bodyweight). A total of 62 500 eggs and 2 850 ml milt were obtained and 23 000 hatching-out larvae were finally produced after 125 h with (19.3 ± 0.2) ℃. The average fertilized rate was 60.1 % with range from 36.3 % to 80.4 %. The elementary research on breeding regulations of brood stocks will contribute to further large-scale production of cultured Chinese sturgeon.

Key words *Acipenser sinensis* Water temperature Feed Ecological regulation Controlled reproduction Gonad development

中华鲟早期发育中17β-雌二醇和睾酮水平变化的初步观察*

吴湘香[1,3,**]　陈细华[1,2,***]　刘鉴毅[1,2]　危起伟[1,2]

(1. 中国水产科学研究院长江水产研究所 农业部淡水鱼类种质资源与生物技术重点开放实验室　湖北荆州　434000
2. 中国水产科学研究院淡水渔业研究中心　江苏无锡　214081　3. 江西农业大学　江西南昌　330045)

摘　要　采用放射免疫分析测定法(RIA)对人工繁殖的中华鲟(*Acipenser sinensis*)早期发育阶段中17β-雌二醇(17β-estradiol,E_2)和睾酮(testosterone,T)的整体水平进行了测定。实验共测定了4批次(分别来自4尾雌鲟)受精卵及仔鱼的E_2和T水平,结果表明:中华鲟胚胎发育期间E_2水平在小范围内波动,T水平除胚胎发育早期出现上升外总体呈下降趋势;仔鱼出膜后E_2水平开始逐渐上升,T水平是先降低后又逐渐升高。其中第1批次的测定结果(单位为pg/g tissue)为,胚胎发育期间E_2水平在1.37~7.38之间波动,仔鱼出膜后3 d为5.21,出膜后9 d为37.26;胚胎发育期间T水平最低为293.58,最高为1420.41,仔鱼出膜后4 d为117.96,出膜后7 d为27.25,出膜后9 d为148.48。对于仔鱼出膜后体内高水平的E_2和T的来源,推测其可能与仔鱼自身合成性类固醇激素有关。另外,本实验中还发现,质量较差的卵子T水平明显低于质量较好者。

关键词　中华鲟(*Acipenser sinensis*)　早期发育　17β-雌二醇　睾酮

中华鲟(*Acipenser sinensis*)隶属于硬骨鱼纲(Osteichthyes)辐鳍亚纲(Actinopterygii)鲟形目(Acipenseriformes)鲟科(Acipenseridae)鲟属(*Acipenser*),是一种大型江海洄游性鱼类,为国家一级保护动物。围绕中华鲟物种保护问题所展开的一系列研究工作中,关于其早期发育的研究主要有中华鲟胚胎发育形态学观察[1]、胚胎耗氧率研究[2]以及早期行为学观察[3],而关于中华鲟早期发育相关生理因子的变化尚未见有研究。

目前,有关鲟形目鱼类的性类固醇激素研究,在施氏鲟(*A. schrenckii*)[4]、闪光鲟(*A. stellatus*)[5]、高首鲟(*A. transmontanus*)[6]中的研究表明,与其他硬骨鱼类一样,性类固醇激素在鲟鱼的生殖调控中发挥着重要作用,但这些研究主要集中在成体和亚成体血清中激素水平的测定分析,还未见有关鲟鱼早期发育阶段整体性类固醇激素水平变化的研究报道。本研究采用放射免疫测定法(RIA)对人工繁殖的中华鲟早期发育阶段的整体17β -雌二醇(17β-estradiol,E_2)和睾酮(testosterone,T)水平进行了测定,了解其变动情况,旨在为中华鲟繁殖生物学积累资料。

1　材料及方法

1.1　实验材料

2005年11月上旬,由渔政部门组织在长江葛洲坝下江段捕获中华鲟成熟亲鱼,运至长江水产研究所水生野生动物驯养繁育与救护中心,经人工催产后采用人工授精获取受精卵。在

---

* 基金项目:国家自然科学基金面上项目(30571411);中国水产科学研究院水产种质资源与养殖技术实验室开放课题(2005A003)。

** 第一作者,女,1981年出生,硕士研究生,专业方向为鱼类生理学,E-mail:wuxiangxiang2007@163.com

*** 通讯作者(Corresponding author),E-mail:chenxh@yfi.ac.cn

孵化(流水孵化,孵化水温为 18.5 ~ 20.9 ℃)车间,参照中华鲟胚胎发育时序[1]分别对 4 批受精卵(分别来自 4 尾雌鲟)进行跟踪取材至仔鱼出膜后 20 d,其中对第一和第二批次受精卵和仔鱼进行了密集取样,第三和第四批次的取样间隔较大。各批次受精卵及仔鱼的取样情况如下:胚胎发育时期每阶段各取受精卵(胚胎)150 颗并称重(湿重,0.06 ~ 0.11 g/ind.);仔鱼孵出后,取全鱼作为测定材料,根据鱼体大小不同,每阶段各取仔鱼 5 ~ 150 尾并称重(湿重,0.03 ~ 0.88 g/ind.)。

采得的受精卵及仔鱼在 −20 ℃下短暂保存之后转入超低温冰箱(−80 ℃)中保存,待进行 E_2 和 T 的抽提。样品制备及激素水平测定工作在中山大学水生经济动物研究所完成。

1.2 样品制备

中华鲟受精卵及仔鱼 E_2 和 T 整体抽提方法参考血清性类固醇激素抽提法[7],略作修改。将已准确称重的各批次不同发育阶段的受精卵及仔鱼分别置于各塑料离心管中,加适当体积磷酸缓冲液(pH 7.6)后进行组织匀浆,随后将组织匀浆液置于 −20 ℃冰箱中过夜保存;次日加无水乙醚于匀浆液中旋转搅拌,之后入液氮中速冻。将上层乙醚倒入另一套塑料离心管中,(52 ± 2)℃水浴蒸发乙醚。待乙醚蒸发完后,将抽提物置于-20 ℃冰箱中保存,待测样品 E_2 和 T 的水平。

1.3 抽提效率实验

取已知浓度 E_2 和 T 标准品(或质控血清),按照上述样品 E_2 和 T 抽提步骤进行抽提效率实验,检查样品抽提效果,E_2 和 T 的抽提效率分别为 56.7 %和 46.7 %。

1.4 样品 E_2 和 T 水平测定

采用 RIA 测定样品 E_2 和 T 水平。E_2 和 T 放射免疫分析试剂盒购自北京北方生物技术研究所。所有操作严格按照试剂盒说明书进行。

2 结果与分析

2.1 中华鲟受精卵及仔鱼整体 E_2 水平变化

由图 1 可知,胚胎发育期间受精卵 E_2 水平变化不大,在小范围内波动,仔鱼出膜后 E_2 水平开始逐渐上升。第一批受精卵胚胎发育期间 E_2 水平在 1.37 pg/g tissue(25. 胚体扭动期) ~ 7.38 pg/g tissue(10. 原肠中期)之间波动,仔鱼出膜后 E_2 水平逐渐上升,出膜 3 d 为 5.21 pg/g tissue,到出膜 9 d 达到 37.26 pg/g tissue。第二批受精卵胚胎发育期间 E_2 水平在 0.51 pg/g tissue (24. 心脏 s 型期) ~ 4.66 pg/g tissue(6. 三十二细胞期)之间波动,仔鱼出膜后,E_2 水平逐步上升,出膜 2 d 为 3.34 pg/g tissue,出膜 8 d 上升到 10.14 pg/g tissue,而到出膜后 20 d 上升到 230.28 pg/g tissue,从成倍上升到成几十倍的上升。第三批受精卵取样间隔较大,胚胎发育期间所测得的数据中,最低为 3.43 pg/g tissue(25. 胚体扭动期),最高为 5.93 pg/g tissue (16. 晚神经胚期),仔鱼出膜后 E_2 水平波动上升,最低水平 0.10 pg/g tissue 出现在出膜 2 d,到出膜 10 d E_2 水平上升到 15.05 pg/g tissue。第四批受精卵胚胎发育期间所测得数据中,最高 12.40 pg/g tissue (19. 尾芽期),最低 1.48 pg/g tissue(20. 尾芽加厚期),仔鱼出膜 5 d

19.33 pg/g tissue，出膜 6 d 为 13.25 pg/g tissue。

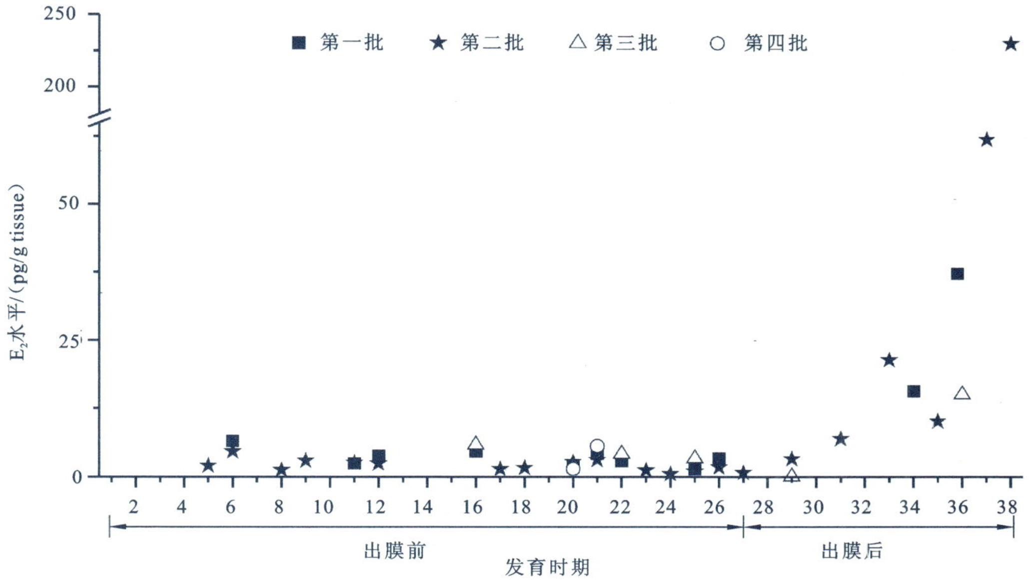

图 1　各批次受精卵及仔鱼整体 E_2 水平变化情况

Fig. 1　Variation in the level of E_2

注：1. 受精后 50 min；2. 受精后 2 h；3. 四细胞期；4. 八细胞期；5. 十六细胞期；6. 三十二细胞期；7. 多细胞期；8. 囊胚期；9. 原肠早期；10. 原肠中期；11. 大卵黄栓期；12. 小卵黄栓期；13. 小卵黄栓晚期；14. 隙状胚孔期；15. 宽神经板期；16. 晚神经胚期；17. 闭合神经管期；18. 眼突形成期；19. 尾芽期；20. 尾芽加厚期；21. 心脏原基期；22. 长心管期；23. 长心管末期；24. 心脏 s 型期；25. 胚体扭动期；26. 尾褶出现期；27. 出膜前期；28. 出膜 1 d；29. 出膜 2 d；30. 出膜 3 d；31. 出膜 4 d；32. 出膜 5 d；33. 出膜 6 d；34. 出膜 7 d；35. 出膜 8 d；36. 出膜 9 d；37. 出膜 10 d；38. 出膜 20 d。

2.2　中华鲟受精卵及仔鱼整体 T 水平变化

由图 2 可知，胚胎发育期间 T 水平总体呈下降趋势，仔鱼出膜后 T 水平是先降低后又逐渐升高。第一批受精卵胚胎发育期间 T 水平最低为 293.58 pg/g tissue（23. 胚体扭动期），最高为 1 420.41 pg/g tissue（4. 八细胞期），仔鱼出膜后 T 水平仍然在下降，出膜 4 d 为 117.96 pg/g tissue，出膜 7 d 下降到 27.25 pg/g tissue，到出膜 9 d 又上升到 148.48 pg/g tissue。进行了密集取样的第二批受精卵胚胎发育期间大部分时期的 T 水平未能测出，所测得的数据中，T 水平最低为 1.08 pg/g tissue（18. 眼突形成期），最高为 20.27 pg/g tissue（20. 尾芽加厚期），仔鱼出膜后 T 水平上升幅度较大，出膜 2 d T 水平为 27.94 pg/g tissue，到出膜 20 d 上升到 5 975.10 pg/g tissue。第三批受精卵胚胎发育期间所测得的数据中，T 最低水平为 158.15 pg/g tissue（25. 出膜前期），最高水平为 1 032.91 pg/g tissue（16. 晚神经胚期），仔鱼出膜后 T 水平先下降后上升，出膜 1 d 为 196.99 pg/g tissue，出膜 2 d 下降到 50.40 pg/g tissue，到出膜 9 d 又上升到 920.75 pg/g tissue。第四批受精卵胚胎发育期间所测得的数据中，T 水平最低为 267.60 pg/g tissue（20. 尾芽加厚期），最高为 360.23 pg/g tissue（21. 心脏原基期），仔鱼出膜后水平下降，出膜 5 d 62.34 pg/g tissue，出膜 6 d 为 43.91 pg/g tissue。

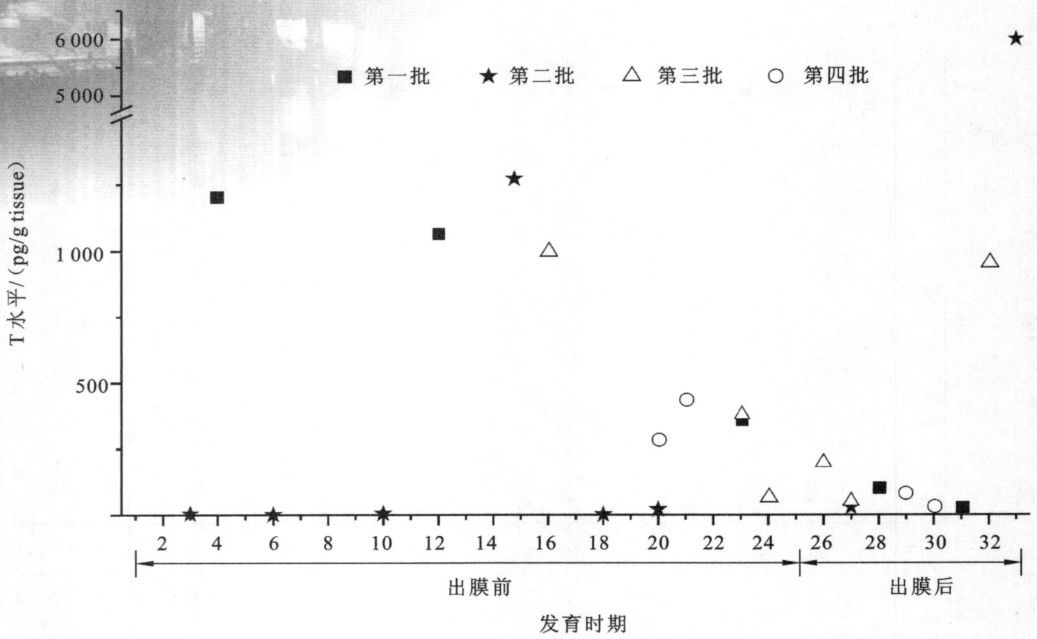

图2　各批次受精卵及仔鱼整体T水平变化情况

Fig. 2　Variation in the level of T

注:1. 受精后50 min;2. 受精后2 h;3. 四细胞期;4. 八细胞期;5. 十六细胞期;6. 三十二细胞期;7. 多细胞期;8. 囊胚期;9. 原肠早期;10. 原肠中期;11. 大卵黄栓期;12. 小卵黄栓期;13. 小卵黄栓晚期;14. 隙状胚孔期;15. 宽神经板期;16. 晚神经胚期;17. 闭合神经管期;18. 眼突形成期;19. 尾芽期;20. 尾芽加厚期;21. 心脏原基期;22. 长心管期;23. 胚体扭动期;24. 尾褶出现期;25. 出膜前期;26. 出膜1 d;27. 出膜2 d;28. 出膜4 d;29. 出膜5 d;30. 出膜6 d;31. 出膜7 d;32. 出膜9 d;33. 出膜20 d。

3　讨论

鱼类卵巢排卵前的卵母细胞双层滤泡膜是产生性类固醇激素的部位[8],未受精的卵子含有从母体带来的性类固醇激素,这些性类固醇激素影响着鱼类的胚胎发育及胚后发育[9],但关于鱼类母源性类固醇激素在其中所发挥的具体作用还不清楚,胚胎发育早期类固醇代谢产物可能对细胞内功能活动起到一定调节作用[10]。已知虹鳟(*Oncorhynchus mykiss*)、罗非鱼(*Oreochromis niloticus*)等硬骨鱼类母源和孵出个体后期自身合成的性类固醇激素在其胚胎及胚后发育中的一般变化情况是,刚受精的卵性类固醇激素水平很高,在胚胎发育期间激素水平逐渐降低,仔鱼出膜后又逐渐上升[9,11,12]。Yeho等[10]认为母源类固醇激素在胚胎发育过程中发生代谢导致这些激素水平下降。本实验测得中华鲟仔鱼出膜前,T水平呈下降趋势,这与上述硬骨鱼类胚胎发育期间的性类固醇激素水平变化相似,但E_2水平只是在一定范围内波动,并没有出现下降趋势。仔鱼出膜后(出膜20 d以前),E_2水平逐渐上升,与上述的仔鱼孵出后的一般变化情况也大致相似,而T水平是先下降再上升。鲟鱼类早期发育过程中性类固醇激素的变化规律是不是与其他硬骨鱼类存在差异,有待进一步研究。

一般认为,鱼类性分化前体内没有类固醇合成细胞[8],但也有学者认为内源性类固醇激素是诱导性腺分化的原因[13]。中华鲟为晚成熟鱼类,出膜1个月后的仔鱼还处于性腺原基的发育阶段,组织学上的两性分化大约发生在9月龄[14]。根据本实验的观察,中华鲟仔鱼出膜后20 d E_2 和T均上升到很高的水平。对于组织学上两性分化之前的中华鲟仔鱼体内高水平的 E_2 和T的来源,作者推测其可能与仔鱼自身合成性类固醇激素有关。

另外,根据本实验取材时孵化车间的生产情况,第二批次的受精卵质量较差,其中存在一些坏死卵(个别时期取样时未将其剔除)。实验结果发现,第一、三、四批次受精卵胚胎发育各时期T水平均为第二批次的百倍甚至千倍,第二批次受精卵胚胎发育的许多时期(包括已经剔除坏死卵的阶段)的T水平都低于检测下限。据此,推测卵子的质量与T水平之间也许存在一定的关系。

致谢

中山大学水生经济动物研究所为样品制备和激素测定提供了实验条件,中山大学张勇博士和华南农业大学赵会宏博士指导了具体实验工作,长江水产研究所博士研究生张辉和长江大学伍卓军同学等协助了实验取材工作,在此一并致谢。

参考文献

[1] 鲁大椿,傅朝君,刘宪亭,等.中华鲟胚胎发育研究[J].淡水渔业,1986,16(4):2-5.

[2] 唐国盘,刘鉴毅,危起伟,等.中华鲟胚胎的耗氧率[J].动物学杂志,2004,39(5):30-34.

[3] Zhuang P, Kynard B, Zhang L, et al. Ontogenetic behavior and migration of Chinese sturgeon, *Acipenser sinensis* [J]. Environ Biol Fish, 2002, 65(1): 83-97.

[4] 孙大江,曲秋芝,王丙乾,等.施氏鲟不同年龄性腺发育与性类固醇激素浓度关系[J].中国水产科学,2004,11(4):317-322.

[5] Semenkova T, Barannikova I, Dimed E, et al. Sex steroid profiles in female and male stellate sturgeon (*Acipenser stellatus* Pallas) during femal maturation induced by hormonal treatment [J] J Appl Ichthyol, 2002, 18: 375-381.

[6] Webb M A H, Feist G W, Trant J M, et al. Ovarian steroidogenes is in white sturgeon (*Acipenser transmontanus*) during oocytem aturation and induced ovulation [J] Gen Comp Endocrinol, 2002, 129: 27-38.

[7] 林浩然,彭纯,梁坚勇,等.鱼类生理学实验技术和方法[M].广州:中山大学出版社,1988:49-53.

[8] 林浩然.鱼类生理学[M].广州:广东高等教育出版社,1999:190-196.

[9] Feist G, S chreck C B, Fitzpatrick M S, et al. Sex steroid profiles of cohosalmon (*Oncorhynchus kisutch*) during early development and sexual differentiation [J] Gen Comp Endocrinol, 1990, 80(2): 299-313.

[10] Yeoh C G, Schreck C B, Fitzpatrick M S, et al. In vivo steroid metabolism inembryonic and newly hatched steelheadtrout (*Oncorhynchus mykiss*) [J] Gen Comp Endocrinol, 1996, 102(2): 197-209.

[11] Hines G A, Boots L R, Wibbels T, et al. Steroid levels and steroid metabolism in relation to early gonadal development in the tilapia Oreochromis niloticus (Teleostei: Cyprinoidei) [J] Gen Comp Endocrinol, 1999, 114(2): 235-248.

[12] Feist G, Schreck C B. Brain-Pituitary-Gonadal Axis during earlydevelopment and sexuald ifferentiation in the rainbow trout, *Oncorhynchus mykiss* [J] Gen Comp Endocrinol, 1996, 102(3): 394-409.

[13] NaKamura M, Bhandari R K, Higa M. The role estrogens play in sex differentiation and sex changes of fish [J] Fish Physiol Biochem, 2003, 28: 113-117.

[14] 陈细华,危起伟,杨德国,等.养殖中华鲟性腺发生与分化的组织学研究[J].水产学报,2004,28(6):633-637.

Variation in the Levels of 17β-Estradiol and Testosterone during Early Development of Chinese Sturgeon(*Acipenser sinensis*)

Wu Xiangxiang[1,3] Chen Xihua[1,2] Liu Jianyi[1,2] Wei Qiwei[1,2]

(1. Key Laboratory of Freshwater Fish Germplasm Resources & Biotechnology, Ministry of Agriculture, Yangtze River Fisheries Research Institute, Chinese Academy of Fishery Science, Jingzhou, Hubei 434000;
2. Freshwater Fisheries Research Center, Chinese Academy of Fishery Science, Wuxi, Jiangsu 214081;
3. Jiangxi Agricultural University, Nanchang 330045)

Abstract The Chinese sturgeon, *Acipenser sinensis* Gray, is a large anadromous fish. The fish is listed in the National Glass I Protected Animals, and the artificial spawning has been performed. In the experiment, the levels of 17β -estradiol (E_2) and testosterone (T) during the early development of Chinese sturgeon were measured by radio immunoassay (RIA). The levels of E_2 and T of four batches of eggs and larvae were measured. The results showed that the level of E_2 during embryonic development changed a little, while the level of T descended all through except for the early embryogenesis stage. After the larvae were hatched out, the level of E_2 increased, but there was a decline of T level before it increased. The levels of E_2 and T for one batch of eggs and larvae were as follows. The unit of levels of E_2 and T were expressed by pg/g tissue. During the embryonic development, the level of E_2 fluctuated between 1.37 ~ 7.38. The level of E_2 was 5.21 at three days after larvae hatching, and 37.26 at nine days after larvae hatching. The lowest and highest levels of T were 293.58 and 1 420.41 respectively during the embryoni development When the larvae were out, the level of T was 117.96 at four days after hatching, 27.25 at seven days after hatching, and 148.48 at nine days afte rhatching. The high levels of E_2 and T in the larvaemay be related with the sex steroids produced by the larvae. In addition, the result showed that the level of T in high-quality eggs was higher than that in the low-quality eggs.

Key words *Acipenser sinensis* Early development 17β-estradiol Testosterone

不同喂食方法对产后野生中华鲟的摄食促进*

张晓雁[1,**]　刘鉴毅[2,3]　危起伟[2,3,***]　杨道明[1]
周文军[1]　蔡经江[1]　杨　静[1]

（1. 北京海洋馆　北京　100081
2. 农业部淡水鱼类种质资源与生物技术重点开放实验室 中国水产科学研究院长江水产研究所　湖北荆州　434000
3. 中国水产科学研究院淡水渔业研究中心　江苏无锡　214081）

摘　要　2006 年 1 月 13 日 ~ 7 月 15 日，研究了不同喂食方法对产后野生雌性中华鲟（*Acipenser sinensis*）的摄食促进效果。结果表明，经人工繁殖的雌性野生中华鲟产后身体虚弱，在人工环境下无主动摄食行为；对投洒的人工配合饲料及天然饲料无摄食反应。水下直接喂食天然饲料的效果依赖于中华鲟主动摄食能力的恢复，而采用导管喂食方法可以实现中华鲟的被动摄食，摄食量由 76 g 逐步增加至 382 g，20 d 后开始出现主动摄食行为，日摄食量逐步上升至 1 822 g，并趋于稳定。导管喂食过程中健康状况也逐步得到改善，胸围和腹围开始提高，主动摄食后健康状况得到彻底改善。表明，产后野生中华鲟有望在水族馆较好的环境中康复。

关键词　中华鲟　产后恢复　水族馆　驯养

世界上现存鲟形目鱼类共 27 种，其中有 17 种处于濒危或极危状态，9 种处于易危状态。如何才能有效地保护这些珍贵鱼类在全世界范围内引起了广泛关注[1]。中华鲟（*Acipenser sinensis*）是一种典型的溯河洄游性鱼类，主要分布于我国的东海、黄海、珠江和长江干流[2]。20 世纪后期，由于过度捕捞、水利工程建设、航运、水体污染等诸多因素，其野生种群资源不断下降，虽然在 1988 年被列为国家一级保护动物，但 1981 ~ 1999 年的 19 年间，中华鲟的幼鲟补充群体和亲鲟补充群体仍然分别减少了 80 %和 90 %左右[3,4]。为了更好地保护中华鲟，需要开展一些移地保护研究。北京海洋馆利用先进的技术和饲养条件，在珍稀濒危水生动物的保护方面进行了一些有益的探索[5]。

野生鲟鱼驯养和摄食行为的研究，国外如美国和俄罗斯有所突破，但是中国仅有一些其他鲟鱼驯养研究的报道[6-9]。2005 年，我们对野生中华鲟在水族馆的驯养方法进行了探索，并取得成功[5-6]，使产后中华鲟在水族馆内健康复原，进而达到主动摄食。这在国内还是首次报道，对于产后中华鲟重复利用和保护中华鲟亲本资源具有重大意义。

根据野生中华鲟的生活习性，在产卵前会停止进食，产卵后一般回到大海时才会进食；但也曾发现，产后中华鲟在长江中游已开始摄食[2]。如何实现产后野生中华鲟在人工环境下进食，恢复体质，进而主动摄食，成为一项重要的研究课题。为了寻找适宜的喂食方法，作者研究

* 基金项目：国家公益研究专项（2004DIB3J099）；国家基础性工作重点专项（2002DEA1004）；国家自然科学基金重大项目（30490231）。

** 第一作者，女，工程师，研究方向为水族生态与管理，E-mail：Zn@ sina. com

*** 通讯作者（Corresponding author），E-mail：Weiqw@ yfi. ac. cn

了不同喂食方法对产后野生中华鲟摄食促进的效果。

1 材料与方法

1.1 实验鱼

实验用野生中华鲟是中国水产科学研究院长江水产研究所于2005年11月9日在宜昌江段科研捕捞的样本。经检查,该鱼为雌性,估计年龄为23龄,全长338 cm,体长280 cm,体重260 kg。在荆州市中华鲟增殖放流基地进行人工催产后,运至北京海洋馆进行康复研究。

1.2 水族馆驯养池

驯养池规格为29.0 m(长)×11.0 m(宽)×4.4 m(高),总水量达1 400 t,展窗20.0 m(长)×3.0 m(高),实验开始前对池中岩石造景进行改造,造景表面覆盖质地柔软的外套,外套使用无毒树脂合成材料制成,避免冲撞或剐蹭造成鱼体受伤。驯养池配备世界先进的生命支撑系统(life support system),可以对水温、水质进行人工监控和调节,还具有淡水、海水转换调控能力。水质监测方法同文献[5]。

1.3 饲料及处理方法

根据中华鲟的食性,准备了天然饲料及人工饲料共20种。天然饲料选用海水及淡水不同种类,饲料处理方法同文献[5]。

导管喂食饲料制作方法:将去掉外壳经过紫外线杀菌的南美白斑虾肉装入去除头部及内脏、经过紫外线杀菌的鱿鱼体内,用医用羊肠线缝合鱿鱼体腔口,将制作完成的饲料穿在长度为90 cm,直径为1 cm的硬塑料管顶部。

1.4 喂食方法

中华鲟进入鲟鱼池后于次日开始喂食,采用4种喂食方法:固定场所喂食人工颗粒饲料;固定场所喂食天然饲料;水下直接喂食天然饲料;水下使用导管插入食道喂食饲料。

在池底中华鲟游动路线位置建立饲料固定喂食场,每天9:00~10:00投喂人工配合饲料,14:00~15:00投喂天然饲料,15:00~16:00由潜水员水下喂食天然饲料(从6月18日到7月7日水下使用导管插入食道喂食天然饲料)。

人工颗粒饲料的投喂采用文献[10]提到的方式。固定场所喂食天然饲料的方式为,潜水员在喂食场洒下天然饵料,其中淡水活饵鱼类剪掉部分尾鳍,以避免鱼饵分散游开。饵料密度为10尾/m^2,每尾饵料鱼体长控制在10~20 cm。

潜水员水下直接喂食的方法为,在产后中华鲟游动路线固定位置等候,待其游至身边,将饲料放于中华鲟嘴边,或者将饲料塞入其口中。潜水员水下使用导管喂食的方法为,通过长90 cm,直径1 cm稍有弧度的硬塑料管将包在顶端的天然饵料送入中华鲟口中,随着中华鲟的吞咽,潜水员将塑料管逐渐深入食道,插入长度70 cm,这样可保证中华鲟不会将饵料吐出。

1.5 身体检查及测量

每日对野生中华鲟体表及鳃进行仔细观察及记录,内容包括体表有无生长水霉、有无淤血及鳃丝的颜色。定期对野生中华鲟进行胸围及腹围的测量并记录。测量方式为潜水员在水下使用皮尺测量胸围、腹围。胸围测量位置是,从头部数第 3 块背骨板最高点,第 2 块侧骨板最高点;腹围测量位置是,从头部数第 5 块背骨板最高点,第 6 块侧骨板。

2 结果及分析

2.1 驯饵过程

产后野生中华鲟对不同喂食方式的行为反应随时间变化的关系见表 1。

表 1 产后中华鲟对于不同喂食方式的行为反应随时间变化的关系

时间(月/日)	喂食方法	行为表现	泳层
1/13 至 3/15	中华鲟游动路线洒喂人工或天然饲料;或等候喂食天然饲料	对饲料无反应,游速缓慢,当潜水员靠近时,加速游开	池体中上层
3/15 至 4 月初	中华鲟游动路线等候喂食或将饲料塞入产后中华鲟口中	不摄食或将塞入口中饲料吐出。对潜水员靠近及抚摸(尾部除外)无受惊反应,游速下降,游姿不平稳	池体中上层
4 月初至 6/18	中华鲟游动路线等候喂食或将饲料塞入产后中华鲟口中	不摄食或将塞入口中饲料吐出。对潜水员触摸尾部无反应,游速继续下降,游姿不平稳幅度加大	池体中上层
6/18 至 7/7	插管喂食	喂食量及喂食次数逐渐增加,7 月 22 日开始对潜水员触摸尾部有躲避反应,泳速稍快	池体中上层
7/8	主动摄食	吸力弱,游动路线摄食第一条鱼后,围绕食场小圈游动	主动摄食第一条鱼后泳层降至池体下层
7/9 日以后	主动摄食	吸力逐渐增强,围绕食场主动摄食	池体下层

从表 1 可知,产后野生中华鲟对水下洒喂人工饲料和天然饲料均无觅食反应。在整个过程中,产后野生中华鲟对投洒的人工配合饲料及天然饲料,既不摄食,也不靠近,基本上无任何反应,其游动路线没有任何变化。在中华鲟非主动摄食阶段,潜水员水下直接喂食天然饲料,均未获得成功,只有在中华鲟恢复了主动摄食能力后,才开始进食潜水员水下直接喂食的天然饲料。

使用导管插入食道喂食饲料的方法,可以较好地实现中华鲟的被动进食。6 月 18 日首次使用即获成功。此后通过导管喂食生鲜饲料的情况越来越好,喂食量及喂食次数逐渐增加,健康状况也逐渐恢复。在接受导管喂食 20 d 后,出现主动摄食行为。从开始进食平均每 2 d 摄食的饲料量随驯养时间变化的关系如图 1 所示。

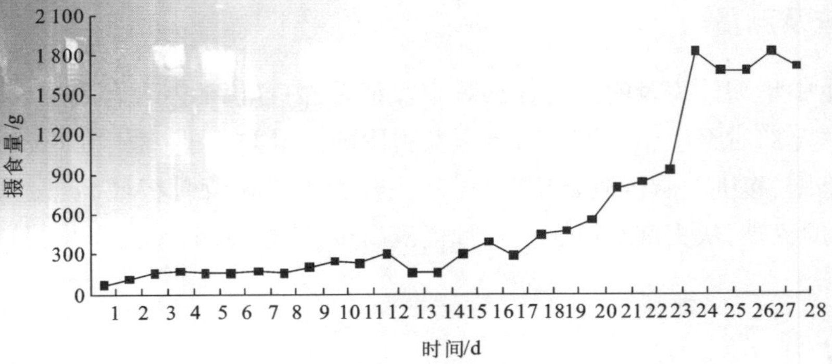

图 1　野生中华鲟平均每 2 d 摄食的饵料量随驯养日期的变化

2.2　中华鲟身体条件的变化

产后中华鲟进入海洋馆驯养池后，长达半年时间一直未摄食，身体逐渐消瘦，胸围及腹围下降，鳃的颜色逐渐变浅，游速缓慢。从 3 月初开始，泳姿不平稳，游动时身体不能完全保持平衡。为避免其身体内伤，体能损耗，从 3 月 14 日开始每 15 d 肌肉注射维生素 C 2 ml:0.5 g×1 支、维生素 B_6 1 ml:0.5 g×1 支、ATP 2 ml:20 mg×2 支加以保护治疗。

从 4 月初开始，游速明显下降，游动时身体晃动幅度增大，潜水员触摸尾部无反应。

6 月 18 日开始进行导管喂食后，产后野生中华鲟身体健康状况明显好转。鳃的颜色逐渐加深，由淡粉红色变成深红色。胸围及腹围开始上升，游动时身姿逐渐平稳，当潜水员触摸尾部时加速游开。平均每 15 d 胸围、腹围随驯养时间变化的关系如图 2 所示。

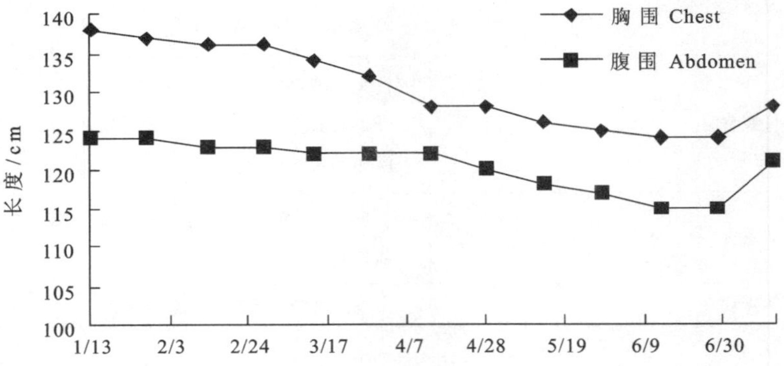

图 2　产后野生中华鲟胸围、腹围随驯养时间变化

3　讨论

插管喂食后，泳速逐渐加快，对潜水员触摸尾部出现躲避行为，说明通过插管喂食使其体能开始上升，对应激能够做出反应。但是泳层仍为中上层，说明产后中华鲟此时仍无摄食欲望。插管喂食 20 d 后，潜水员在游动路线等候投喂，产后中华鲟摄食第一条鱼后，游动路线及泳层发生改变，主动觅食，随后，产后中华鲟摄食行为稳定，在食场摄食。说明产后中华鲟已经

彻底恢复摄食。从表 1 产后中华鲟对于不同喂食方法的行为变化的全部过程看出，插管喂食是导致产后中华鲟体质恢复的关键因子。产后中华鲟体能开始恢复后，进而主动摄食，说明插管辅助喂食对此尾产后中华鲟主动摄食具有促进作用。另外，从图 2 看出，从 1 月至 6 月 18 日插管喂食前，产后中华鲟胸围及腹围明显下降。于 3 月 14 日开始进行肌肉注射维生素 C 2 ml:0.5 g×1 支、维生素 B_6 1 ml:0.5 g×1 支、ATP 2 ml:20 mg×2 支加以保护治疗。随着食道插入导管喂食，产后野生中华鲟胸围及腹围下降趋势停止，并维持稳定，说明通过导管喂食的食量已经基本可以维持产后中华鲟自身的能量消耗。因此，产后野生中华鲟停止消瘦。在产后野生中华鲟开始主动摄食后，由于食量上升较大，胸围及腹围上升速度也随之明显加大，说明产后野生中华鲟通过摄食得到的能量已经超过自身基础代谢量，身体开始储存多余的能量。研究表明，产后野生中华鲟在进食后，身体健康状况日渐复原，随着进食量加大，体质恢复速度明显加快。

另外，使用导管喂食的优点在于，通过导管可以将饲料送入产后野生中华鲟的食道下段，由于导管插入食道较深（插入食道 70 cm），致使其不能将饲料吐出，可确保其进食。此种喂食方法便于控制食量，对于救治野生中华鲟提供了新的思路及方法。

此尾产后野生雌性中华鲟在进入驯养池初期，对潜水员的行为反应强烈。当潜水员靠近时，表现出强烈的受惊躲避行为，使潜水员无法靠近喂食。经过 2 个月的水下接触训练，产后野生雌性中华鲟才可以配合潜水员进行一些动作，为后期潜水员使用导管喂食打下了基础。水下接触训练成功将大大推动对野生中华鲟驯养的研究。另外，野生中华鲟对潜水员的受惊躲避行为是否与人工繁殖有关还需要进一步研究。

产后野生中华鲟至今对人工配合饲料没有任何摄食反应。虽然我们 2005 年驯养成功的野生中华鲟也不摄食人工合成饲料，但是对于野生中华鲟究竟能不能摄食人工配合饲料的问题还需要进一步研究[10,11]。

参考文献

[1] Billard R, Lecointre G. Biology and eonservation of sturgeon and paddlefish, Reviews in Fish Biology and Fisheries, 2001, 10(4): 355-392.
[2] 四川省长江水产资源调查组. 长江鲟鱼类生物学及人工繁殖研究. 成都：四川科学技术出版社，1988：88.
[3] 柯福恩. 论中华鲟的保护与开发. 淡水渔业，1999，29(9)：4-7.
[4] 柯福恩，危起伟，罗俊德，等. 论长江珍稀水生野生动物的保护与补救//李渤生，詹志勇. 绿满东亚：第一届东亚地区国家公园与保护区会议暨 CNPPA/IUCN 第 41 届会议论文集. 北京：中国环境科学出版社，1994：780-785.
[5] 刘鉴毅，张晓雁，危起伟，等. 野生中华鲟水族馆驯养观察. 动物学杂志，2006，41(3)：48-53.
[6] 刘鉴毅，张晓雁，危起伟，等. 中华鲟水族馆驯养研究. 淡水渔业，2006，36(5)：52-56.
[7] Wei Qiwei, Ke F, Zhang J, et al. Biology, fisheries, and conservation of sturgeons and paddlefish in China. Environmental Biology of Fishes, 1997, 48(1/4): 241-255.
[8] 陈凌云，章秋虎，项卫民，等. 匙吻鲟孵化驯养与无公害网箱养殖新技术. 渔业致富指南，2005(7)：33-34.
[9] 张彤晴，葛家春. 小体鲟幼鱼水库坝下流水池塘驯养技术. 江苏农业科学，2002(5)：59-61.
[10] 罗晓松，曾令兵. 工厂化驯养鲟鱼病害防治技术. 科学养鱼，2000(11)：32-33.
[11] 颜远义. 中华鲟生物学特性及养殖方法. 水产科技，2003(5)：14-16.

Recovery of Forging Ability of Wild Female Chinese Sturgeon after Artificial Propagation in Aquarium by Different Feeding Methods

Zhang Xiaoyan[1] Liu Jianyi[2,3] Wei Qiwei[2,3] Yang Daoming[2]
Zhou Wenjun[1] Cai Jingjiang[1] Yang Jing[1]

(1. Beijing Aquarium, Beijing 100081, China;
2. Key Lab of Freshwater Fish Gerplasm Resources and Biotechnoligy, Ministry of Agriculture of China, Yangtze River Fisheries Research Institute, Chinese Academy of Fishery Sciences, Jingzhou 434000, China;
3. Freshwater Fisheries Research Center, Chinese Academy of Fisheries Science, Wuxi 214081, China)

Abstract A wild female Chinese sturgeon (*Acipenser sinensis*), caught from the Yichang section of Yangtze River was domesticated in a large public aquarium (Beijing Aquarium) from January 13 to July 15 in 2006 after artificial propagation. Different feeding methods, such as the artificial baits were sprinkled on the bottom of the aquarium or the natural baits were sent to the sturgeon mouth by a diver directly, were conducted for the recovery of forging ability of Chinese sturgeon which was so weakness to eat any food intimately. Of the methods, a successful way was established through introducing a plastic tube by a diver to the end of esophagus to transport the food to the stomach of the fish after two months test. The Chinese sturgeon gradually adapted the diver and the feeding way and daily food consumption increased from 76 g to 1 822 g. With the energy supplying, the Chinese sturgeon became strong enough to find the food initiatively after 20 d. The condition of the sturgeon improved largely with perimeter of the section of chest and abdomen becoming larger. The results indicated challenges could be realized in the good conditions as aquarium for recovery of the wild Chinese sturgeon after artificial propagation.

Key words Chinese sturgeon (*Acipenser sinensis*) Recovery Aquarium Domestication

养殖密度对中华鲟行为、免疫力和养殖环境水质的影响*

张晓雁[1,2,3,5,**]　李罗新[4]　危起伟[1,3,4,***]　张先锋[1,3,****]

沈　丽[4]　张艳珍[5]　杨道明[5]　栾　刚[5]　刘建伟[6]

（1. 中国科学院水生生物研究所　湖北武汉　430072　2. 中国科学院研究生院　北京　100049

3. 中国科学院水生生物多样性与保护重点实验室　湖北武汉　430072

4. 农业部淡水生物多样性保护重点开放实验室 中国水产科学研究院长江水产研究所　湖北武汉　430223

5. 北京信沃达海洋科技有限公司（北京海洋馆）　北京　100081

6. 中国水产科学研究院　北京　100141）

摘　要　以北京海洋馆驯养的野生和子一代中华鲟为研究对象，研究了养殖密度对中华鲟行为、免疫力和养殖环境水质的影响。结果表明，不同养殖密度（2.7 kg/m^3，2.4 kg/m^3，2.0 kg/m^3）影响中华鲟的泳层分布，而对中华鲟食欲，对野生和 F1-1998 的泳速和呼吸频率影响不明显。2.7 kg/m^3，2.4 kg/m^{3}密度使中华鲟抗氧化力及多项水质指标变化，高年龄的中华鲟，包括野生中华鲟和 F1-1998 中华鲟细菌感染。实验表明，使用循环水体养殖中华鲟，环境负荷对水体微生物和养殖鱼免疫力的影响是极为显著的，本文最佳养殖密度为 2.0 kg/m^3。

关键词　中华鲟　养殖密度　循环水　水质　行为　免疫力

高养殖密度可以高效利用水资源，提高养殖效益；但是高密度养殖容易导致系统负载率增高，系统不能及时清除代谢物，造成水质量下降[1]。高养殖密度导致饲养个体间食物和空间竞争，作为一种慢性环境胁迫因子，也可导致鱼类行为和免疫力等发生改变[2-4]。高密度养殖引起的鱼类应激通过下丘脑-垂体-肾间组织轴，交感神经-嗜铬组织轴，使得鱼生理改变，对应激产生生理变化[5-6]，增大鱼类疾病发生的可能性。低 pH 可使草鱼的呼吸生理功能受到干扰，氧代谢紊乱，使肝脏的抗氧化力下降[7-8]；高氨氮使鲤鱼等鱼类鳃变性和肝细胞破坏[9]；高亚硝酸盐也可成为影响鱼生长的胁迫因子[10]；鱼的消化酶活力也直接受到生态因子的影响，进而影响鱼的健康和生长[11]。中华鲟（*Acipenser sinensis*）是我国的一级保护动物，近年来资源量持续衰退[12]，已经被列为 IUCN 的极危物种。对其进行人工圈养是实施保护和人工增殖研究的有效措施之一[13]。研究循环水养殖中华鲟，以及养殖密度对水质和养殖鱼免疫力及行为的的关系，对规范中华鲟的养殖技术具有指导意义。2008 年 12 月 ~2009 年 8 月，作者在北京海洋馆就养殖密度对水质和中华鲟行为及免疫力的影响进行了实验。

---

* 基金项目：中国水产科学研究院研究型业务专项（2011C002）；农业公益性行业科研专项淡水水生生物资源增殖放流及生态修复技术研究（200903048）；三峡工程环境保护补偿经费生态建设类（施工区外）三峡工程珍稀鱼类增殖放流项目（2005-2009）。

** 第一作者，女，山东省栖霞人，高级工程师，博士研究生，主要从事鱼类生理生态学和水族馆科学与技术方向的研究，E-mail：zn7@ sina. com

*** 通讯作者（Corresponding author），E-mail：weiqw@ yfi. ac. cn

**** 通讯作者（Corresponding author），E-mail：Zhangx@ ihb. ac. cn

1　材料与方法

1.1　实验鱼

实验鱼是在2005～2008年，先后从长江水产研究所中华鲟实验基地迁至北京海洋馆淡水饲养，共68尾，包括6尾野生中华鲟和3个年龄组若干子一代养殖中华鲟。其中野生中华鲟为人工产后亲鱼，在北京海洋馆达到康复（简写为W-N），子一代中华鲟分别为1998年、2001年、2005年人工繁殖个体（编号为F1-1998，F1-2001，F1-2005），实验初始体重分别为（227±39.7）kg（n=6），（54.3±7.4）kg（n=7），（36.1±10.6）kg（n=18），（24.7±4.7）kg（n=37）。

1.2　养殖设施和日常管理

饲养池为长方形水泥池，规格为29.0 m（长）×11.0 m（宽）×4.4 m（高），观察窗为20.0 m（长）×3.0 m（高），总水量达1 200 m^3。饲养池循环率为50 min/次，可以对水温及其他水质指标进行实时人工监控和调节。照明恒定为8:30～18:30。试验温度控制在（22.3±0.4）℃。

每周定期吸出池底污物，冲洗砂滤罐。

饲料主要以鲜饲料和人工合成饲料为主，每周喂5次，每次投饵率为（体重比）：W-N（0.8%），F1-1998（1.3%），F1-2001（3%）和F1-2005（3%）。W-N和F1-1998采用潜水员水下喂食鲜饲料，F1-2001和F1-2005采用撒食鲜饲料和人工合成饲料。

1.3　试验设计

试验共3个密度组，分别为2.7 kg/m^3，2.4 kg/m^3和2.0 kg/m^3，也即分别为3 300 kg，2 900 kg和2 400 kg养殖总重量，通过移出子一代个体以调节养殖密度，养殖数量分别为68尾、46尾、25尾。每个密度组连续养殖3个月。

1.4　水质参数

实验期间对水体溶解氧（DO）、酸碱度（pH）每日测定1次，对氨/铵（NH_3/NH_4+）、亚硝酸盐（NO_2^-）、硝酸盐（NO_3^-）、磷酸盐（PO_4^{3-}）、浊度、总细菌6个指标每周检测2次以上。

pH采用HANNA便携pH计（复合pH电极，带自动温度补偿ATC，产自意大利）测量。DO采用HACH溶氧仪，渗透膜法测量。NH_3/NH_4^+，NO_2^-，NO_3^-和PO_4^{3-}含量采用HACH分光光度计测量，检测值低于0.01 mg/L时，以0.01记录。总细菌数量采用HACH试剂，Millipore膜过滤法检测。

1.5　行为与生理观察

观察项目包括：泳层（下层，距池底0～1.5 m；中层，距池底1.5～3.0 m；上层，距池底3.0～4.4 m），平衡性，摄食主动性，泳速，呼吸频率等。前三项每天定时观察3次，每次观察40 min。后两项采用随机采样，分别随机选取4尾健康的W-N和F1-1998观察。固定每日监测发病鱼的呼吸频率，并增加B超（Sonosite Macromaxx）检查消化道和腹腔项目。对病亡鱼类进行病理组织取样，病理组织标本经常规石蜡包埋切片法光镜观察。

在实验末期测定血浆超氧化物歧化酶（SOD）和谷胱甘肽过氧化物酶（GSH-PX）活性、丙二醛（MDA）含量。测定方法：尾静脉采血后用肝素钠抗凝，3 000 r/min 离心 20 min 后取上清，－4℃保存备用。SOD 含量检测用黄嘌呤氧化酶法，MDA 含量测定用硫代巴比妥酸法，GSH-PX 含量测定采用比色测定法。试剂盒由南京建成生物工程研究所提供。

1.6 数据统计与参数计算

数据通过统计软件 SPSS 13.0 进行分析处理。结果为平均值 ± 标准差。水质数据，鱼的泳速和呼吸频率利用方差分析（one-way ANOVA）分别检验各实验组的差异显著性，利用 S-N-K 法进行多重比较。使用 Pearson Correlations 分析养殖密度和水质数据的相关性。不同组的血清抗氧化酶活力利用 Independent Two-Sample *T* Test 检验显著性差异。

2 结果

2.1 游泳行为

实验表明，随养殖密度下降，不同体型的中华鲟泳层分布会发生变化（表 1）。高密度时，体型较大的中华鲟，包括 W-N 和 F1-1998，主要分布在鱼池上层；体型较小的中华鲟，包括 F1-2001 和 F1-2005，分布在鱼池下层。低密度时，除体型最大的 W-N 在上层活动居多外，其余体型的中华鲟在各水层活动。高密度环境时下层鱼的碰撞频率明显高于低密度时的频率。

表 1 不同密度组中华鲟的泳层分化

Tab. 1 The swimming-level distribution at different densities

项目 Item	养殖密度/(kg/m^3) Stocking density		
	2.7	2.4	2.0
W-N	上层	上层	上层，偶尔下层
F1-1998	上层，中层	上层，中层	中层，下层，上层
F1-2001	下层	下层	中层，下层，上层
F1-2005	下层	中层，下层	中层，下层，偶尔上层

各实验组，健康的中华鲟均积极主动摄食，游泳无统一的方向，平衡性较好；泳速和呼吸基本稳定，各组间无显著性差异（$p > 0.05$）。在同一组内，W-N 的泳速低于 F1-1998，而呼吸频率反而较快（表 2）。

表 2 各密度组中 W-N 和 F1-1998 的泳速和呼吸频率

Tab. 2 The swimming velocity and respiratory frequency of the wild ($n=4$) and the F1-1998 Sturgeon ($n=4$) at different densities

养殖密度/(kg/m^3) Stocking density	W-N ($n=4$)		F1-1998 ($n=4$)	
	泳速/(s/全长*) Swimming velocity	呼吸频率/(次/min) Respiratory frequency	泳速/(s/全长*) Swimming velocity	呼吸频率/(次/min) Respiratory frequency
2.7	6.25 ± 1.31	23 ± 7	4.18 ± 0.85	20 ± 8
2.4	6.93 ± 1.25	22 ± 7	4.60 ± 1.40	18 ± 7
2.0	6.98 ± 2.25	25 ± 5	4.24 ± 1.24	20 ± 5

* 全长指鱼全长。

2.2 生理指标

在 2.7 和 2.0 kg/m^3 实验组，F1-1998 的 SOD，GSH-PX 和 MDA 检测结果显示（图 1），在

2.0 kg/m³ 养殖密度下 SOD 活性显著升高，两组间呈现显著性差异($p<0.05$)，MDA 水平和 GSH-PX 活性下降，两组间无显著性差异($p>0.05$)，SOD/MDA 值显著上升，增幅近 1 倍。

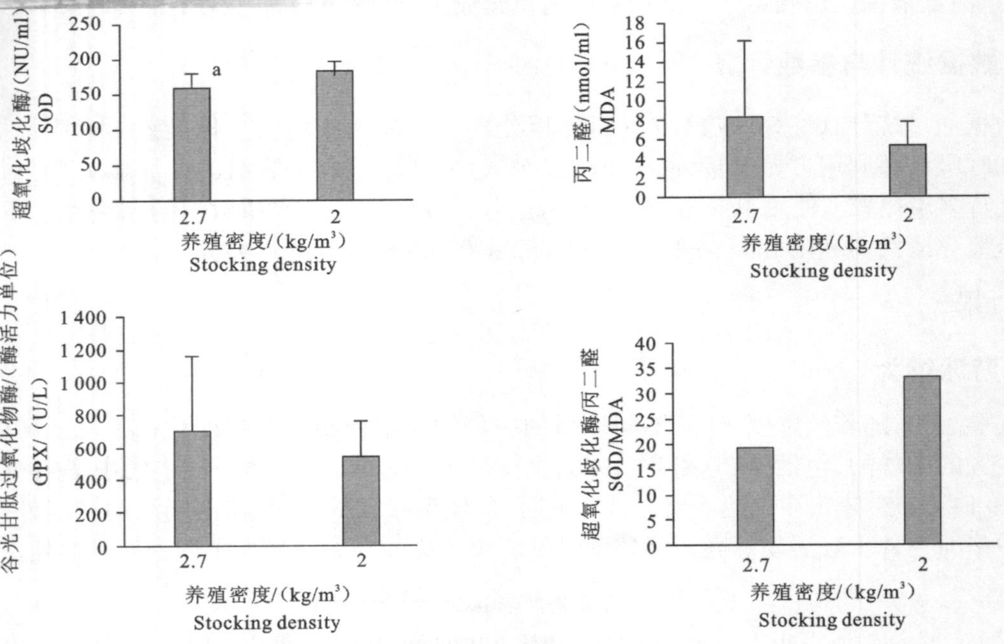

图 1　不同实验组 F1-1998 的抗氧化指标(SOD，MDA，GPX，SOD/MDA)比较

$n=6$；a 表示在 5 %水平上的显著差异

Fig. 1　SOD, MDA, GPX, SOD/MDA in serum of F1-1998 at different densities.

$n=6$; The letter a stand for significant difference($p<0.05$)

2.3　病亡鱼的行为学异常及组织学与微生物诊断

在养殖末期，高、中密度组中，有 5 尾中华鲟相继发病，发病鱼的异常行为和检查结果见表 3。

表 3　各实验组发病鱼的异常行为和检查结果

Table 3　The information of diseased fish at different stocking densities

养殖密度/(kg/m³) Stocking density	发病鱼 Diseased fish	发病鱼的异常行为和检查结果 The information of diseased fish
2.7	1 尾 W-N	拒食。嘴红肿，明显外突
2.4	2 尾 W-N	第一尾：右侧鳃盖张开 6 cm，不能随呼吸而闭合 第二尾：突然拒食。次日腹部肿胀，泄殖孔不能完全闭合
2.4	2 尾 F1-1998	第一尾：突然拒食，正常游泳时呼吸频率加快，于正常 2 倍。体表黏液分泌增多 第二尾：呼吸急促，游动时头向上呈 45°。泄殖孔微红

经使用抗生素对疾病中华鲟医疗，4 尾痊愈，1 尾 F1-1998 死亡。死亡鱼病理诊断结果显示，多处组织器官出现炎症病变，肠道和鳃炎症严重，咽、食道和胃炎症较轻，肝脏局灶性肝炎，冠状室、性腺和脾脏炎，肾小球肾炎，肾管型(图 2)。对死亡鱼的组织和血液微生物培养的结果为，鳃受多种菌感染，一种为优势菌，肝脏一种菌；肾脏两种菌，第一种菌多于第二种菌，血培养结果为嗜水气单胞菌。对疾病和死亡鱼的诊断结论为细菌性感染。

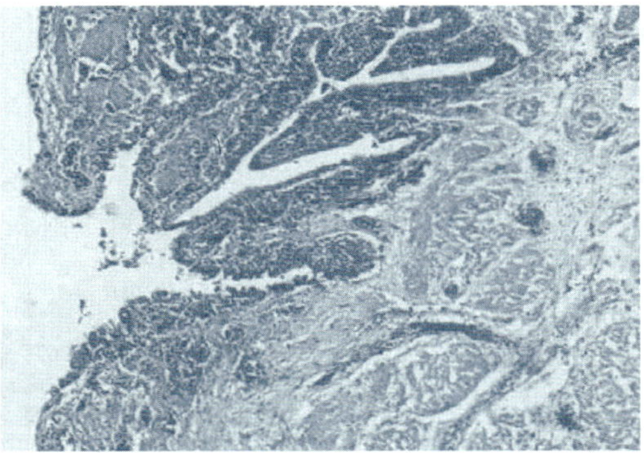
直肠 HE 100×

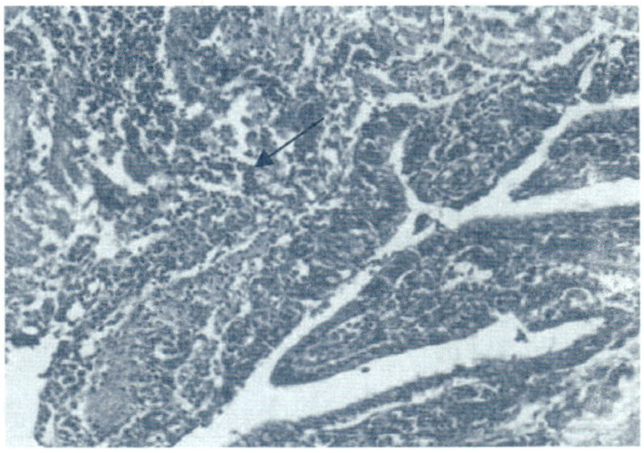
直肠 HE 200×

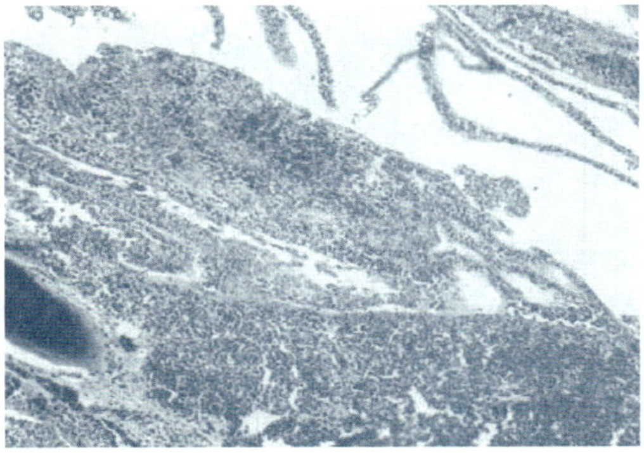
鳃丝 HE 100×

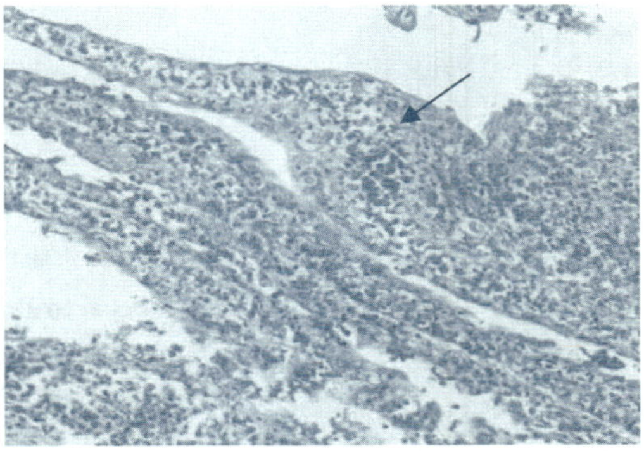
鳃丝 HE 200×

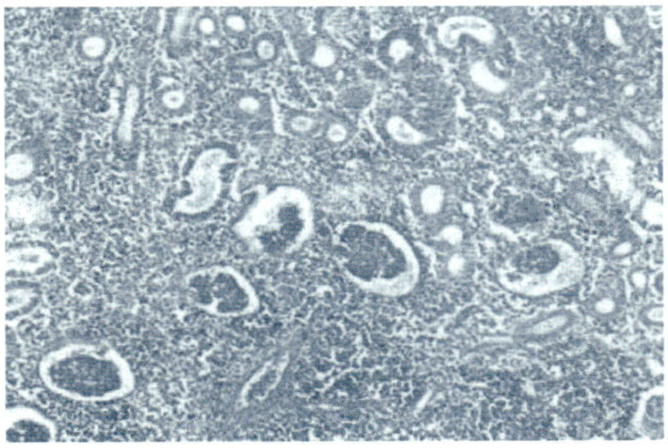
肾脏 HE 100×

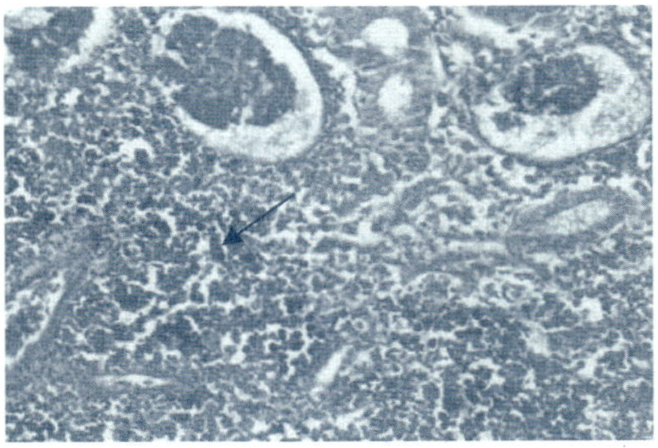
肾脏 HE 200×

图 2　直肠、鳃和肾脏的病理组织切片图

箭头所指为炎性细胞

Fig. 2　Histology of the rectum, the gill lamellae and the kidney.

The inflammation cell is marked by arrow

2.4　水质指标

在监测的 8 项水指标中(表 4),NH_3/NH_4^+,NO_2^- 和浊度值基本保持平稳,各实验组间无显著性差异。NO_3^-,PO_4^{3-},总细菌值随着密度的降低显著下降($p<0.05$),低密度组的 NO_3^- 和 PO_4^{3-} 值比中、高密度组降幅分别达 50 %和 40 %左右,低、中、高密度组总细菌值依次下降近 30 %和 50 %。DO 随密度降低而升高($p<0.05$),各实验组依次增高不足 3 %。

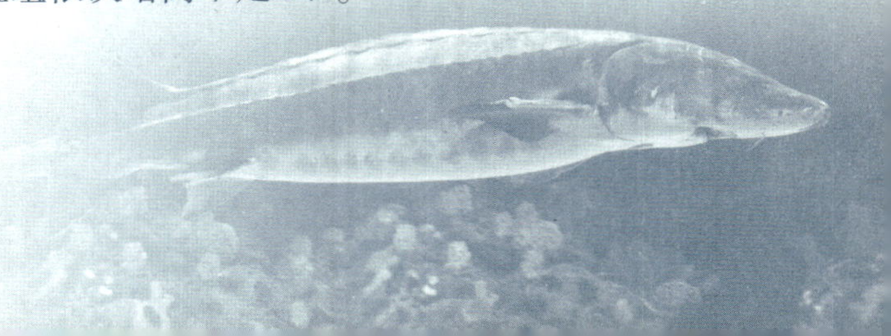

表 4　水质指标的变化

Table 4　The physichemical and microbiological parameters at different stocking densities

养殖密度 /(kg/m^3) Stocking density	酸碱度 pH	溶解氧 /(mg/L) DO	氨/铵 /(mg/L) NH^3/NH_4^-	亚硝酸盐 /(mg/L) NO_2^-	硝酸盐 /(mg/L) NO_3^-	磷酸盐 /(mg/L) PO_4^{3-}	浊度(NTU) Turbidity	总细菌 /(cfu/100 ml) Total bacteria
2.7	7.87 ± 0.05^a	7.56 ± 0.34^a	0.01	0.01	54.89 ± 6.18^a	4.08 ± 0.57^a	0.06 ± 0.010	$10\,908 \pm 6\,639^a$
2.4	7.85 ± 0.02^b	7.72 ± 0.33^b	0.01	0.01	50.40 ± 10.10^a	3.89 ± 0.70^a	0.057 ± 0.005	$7\,672 \pm 5\,208^b$
2.0	7.80 ± 0.06^c	7.80 ± 0.45^c	0.01	0.01	26.44 ± 7.50^b	2.40 ± 0.45^b	0.058 ± 0.005	$3\,732 \pm 2\,447^c$

注：同一列不同字母表示在 5 %水平上的显著差异。

Note: The different letters in one line stand for significant difference($p < 0.05$), otherwise, the same ones stand for no significant difference.

对以上检测的各项水指标结果使用线性相关分析显示(表 5)，NO_3^-，PO_4^{3-}，pH 和总细菌与养殖密度之间显著正相关($p < 0.01$)，DO 与养殖密度呈显著负相关($p < 0.01$)，而浊度与养殖密度无显著相关性($p > 0.01$)。

表 5　养殖密度和水质指标的相关性(以 r 值表示)

Table 5　Correlations between the physiochemical and microbiological parameters and stocking density(r value shown)

水质指标 Water indices	酸碱度 pH	溶解氧 DO	硝酸盐 NO_3^-	磷酸盐 PO_4^{3-}	总细菌 Total bacteria	浊度 Turbidity
r 值	0.514 *	-0.450 *	0.810 *	0.766 *	0.485 *	0.092

* 表示在 0.01 水平上的显著差异。

* stand for significant difference($p < 0.01$).

3　讨论

3.1　养殖密度对行为的影响

Papoutsoglou[4] 曾报道海鲈在不同养殖密度下行为发生变化，认为高养殖密度时摄食集群行为可能明显，而其他学者对史氏鲟稚鱼研究时，并未发现随密度改变出现集群行为[14]。本实验的各密度环境中，中华鲟没有出现明显的集群行为。然而随养殖密度下降，不同体型的中华鲟泳层发生不同的分化，体型最大的 W-N 泳层变化不大，体型最小的 F1-2005 泳层变化最大。这说明中华鲟对空间竞争优势与其体型成正比。由于密度降低，鱼对生活空间竞争下降，体型小的鱼可以在更大的空间活动。

水体溶解氧、酸碱度影响鱼的呼吸频率[15]，水的流速将干扰鱼的游泳行为[16]。本文实验期间，养殖水体的溶解氧、酸碱度和水流比较平稳，因而不会影响鱼的呼吸频率和泳速。另外，W-N 和 F1-1998 体型大，具有水域空间竞争优势，在高密度时“情绪”也能保持稳定，因此游泳行为可能受养殖密度影响不大。

3.2　养殖密度对免疫力的影响

有学者报道，SOD 活性对水生动物的免疫功能有重要作用，MDA 的含量间接反映机体细胞损伤的严重程度[17,18]。刘松岩[19] 报道，在高密度环境下，中华鲟的 SOD 活性下降，MDA 水平上升。本试验中,SOD 活性和 MDA 含量在高密度环境时的变化趋势与其结果一致,然而 SOD 活性低于其报道的结果,这可能是由于实验鱼体型差别大,致使结果存在差异。高密度环境时,SOD/MDA 值显著低于低密度环境,这个结果也提示高密度环境使机体的抗氧化能力下降,脂质过氧化反应增强[19]。吴长珠等曾报道[20],GSH-PX 具有抗氧化功能,当受到过氧化损伤时,测定值升高。本文中高密度环境时鱼脂质过氧化增强,MDA 含量上升,使 GSH-PX 值升高,与上述报道一致。

本实验中,W-N 和 F1-1998 发病率高,分别为 50 %和 30 %左右,而 F1-2001 和 F1-2005 为 0。分析其原因,一方面,馆养 W-N 都在 25 龄左右,趋于高龄[21],而高龄鱼的生理机能退化,适应能力呈下降趋势,因而其受到环境因子胁迫时更易于被淘汰;另一方面,野生中华鲟在纯人工环境养殖,其营养和生态条件不同于其天然生境,可能对人工环境的敏感度较高,如果环境条件不适宜,发病率将会增加。总之,虽然 W-N 体型大,具有水域空间竞争优势,但是高密度环境的制约仍然可能导致其抗病力下降,易感性增强。

子一代中华鲟长期生活在人工环境中,对空间的占有率明显低于自然生境,更容易适应高密度环境,适应力高于野生鱼,发病率相对较低。然而,由于人工投喂的饵料较单一,提供的营养不能完全满足其生长发育变化的需求,就表现为养殖时间长的大龄子一代中华鲟对不适环境表现得更敏感。因此可知,养殖大龄的野生和子一代中华鲟,通过改良水质条件,也许能减少营养单一对其长期养殖的不利影响,使其保持健康生长。

3.3 养殖密度对水质的影响

已有的资料表明,封闭循环水养殖水处理系统对各种物质的处理效率不同[1,22]。本实验中,在不同养殖密度下,对鱼有毒性的 NH_3/NH_4^+ 一直保持在 0.012 5 mg/L 的养殖安全水平[23]以下,NO_2^- 也保持 0.01 mg/L,其余 6 个水指标随着养殖密度下降分别不同程度降低或者升高。这说明循环水养殖环境中,系统处理各指标的效率存在差别,因此这些波动的水指标及波动程度也是控制环境负荷的关键指示指标。

本文结果显示,NO_3^- 和 PO_4^{3-} 与养殖密度的相关性极高。Thomas 等[24]对高首鲟排泄氨的研究表明,全封闭水环境中硝酸盐的累积主要受到鱼摄食蛋白质总量的影响。本实验中,随着养殖密度下降,日投饵总量减少,硝酸盐累积量随着下降。鱼主要从食物中摄取磷,经代谢后将磷主要以磷酸盐形式排出体外[24]。当养殖密度降低时,总排泄物减少,水中磷酸盐的累积量也相应减少。

对水中总细菌检测表明,其密度随着养殖密度的降低而显著下降($p<0.05$)。这主要因为高养殖密度和投饵率使水体和底质的有机物含量升高,细菌大量繁殖。这种情况将导致水体中原有正常菌群失衡,条件致病菌快速增殖[25]。本文中,高、中、低密度组的水细菌数依次降低 30 %和 50 %,并且在密度从 2.4 kg/m^{3}降至 2.0 kg/m^{3}时,优势菌从淡水鱼致病力较强的嗜水气单胞菌和维罗纳气单胞菌温和生物变种,转变为致病力较弱的鲁氏不动杆菌和鲍曼不动杆菌等[25-27]。这显然说明,水中优势菌的种类与水体细菌总数相关联。另外,在密度 2.7 和 2.4 kg/m^{3}时,水细菌总数基本大于 8 000 cfu/100 ml,鱼的免疫力下降,鱼陆续出现细菌感染性疾病,而且病亡鱼血液中的感染菌与水体优势菌相同,而密度为 2.0 kg/m^{3}时,细菌总数低于 8 000 cfu/100 ml,鱼不发病,这说明水细菌总量和优势菌种类及鱼的免疫力与养殖密度相关联,2.0 kg/m^{3}基本是健康养殖的最大养殖密度。

随着养殖密度降低，池水溶解氧上升，理论上 pH 也应上升，实际却略有下降。由于使用水净化工艺处理循环使用的养殖水，pH 的变化除受到生物代谢和有机物分解等影响以外，可能还受到其他处理工艺的影响，这有待进一步研究。

参考文献

[1] Syresha V, Lin C K. Effcet of stocking density on waterquality and production of red tilapia in a recirculated watersystem[J]. Aquacultural Engineering, 1992, 11(1): 1-22.

[2] 王文博，李爱华. 环境胁迫对鱼类免疫系统影响的研究概况[J]. 水产学报，2002，26(4)：368-374.

[3] 洪磊，张秀梅. 环境胁迫对鱼类生理机能的影响[J]. 海洋科学进展，2004，22(1)：113-121.

[4] Papoutsogloi S E, Taiha G, Vrettos X, et al. Effects of stocking density on behavior and growth rate of European sea bass(*Dicentrarchus labrax*) juveniles reared in aclosed circulated system[J]. Aquacultural Engineering, 1998, 18(2): 135-144.

[5] Ruane N M, Huisman E A, Komen J. Plasma cortisoland metabolite level profiles in two isogenic strains of common carp during confinement[J]. Journal of Fish Biology, 2001, 59(1): 1-12.

[6] Bonga W. The stress response in fish[J]. Physiological Reviews, 1997, 77(3): 591-625.

[7] 余日清，贺锡勤. 低 pH 对草鱼呼吸活动和耗氧代谢的影响[J]. 环境科学学报，1992，12(1)：105-111.

[8] 马广智，唐玫，徐军. 低 pH 值对草鱼鳃和肝组织超氧化物歧化酶(SOD)活性的影响[J]. 中国水产科学，2001，8(1)：23-25.

[9] 徐镜波，马逊风. 温度、氨对鲢、鳙、草、鲤草的影响[J]. 中国环境科学，1994，14(3)，214-219.

[10] 余瑞兰，聂湘平，魏泰莉，等. 分子氨和亚硝酸盐对鱼类的危害及其对策[J]. 中国水产科学，1999，6(3)，73-77.

[11] 田宏杰，庄平，高露姣. 生态因子对鱼类消化酶活力影响的研究进展[J]. 海洋渔业，2006，28(2)：158-162.

[12] 张慧杰，杨德国，危起伟，等. 葛洲坝至古老背江段鱼类的水声学调查[J]. 长江流域资源与环境，2007，16(1)：86-91.

[13] Wei Qiwei, He J, Yang D, et al. Status of sturgeon aquacultureand sturgeon trade in China: a review based on two recent nationwide surveys[J]. Journal of Applied Ichthyology, 2004, 20(5): 321-332.

[14] 庄平，李大鹏，王明学，等. 养殖密度对史氏鲟稚鱼生长的影响[J]. 应用生态学报，2002，13(6)：735-738.

[15] 林浩然. 鱼类生理学[M]. 广州：广东高等教育出版社，2004：146-148.

[16] 李大鹏，庄平，严安生，等. 光照、水流和养殖密度对史氏鲟稚鱼摄食、行为和生长的影响[J]. 水产学报，2004，28(1)：54-61.

[17] 姚翠鸾，王维娜，王安利. 水生动物体内超氧化物歧化酶的研究进展[J]. 海洋科学，2003，27(10)：18-21.

[18] 李东洋，张天成，让慰清，等. 烹调油烟对大鼠肺组织 SOD 和 MDA 及骨髓 MN 变化的研究[J]. 卫生研究，1998，27(6)：382-384.

[19] 刘松岩. 环境胁迫对中华鲟体内自由基水平和抗氧化酶活力的影响[D]. 武汉：华中农业大学，2006.

[20] 吴长珠，刘友尧，林建华，等. 骨瘤患者血液 LPO、SOD、GSHPX 与 GSH 临床价值的初步探讨[J]. 肿瘤防治研究，1994，21(1)：38-40.

[21] 柯副恩. 论中华鲟的保护与开发[J]. 淡水渔业，1999，29(9)：4-7.

[22] 祁真. 对虾生态工程化养殖系统水环境变化及影响因子研究[D]. 浙江：浙江大学，2004.

[23] Meade J W. Allowable ammonia for fish culture[J]. Progressive Fish-Culturist, 1985, 47(3): 135-145.

[24] Thomas S L, Piedrahita R H. Apparent ammonia-nitrogen production rates of white sturgeon (*Acipensertransmontanus*) in commercial aquaculture systems[J]. Aquacultural Engineering, 1998, 17(1): 45-55.

[25] 战文斌. 水产动物病害学[M]. 北京：中国农业出版社，2004：8-12.

[26] 黄志坚，何建国，翁少萍，等. 鳜鱼细菌性病原的分离鉴定及致病性初步研究[J]. 微生物学报，1999，26(4)：241-246.

[27] 陆文浩，陈辉，邹勇，等. 银鲫不动杆菌病原菌鉴定及药敏试验[J]. 水产科学，2010，29(3)：156-161.

Effects of Culturing Density on Behavior and Immunity of Captive *Acipenser sinensis* and That on the Water Quality of Their Living Pool

Zhang Xiaoyan[1,2,3,5] Li Luoxin[4] Wei Qiwei[1,3,4]
Zhang Xianfeng[1,3] Shen Li[4] Zhang Yanzhen[5]
Yang Daoming[5] Luan Gang[5] Liu Jianwei[6]

(1. Institute of Hydrobiology, Chinese Academy of Sciences, Wuhan 430072, China;
2. Graduate University of Chinese Academy of Science, Beijing 100049, China;
3. Key Lab of Aquatic Biodiversity and Conservation, Chinese Academy of Sciences, Wuhan 430072, China;
4. Key Laboratory of Freshwater Biodiversity Conservation and Utilization, Ministry of Agriculture; Yangtze River Fisheries Research Institute, Chinese Academy of Fisheries Sciences, Jingzhou 434000, Hubei Province, China;
5. Beijing Aquarium, Beijing, 100081, China; 6. Chinese Academy of Fisheries Science, Beijing 100141, China)

Abstract Studies on effects of culturing density on behaviour and immunity of Chinese sturgeon *Acipenser sinensis* and the effects on the water quality of their living pool were reported in this paper. Six wild Chinese sturgeons and the 62 offsprings were cultured in three living densities of 2.7 kg/m^3, 2.4 kg/m^3, 2.0 kg/m^3 respectively in Beijing Aquarium between 2008 and 2009. Behaviour observation showed that the distribution of swimming-level of fish at different culturing densities, and the behaviour of younger fish changed more frequently. The sturgeon's appetite didn't show difference in different densities. Swimming speed and respiratory frequency of wild sturgeon and F1-1998 didn't change either. On the other hand, serum SOD, SOD/MDA and GSH-PX of F1-1998 decreased, while MDA increased at the density of 2.7 kg/m^3. The older fish including wild and F1-1998 were infected by bacteria at the densities of 2.7 kg/m^3 and 2.4 kg/m^3 when the total bacteria numbers were continuously high than 8 000 cfu/100ml. With the decrease of density, NH_3/NH_4^-, NO_2^- and turbidity did not fluctuate significantly, NO_3^-, PO_4^{3-} and total bacteria numbers were positively correlated significantly with the density. These results indicate that the effect of environmental capacity on aquatic microbial communities and the fish immunity is significant when the sturgeons are cultured in recalculated water system. An optimal culturing density is drawn at 2.0 kg/m^3 from this study.

Key words *Acipenser sinensis* Stocking density Circular tank Water quality Behaviour Immunity

性腺发育及年龄对养殖中华鲟抗氧化力的影响*

张晓雁[1,3,4,5,**]　李罗新[2]　张燕珍[5]
危起伟[1,2,4,***]　张先锋[1,4,****]　刘建伟[6]

(1. 中国科学院水生生物研究所　湖北武汉　430072
2. 农业部淡水生物多样性保护重点实验室　中国水产科学研究院长江水产研究所　湖北武汉　434000
3. 中国科学院大学　北京　100049　4. 中国科学院水生生物多样性与保护重点实验室　湖北武汉　430072
5. 北京海洋馆　北京　100081　6. 中国水产科学研究院　北京　100141)

摘　要　以养殖的子一代中华鲟为对象,研究了性腺发育及年龄对其抗氧化力的影响。结果表明,对于不同发育阶段的F1-1998雌性、雄性,睾酮和雌二醇/睾酮值,抗氧化力指标差别明显,且雌二醇/睾酮值与不同指标呈负相关、正相关或无显著相关性。不同年龄个体间的部分抗氧化力指标存在显著性差异,且与年龄存在一定相关性。表明进入性腺快速发育阶段的雄性氧化应激水平明显高于处于性腺慢速发育阶段的同龄雌性;随着年龄增长,未达性成熟中华鲟的抗氧化力水平逐渐增强;使用以上指标作为亚健康群体的筛选指标和氧化应激的生物标志物时,应充分考虑监测群体性腺发育和年龄因素,以使参考指标更准确。

关键词　中华鲟　性腺发育　年龄　超氧化物歧化酶　谷胱甘肽过氧化酶　丙二醛　超氧化物歧化酶/丙二醛

中华鲟是国家一级重点保护动物,由于葛洲坝和三峡大坝的建设,中华鲟产卵场变迁,新产卵场环境和水文条件的变化,使自然种群的繁殖受到影响,亲鲟年龄上升,体质下降,精子、受精卵和仔鱼质量下降[1,2]。由于近年水域环境污染,中华鲟种群性比结构改变,种群也持续衰退[3]。中华鲟资源量持续下降[4],在2009年被IUCN列为极危物种。实施人工繁殖和易地保护措施是开展中华鲟物种保护工作的重要途径之一[5]。然而,中华鲟性成熟时间长,在养殖过程中,中华鲟的健康水平将受到营养和环境因子等多方面因素的影响,易于引发疾病[6]。

需氧生物在长期进化中建立了抗氧化系统,以及时清除氧代谢过程产生的氧自由基,使机体保持在一定的稳态。当氧化-抗氧化的调控体系失衡时,生物将发生疾病[7]。超氧化物歧化酶(SOD)和谷胱甘肽过氧化酶(GPX)是硬骨鱼类主要的抗氧化酶,丙二醛(MDA)是脂质过氧化的终产物,代表机体或组织的氧化水平[8,9],超氧化物歧化酶(SOD)/丙二醛(MDA)指示抗脂质过氧化的潜在能力,这些指标反映生物体氧化损伤程度和调节能力,可以作为监测环境和生物健康情况的生物标志物[7-10]。一般对营养、摄食、环境因子等影响抗氧化系统的研究较多,而关于性腺发育、年龄等生理情况对抗氧化力影响的报道较少[11]。但是氧化应激产物会随着年龄积累,抗氧化酶活力随年龄增高而下降[12];在性成熟过程的不同阶段,抗氧化酶水平

---

*基金项目:国家自然科学基金项目(31172413);中国水产科学院研究型业务专项(2011C002);公益性行业(农业)科研专项(201203086)。

**第一作者,女,山东省栖霞人,高级工程师,博士研究生,主要从事鱼类生理生态学和水族馆科学与技术方向的研究,E-mail:zn7@ sina. com

***通讯作者(Corresponding author),E-mail:weiqw@ yfi. ac. cn

****通讯作者(Corresponding author),E-mail:Zhangx@ ihb. ac. cn

与性别的相关性也存在不同[11,13,14]。因此,可以通过分析性成熟过程和年龄对中华鲟抗氧化力的影响,评价鱼体和环境的健康情况。

2009 年 10 月我们进行了性腺发育及年龄对中华鲟抗氧化力的影响实验,以期为筛选亚健康中华鲟,调控养殖环境,建立疾病控制的有效技术手段等提供理论依据,也可为野生中华鲟繁殖种群的生理监测提供基础材料。

1 材料与方法

1.1 研究材料

实验用中华鲟为人工繁殖培育所得。性腺发育的影响实验在厦门中华鲟繁育保护基地进行。1998 年繁殖的中华鲟同一养殖条件下饲养,投喂相同的冰冻鲜饲料。养殖池为 15 m × 30 m × 1.5 m的长方形水泥池,水池正中有一排污出水口,养殖水位保持在 1.2 m。养殖用水为汀溪水库排出水,每日换水 1 次,日换水量为池中水量的 50 %,实验期间池中水温 21 ~ 22 ℃,溶氧 6.0 ~ 9.0 mg/L,酸碱度 7.5。试验样本共 19 尾,身体健康,体重为(51.1 ± 6.0) kg,体长为(167.9 ± 10.1) cm。

年龄实验在北京海洋馆进行。2005 年(4 龄,$n = 6$)、2001 年(8 龄,$n = 7$)、1998 年(11 龄,$n = 4$)和 1997 年(12 龄,$n = 4$)繁殖的中华鲟(F1-2005,F1-2001,F1-1998,F1-1997)在同一条件下饲养,投喂相同的冰冻鲜饲料。养殖池为 29.0 m × 11.0 m × 4.4 m 的长方形水泥池,养殖水深保持在 3.5 m,循环率为 50 min/次。每周定期吸出池底污物,冲洗砂滤罐。实验期间池中水温 21 ~ 22 ℃,溶氧 7.0 ~ 7.8 mg/L,酸碱度 7.6 ~ 7.8。4 个年龄组的体重分别为(25.2 ± 4.9) kg,(37.9 ± 11.6) kg,(54.3 ± 7.5) kg,(77.9 ± 11.3) kg;体长分别为(142.2 ± 11.0) cm,(158.4 ± 9.1) cm,(184.8 ± 10.5) cm,(190.8 ± 17.2) cm。

1.2 样品的采集与测定

采用微创手术活体检查的方法,对中华鲟进行活体取样,利用肉眼和显微镜进行性别和性腺发育时期的鉴定,性腺分期方法按陈细华等[15]对中华鲟的性腺分期方法。性腺发育实验中雌性卵巢发育大约至 II 期初至 II 期中,雄性精巢处于 II 期末到 III 期初。年龄实验中实验鱼未达性成熟,性腺发育大约至 II 期初至 II 期中。

为降低应激反应,实验鱼的捕捞时间尽量缩短,每尾鱼的捕捞控制在 5 min 内完成。尾静脉采血后用肝素钠抗凝,3 000 r/min 离心 20 min 后取上清,-4 ℃保存备用。超氧化物歧化酶(SOD)含量检测用黄嘌呤氧化酶法,丙二醛(MDA)含量测定用硫代巴比妥酸法,谷胱甘肽过氧化酶(GPX)含量测定采用比色测定法。使用南京建成生物工程研究所的试剂盒测定。雌二醇(E2)和睾酮(T)用放射免疫法测定,试剂盒由北京普尔伟业生物科技有限公司提供。

1.3 统计分析

数据通过统计软件 SPSS 13.0 进行分析处理。结果为平均值 ± 标准差。不同性别的血清抗氧化力利用 Independent Two-Sample *T* Test 检验显著性差异。利用方差分析(One-way ANOVA)分别检验各年龄组的差异显著性,利用 S-N-K 法进行多重比较。使用 Pearson Correlations 分析性腺发育及年龄与抗氧化力水平的相关性[11,12]。显著性水平设为 0.05。

2 结果与分析

2.1 性腺发育与中华鲟的抗氧化力

不同性别 F1-1998 的性类固醇激素水平和抗氧化力结果见表 1。雌、雄实验鱼的雌二醇(E2)无显著性差异,而睾酮(T)和 E2/T 差异显著,雌性 T 仅为雄性 1/10,而 E2/T 则约 6 倍于雄性。雌性中华鲟的超氧化物歧化酶(SOD)、谷胱甘肽过氧化酶(GPX),超氧化物歧化酶(SOD)/丙二醛(MDA)水平均高于雄性,但差异不显著,而 MDA 显著低于雄性,均数约为雄性的 76 %。雌性和雄性的抗氧化力指标间只有 MDA 与 SOD/MDA 之间相关性极显著($r = -0.915$, $r = -0.818$, $P < 0.01$),其余指标间没有显著相关性。E2,T 与抗氧化力指标间无相关性,E2/T 与 SOD,GPX 无显著相关性,而与 MDA 显著负相关,与 SOD/MDA 显著正相关,r 分别达到 -0.635和 0.709。

表 1 厦门基地不同性别 F1-1998 的养殖中华鲟性类固醇激素和抗氧化力水平

Tab. 1 Sex steroids and activity of antioxidant enzymes in serum of different sex Chinese sturgeon cultured at Xiamen

性别 Gender	E2 /(Ng/ml)	T /(Ng/ml)	E2/T	SOD /(NU/ml)	GPX /酶活力单位	MDA /(nmol/ml)	SOD/MDA
雌性 females, $n=9$	3.72 ±0.23	0.19 ±0.06[a]	20.68 ±7.62[a]	173.5 ±16.7	419.7 ±236.1	5.8 ±1.8[a]	32.2 ±9.9
雄性 males, $n=10$	3.07 ±0.99	1.94 ±1.58	3.36 ±3.02	167.5 ±17.5	396.1 ±187.0	7.6 ±2.0	23.5 ±8.0

注:同一列中参数上方字母不同代表有显著性差异($P<0.05$),相同则无显著性差异($P>0.05$).

Note: The different letters on the parameters in one row mean significant differences ($P < 0.05$); the same ones mean no significant differences($P > 0.05$).

2.2 年龄与中华鲟的抗氧化力

2.2.1 SOD 的活力水平

4 个年龄组的 SOD 无显著性差异,说明各年龄组间 SOD 差异不大,活力水平较为稳定(图 1)。随着年龄的增长,中华鲟 SOD 呈逐渐上升趋势,其中 4 龄到 8 龄上升的比例最大,达到 15 %,而 8 龄到 11 龄和 12 龄,依次分别上升 2 %左右。SOD 与年龄间无显著相关性。

2.2.2 GPX 的活力水平

随年龄增长,GPX 呈下降趋势。从 4 龄到 8 龄降幅较小,至 11 龄时稍有回升,但是在 12 龄又快速下降,降幅达 62 %(图 2)。4 龄、8 龄和 11 龄间差异不显著,但 12 龄组的 GPX 明显小于其余三组,且有显著性差异。GPX 与年龄之间相关性极显著($r = -0.547$, $P < 0.01$)。

2.2.3 MDA 的水平

MDA 在 8 龄时稍微增长,然后呈逐渐下降趋势(图 3)。从 8 龄经 11 龄到 12 龄组,降幅依次达 20 %和 25 %。4 龄和 12 龄的组内变异系数明显大于 8 龄和 11 龄组(CV = 49.12 %,18.82%,27.12%,34.78 %)。各年龄组间差异不显著。MDA 与年龄之间相关性显著($r = -0.519$, $P < 0.05$)。

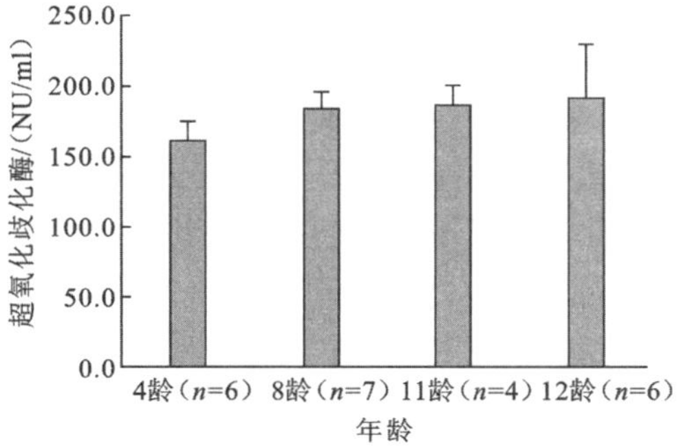

图1 不同年龄中华鲟的超氧化歧化酶活力

Fig. 1 Activity of SOD in serum of cultured Chinese sturgeon examined in different age Class

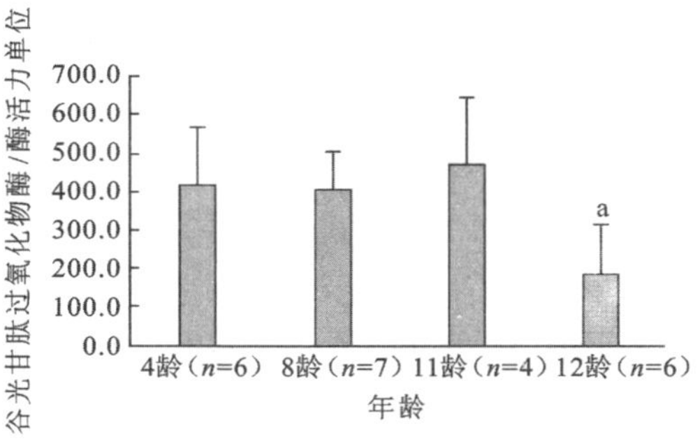

图2 不同年龄中华鲟的谷胱甘肽过氧化物酶活力

Fig. 2 Activity of GPX in serum of cultured Chinese sturgeon examined in different age class

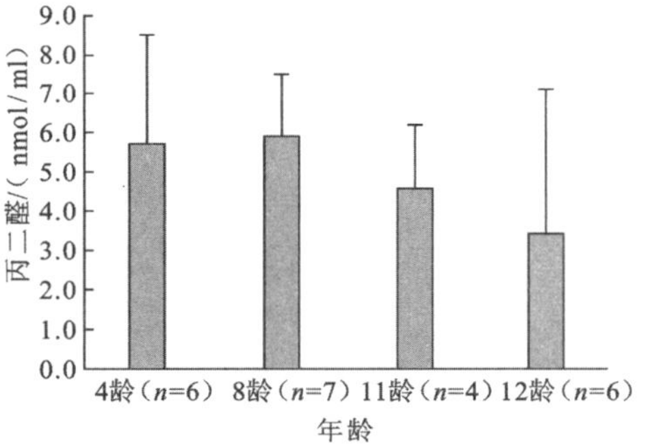

图3 不同年龄中华鲟的丙二醛活力

Fig. 3 Level of MDA in serum of cultured Chinese sturgeon examined in different age class

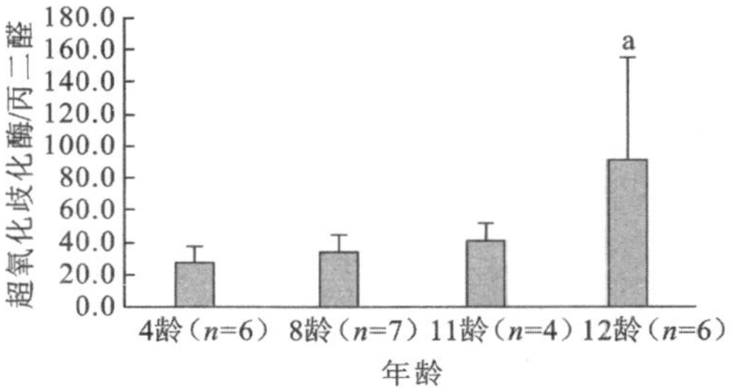

图4 不同年龄中华鲟超氧化歧化酶活力/丙二醛的水平

Fig. 4 Level of SOD/MDA in serum of cultured Chinese sturgeon examined in different age class

2.2.4 SOD/MDA 的水平

中华鲟 SOD/MDA 值随年龄的增长而上升(图4),4 龄、8 龄和 11 龄组间无显著性差异,12 龄组的 SOD/MDA 值明显高于其余3组,并且有显著性差异,组内变异系数最大($CV=71.83\%$)。SOD/MDA 与年龄之间相关性极显著($r=0.569$, $P<0.01$)。

3 讨论

3.1 性腺发育对抗氧化力的影响

以往实验结果显示,处于性腺发育缓慢过程的中华鲟(性腺发育至 II 期初期到中期),雌雄抗氧化酶水平没有显著性差异(张晓雁等,数据未公开发表)。本文实验中,雌性 F1-1998 未达性成熟,性腺尚处于缓慢的发育过程;而雄性已到发育加快的阶段(精巢发育达到 II 期末到 III 期初),SOD,GPX 和 SOD/MDA 值在性别间不存在显著差异,而 MDA 水平雄性显著高于雌性 25 %左右,说明雄性性腺进入发育加快阶段时引起 MDA 水平升高。

已有研究表明西伯利亚鲟 E2 和 T 在性腺发育的不同时期存在规律性变化,因此 T 和 E2 二个激素的比值可以比较灵敏的判别性别和性腺发育水平,且雄性性腺发育进入 III 期,T 明显上升[16]。本文雄性的 T 显著高于雌性,而 E2 稍低于雌性,E2/T 显著低于雌性,结合组织学检查结果说明,这主要是雄性性腺发育加快引起的差异。有研究表明,性成熟的雌雄性腺中的抗氧化力水平不同,这可能与性激素水平相关[14,17]。而在本研究中,E2/T 与 SOD/MDA 正相

关,与 MDA 负相关,说明随着雄性性腺发育加快,性类固醇激素水平(特别是 T)变化,雄性实验鱼的氧化应激水平改变。性腺发育加快的雄性 MDA 水平高,这指示在代谢过程中产生了更多的脂质代谢物。脂氧化应激产物是非常有价值的生物氧化损伤标志物[10],这也表明,一旦进入性腺快速发育过程,雄性的氧化应激水平也随之升高。一般来讲,氧化-抗氧化系统处于动态平衡,当氧化应激水平高,将激发抗氧化水平上升,使生物体处于氧化应激状态。本文实验鱼 MDA 和 SOD/MDA 相关,雄性 MDA 水平高于雌性,而 SOD/MDA 值却低于雌性,这有可能使雄性在环境不适时,其氧化-抗氧化系统易于失衡,机体免疫力下降。进而推知,进入性腺快速发育阶段的雄性中华鲟可能对养殖环境更敏感,当环境发生变化时易于引发疾病。

3.2 年龄对抗氧化力的影响

对人的研究表明年龄与抗氧化力存在相关性,Harman[18]早在 1956 年就提出了随年龄增长,活性氧产物增加,而抗氧化力将减弱的衰老与自由基学说。后期很多研究结果证实[19,20],人衰老后抗氧化酶活力下降,氧化应激产物增加,或者各种原因引发氧自由基产物增加,清除率下降时也将引发衰老等。而年龄对鱼抗氧化力的影响却鲜于报道,且研究结果存在不同。已有研究表明鱼年龄对抗氧化力活性影响的差别较大,各种鱼抗氧化力水平随年龄增加升高、降低或不变化[12,21]。鱼的生理和生态特点都会影响抗氧化力活性,但是底栖鱼类由于生活环境和食物来源及食物种类稳定,对抗氧化力的干扰相对较弱,可以更准确地说明年龄对抗氧化力的影响程度[12]。水族馆养殖环境稳定,鱼类受到的干扰小,因此,可以比较准确的表现抗氧化力在不同年龄的活性。

实验中,随年龄增长,SOD 比较稳定,随年龄小幅度增长,各年龄组间没有显著性差异。Sanz 等[22]对 0,1,2,3,5 + 龄的纳氏鲟研究发现,仅血浆 SOD 在 5 + 时升高。而 Rudneva 等[12]发现游泳慢的底栖鱼类血细胞 SOD 随年龄升高而降低,可是活动性强的鱼类多保持稳定。活动性强的鱼类为了抵抗大量的活性氧产物,有可能在一生中保持比较高且稳定的 SOD 活性[7]。随年龄增长,中华鲟的活动力增加,也有可能保持较高的 SOD 水平以抵抗氧自由基的损害。性成熟期间抗氧化力水平一般也会受到影响[7,23]。而此实验鱼为 4 ~ 12 龄,尚未达性成熟。因此,不存在性成熟的影响问题。

随年龄增长,实验鱼 MDA 水平降低,表明脂质过氧化反应的程度逐渐下降。这可能是由于随年龄增加 SOD 少量升高的结果。另外,SOD/MDA 明显增加也表明随着年龄的增加,处于旺盛生长期的中华鲟抗氧化力在逐渐增强。

SOD 将氧自由基催化生成氧分子和过氧化氢,一部分过氧化氢以及脂质过氧化物将被 GPX 还原成相应的水和醇。实验鱼 SOD 活性变化不显著,而 MDA 水平随增龄明显降低,因此,GPX 将随增龄降低。这也说明抗氧化酶之间协同作用,共同保持机体的稳定,而 SOD 在抗氧化过程中起主要作用,主要决定机体抗氧化力强弱。

4 结论

未达性成熟中华鲟 GPX 和 SOD/MDA 随年龄增长而上升。接近性成熟的雄鱼性腺发育加快,MDA 的水平显著升高。因此,使用这些指标作为亚健康群体的筛选指标和氧化应激的生物标志物时,应充分考虑监测群体年龄及性腺发育的影响,使参考指标更准确。

致谢

厦门中华鲟繁育保护基地李奕慰先生为本实验提供了大力支持,在此表示感谢!

参考文献

[1] 张辉,危起伟,杨德国,等. 葛洲坝下游中华鲟(*Acipenser siensis*)产卵场地形分析[J]. 生态学报,2007,27(10):3945-3957.

[2] 杨德国,危起伟,陈细华,等. 葛洲坝下游中华鲟产卵场的水文状况及其与繁殖活动的关系[J]. 生态学报,2007,27(3):863-869.

[3] 危起伟,陈细华,杨德国,等. 葛洲坝截流24年来中华鲟产卵群体结构的变化[J]. 中国水产科学,2005,12(4):452-457.

[4] 张慧杰,杨德国,危起伟,等. 葛洲坝至古老背江段鱼类的水声学调查[J]. 长江流域资源与环境,2007,16(1):86-91.

[5] Wei Qiwei, He J, Yang D, et al. Status of sturgeon aquaculture and sturgeon trade in China: a review based on two recent nationwide surveys[J]. Journal of Applied Ichthyology, 2004, 20 (5): 321-332.

[6] 潘连德,刘健,陈锦辉,等. 中华鲟主要病害临床诊断及控制技术[J]. 渔业现代化,2009,36(6):29-33.

[7] Martinea-Alvarez R M, Morales A E, Sana A. Antioxidant defenses in fish: biotic and abiotic factors[J]. Reviews in Fish Biology and Fisheries, 1997, 15(1/2): 75-88.

[8] Lepage G, Munoz G, Champagne J, et al. Preparative steps for the accurate measurement of malondialdehyde by high-performance liquid chromatography[J]. Analytical Biochemistry, 1991, 197: 277-283.

[9] Toguyeni A, Fauconneau B, Boujard T, et al. Feeding behavior and food utilization in tilapia, *Oreochromis niloticus*: Effect of sex ratio and relationship with the endocrine status[J]. Physiology &Behavior, 1997, 62: 273-279.

[10] Vega-Lopez A, Galar-Martinez, Jimeneaorozco F A, et al. Gender related differences in the oxidative stress response to PCB exposure in an endangered goodeid fish (*Girardinichthys viviparous*) [J]. Comparative Biochemistry and Physiology-Part A: Molecular &Integrative Physiology, 2007, 146: 672-678.

[11] Rudneva I I, Skuratovskaya E N. Gender peculiarities of blood antioxidant enzyme activity of some Black Sea coastal fish species[J]. Journal of Applied Ichthyology, 2009, 49(1): 125-128.

[12] Rudneva I I, Skuratovakaya E N, Kuzminova N S, et al. Age composition and antioxidant enzyme activities in blood of Blank SEA teleosts[J]. Comparative Biochemistry and Physiology Part C: Toxicology & Pharmacology, 2010, 151: 229-239.

[13] Winzer K, Winston G W, Becker, et al. Sex-related responses to oxidative stress in primary cultured hepatocytes of European Flounder(*Platichthys flesus* L.)[J]. Aquatic Toxicology, 2001, 52: 143-155.

[14] Wei R, Zhang S, Wang C, et al. Antioxidant enzyme activities in different genders and tissues of amphioxus *Branchiostoma belcheri tsingtauense*[J]. Chinese Journal of Oceanology and Limnology, 2007, 25(1): 73-77.

[15] 陈细华,危起伟,杨德国,等. 养殖中华鲟性腺发生与分化的组织学研究[J]. 水产学报,2004,28(6):633-639.

[16] 张涛,章龙珍,赵峰,等. 西伯利亚鲟不同性别与卵巢发育时期血液性类固醇激素的差异与判别分析[J]. 海洋渔业, 2008,30(1):19-25.

[17] Rudneva I I, Skuratovakaya E N. Relationship between processes of peroxide oxidation of Lipids in gonads of Red Mullet *Mullus barbatus ponticus* [J]. Ukrainskii Biokhimicheskii Zhurnal, 1995, 67(5): 72-79.

[18] Harman D. A theory based on free radical and radiation chemistry[J]. Journal of Gerontology, 1956, 11: 298-300.

[19] 周强,曹春燕,陈世华. 年龄对血中抗氧化酶活性及病二醛水平的影响[J]. 哈尔滨医科大学学报,2002,36(6):485-486.

[20] 孙存普,张建中,段绍瑾. 自由基生物学导论[M]. 合肥:中国 科学技术大学出版社,1999:2-3.

[21] Wdzieczak J, Zalesna, Wujec, et al. Comparative studies on superoxide dismutase, catalase, and peroxidase levels in erythrocytes and livers of different freshwater and marine fish species[J]. Comparative Biochemistry and Physiology - Part B:. Biochemistry and Molecular Biology, 1982, 73: 363-365.

[22] Sanz A, Hidalgo M C, Morales A E, et al. Evolution of antioxidant defenses and lipid peroxidation with age in the sturgeon Acipenser naccaii[C]//Proceeding of the 4th International Symposium on Sturgeon, July 8-13, Oshkosh, WI, 2001.

[23] Otto D M E, Moon T W. Endogenous antioxidant systems of two teleost fish, the rainbow trout and the black bullhead, and the effect of age[J]. Fish Physiology and Biochemistry, 1996, 15(4): 349-358.

Effects of Gonad Development and Age on the Antioxidant Defenses of Chinese Sturgeon(*Acipenser sinensis*)

ZhangXiaoyan[1,3,4,5] Li Luoxin[2] Zhang Yanzhen[5]
Wei Qiwei[1,2,4] Zhang Xianfeng[1,4] Liu Jianwei[6]

(1. Institute of Hydrobiology, Chinese Academy of Sciences, Wuhan 430072, China;
2. Key Laboratory of Freshwater Biodiversity Conservation, Ministry of Agriculture; Yangtze River Fisheries Research Institute, Chinese Academy of Fisheries Sciences, Wuhan 434000, China;
3. Graduate University of Chinese Academy of Science, Beijing 100049, China;
4. Key Lab of Aquatic Biodiversity and Conservation, Chinese Academy of Sciences, Wuhan 430072, China;
5. Beijing Aquarium, Beijing 100081, China; 6. Chinese Academy of Fishery Sciences, Beijing 100141, China)

Abstract The effects of gonad development and age on the serum antioxidant enzyme activities of captive F1 Chinese sturgeon, *Acipenser sinensis*, were studied. When females in the period of slow gonadal development and males in the period of fast gonadal development were compared, testosterone (T) and the ratio of estrogen (E2): testosterone levels were significantly different. Compared to males, the females had a significantly lower malondialdehyde (MDA) level ($P < 0.05$), but higher activity levels of superoxide dismutase (SOD), glutathione peroxidase (GPX) and SOD/ MDA, although these differences were not significant ($P > 0.05$). For MDA and SOD/MDA in the serum of males and females, high values were found for the correlation coefficients ($r = -0.915$, $r = -0.818$, $P < 0.01$, respectively). Serum E2/T levels correlated with MDA, SOD/MDA ($r = -0.635$, $r = 0.709$, $P < 0.05$, respectively), while not correlated with SOD or GPX. No significant differences in the activity of serum SOD or MDA were observed in all four different ages (4, 8, 11 and 12 years old) during the period of slow gonad development. The lowest level of GPX activity was found in the 12-year old age group ($P < 0.05$), whereas the highest level of the SOD/MDA ratio was observed in this group ($P < 0.05$). The correlation between the age and the activity of GPX, MDA was negative ($r = -0.547$, $P < 0.01$, $r = -0.519$, $P < 0.05$, respectively), while the correlation between the age and the level of the SOD/MDA ratio was positive ($r = 0.569$, $P < 0.01$). These results suggest that the level of oxidant stress in male sturgeons during the period of fast gonad development is higher than those of females during the period of slow gonad development. The antioxidant defense mechanisms of sturgeons gradually increase with increasing age. Therefore, the gonad development and age of the fish should be considered when indexes of antioxidant enzyme activities are used to assess status of fish health. Understanding the biological signaling that underlies these changes will make the use of these parameters even more accurate.

Key words *Acipenser sinensis* Gonad development Age SOD GPX MDA SOD/MDA

降盐度梯度对中华鲟血液生化指标的影响

张晓雁[1,2,4]　危起伟[1,3,*]　张先锋[1,**]　吴香香[4]
王彦鹏[3]　李罗新[3]　岳　欣[5]　刘建伟[6]

(1. 中国科学院水生生物研究所　湖北武汉　430072　2. 中国科学院研究生院　北京　100049
3. 农业部淡水生物多样性保护重点开放实验室 中国水产科学研究院长江水产研究所　湖北武汉　430223
4. 北京信沃达海洋科技有限公司(北京海洋馆)　北京　100081
5. 中国农业大学　北京　100193　6. 中国水产科学研究院　北京　100141)

摘　要　分析了3龄中华鲟27天盐度驯化(盐度26.7‰~2.5‰)及30天淡水适应时24项血液生化指标的变化,并与淡水对照组进行了比较。随盐度下降,血清K^+,Na^+,Ca^{2+}水平减低,尿素氮、肌苷、尿酸、丙氨酸氨基转移酶、γ-谷氨酰转肽酶(GGT)、血清总胆汁酸值也呈下降趋势,其他指标短时波动或变化不明显。在淡水适应30天后,4项指标恢复至海水初始水平,表明中华鲟生理有一定稳定调节能力。6项指标没有恢复至初始,且3项略高于淡水对照组,表明中华鲟的生理随盐度下降发生了变化。盐度21‰时,多项生化指标明显波动,提示,这可能是中华鲟生理调节的关键盐度,渗透压调节过程也需要消耗更多的能量。

关键词　中华鲟　盐度降低　血液生化指标

Changes in Serum Biochemical Parameters of *Acipenser sinensis* Gray 1835 Caused by Decreasing Environmental Salinity

Zhang Xiaoyan[1,2,4]　Wei Qiwei[1,3,*]　Zhang Xianfeng[1,**]　Wu Xiangxiang[4]
Wang Yanpeng[4]　Li Luoxin[3]　Yue Xin[5]　Liu Jianwei[6]

(1. Institute of Hydrobiology, Chinese Academy of Sciences, Wuhan 430072, China;
2. Graduate University of Chinese Academy of Science, Beijing 100049, China;
3. Key Laboratory of Freshwater Biodiversity Conservation and Utilization, Ministry of Agriculture; Yangtze River Fisheries Research Institute, Chinese Academy of Fisheries Sciences, Jingzhou 434000, Hubei Province, China;
4. Beijing Aquarium, Beijing, 100081, China;　5. China agricultural University, Beijing 100193, China;
6. Chinese Academy of Fisheries Science, Beijin 100141, China)

Abstract　Blood serum parameters were Chinese sturgeons (*Acipenser sinensis*) were examined during a 27 day gradual acclimation period from seawater (26.7 ‰ salinity) to slightly brackish water (2.5 ‰ salinity) and subsequent transfer to freshwater (0 ‰ salinity) for a further 30 day period. The results were compared against serum samples from a control group reared throughout in freshwater. For each fish, the levels of 24 serum

* 通讯作者(Corresponding author), E-mail: Weiqw@ yfi. ac. cn
** 通讯作者(Corresponding author), E-mail: Zhangx@ ihb. ac. cn

biochemical parameters were examined. The results indicated that gradual transfer from seawater to brackish water caused the serum concentrations for potassium, sodium, and calcium ions to decrease. This trend was also observed for blood urea-nitrogen (BUN), inosine (CR), uric acid (UA), alanine aminotransferase (ALT), gamma glutamyl transpeptidase (GGT), and total bile acid (TBA). All other serum parameters showed only transient changes or no changes at all. After being held subsequently for 30 days in freshwater, the four ion concentrations returned to values considered normal for seawater, indicating that Chinese sturgeon are capable of homeostasis. However, the other six parameters did not return to initial values. In fact, ALP, ALT and Fe^{2+} levels were raised in comparison to the freshwater control group. Theses observations indicate that major physiological changes occur during the acclimation process. Specially, at 21 ‰ salinity a number of biochemical parameters were subject to noticeable fluctuations, suggesting this level may represent the pivotal condition of physiological regulation. Hence, it may be energetically costly for osmoregulatory process to return to normal in Chinese sturgeon, due to metabolic and nutrient related functions taking much longer to recover.

Key words *Acipenser sinensis* Gray　Decreasing salinity　Serum biochemical indices

1　Introduction

All sturgeon species have complex life histories (Kynard, 1997), requiring tolerance to a large range of environmental conditions at different times of the life cycle, which may be facilitated by physiological mechanisms (Burggren et al., 1978). As a result, physiological research on sturgeon has primarily been focused on osmoregulatory capacity (Krayushkina, 1998; Lebreton et al., 1998) and on respiratory parameters (McKenzie et al., 1997; Maxime et al., 1995). Blood parameters have also been used to monitor the effects of water quality and other environmental parameters (Asadi et al., 2006a, b).

The Chinese sturgeon is a typical anadromous fish, which migrates upstream to spawn in the Yangtze River and downstream of the Jinsha Jiang River (Chen, 2007). Due to the construction of the Gezhouba Dam on the Yangtze River, the alternative natural spawning grounds below Gezhouba Dam have been adopted by the species, however the new grounds have also been critically reduced. Furthermore, the geo-topographic environment and the water quality of these new locations are different from the historical spawning grounds (Wei et al., 2005), which may impact on natural reproductive parameters and on the successful recruitment of larvae into the recruiting population, further endangering the survival of this species. Therefore, catching adults already at the estuary and bringing them into a captive environment for propagation before the fish has started to go upstream on its spawning migration, may be an effective measure to allow sturgeons to fully spawn. Moreover, wild sturgeons that are accidentally caught and injured in the estuary require rehabilitation in artificial environments, which are usually located in freshwater. While the serum biochemistry of juvenile Chinese sturgeon cultured in fresh water has been described by Zheng et al. (2007), the impact of changes in environmental salinity on serum biochemical parameters has not been examined in specimens caught in the wild.

This study had, therefore, two major objectives: (i) to provide baseline data on serum biochemical parameters, and (ii) to assess how serum biochemistry is modified following salinity stress. The response of sturgeon serum biochemistry to environmental conditions may be used to evaluate the overall health and nutrition status, as well as the tolerance to environmental change.

Consequently, the conditions to maintain the optimal physiological state of this species may be delineated for application to management schemes in captive breeding/rehabilitation centers and may help to understand how optimal sites for reintroduction to natural habitats can be selected.

2 Materials and methods

2.1 Study animals

The fish selected for this experiment were from a 3 year old filial generation. The fish were raised in freshwater holding tanks at the Beijing Aquarium from 2 months onward to year class two. Thereafter, for a 1 year period prior to the present study, a proportion of the raised individuals were transferred to a holding tank containing artificial seawater (after gradual acclimation; i. e. 3 % increase in salinity per 3 days over a 30 day period), while the remainder of the cultured population were transferred to the Chinese sturgeon public aquarium, and raised in fresh water also for a full year. Prior to the start of the trials, the average fork lengths and average weights of experimental sturgeon ($n=6$) raised in artificial seawater was (84.3 ± 2.4) cm and (5.1 ± 0.4) kg, respectively. The control group ($n=6$) raised in freshwater had at that time an average fork length and average weight of (110.3 ± 2.9) cm and (10.3 ± 0.5) kg, respectively.

2.2 Experimental tanks

The seawater holding tank is circular and of dimensions 10.0 m (diameter) × 0.9 m (height), with a maximum water volume of 72 m^3. The freshwater display aquarium is rectangular and of dimensions 29.0 m (length) × 11.0 m (width) × 4.4 m (height), with a maximum water volume of 1 200 m^3. It has a public exhibition window of 20.0 m (length) × 3.0 m (width). The bottom of the aquarium is covered by sand, and the edges of some rocks are covered with soft, non-toxin synthetic resin material to protect the fish from accidental injury. The aquarium is equipped with an advanced life support system, which controls the water quality artificially. The stocking density of the seawater and freshwater tanks are 2.2 and 2.12 kg/m^3 respectively.

2.3 Experimental design

The experiment included two sturgeon groups: (i) a group continuously reared in fresh water, (ii) a group subjected to gradual acclimation from full-strength salinity to freshwater. Before the start of the trials, tests were run to detect any variation in biochemical blood parameters in fish of different size ranges (5-11 kg). No significant differences were found for any of the indicators. As a result 12 fish were selected at random for the two experimental groups.

The experiment on salinity acclimation was run for 57 days. During the first 27 days, the salinity levels were progressively reduced by about 3 ‰ (mean: 3.0; range: 2.4-3.9; SD: ± 0.7) every 3 days (equal amount of freshwater added to the control). Hence, the gradual reduction from the initial salinity level at 26.7 ‰, passed steps at 24.1, 21.3, 17.5, 13.6, 9.8, 7.4, 5.0, and 2.5 salinity. After a further 3 days, the acclimated sturgeons were transferred to the Chinese sturgeon public

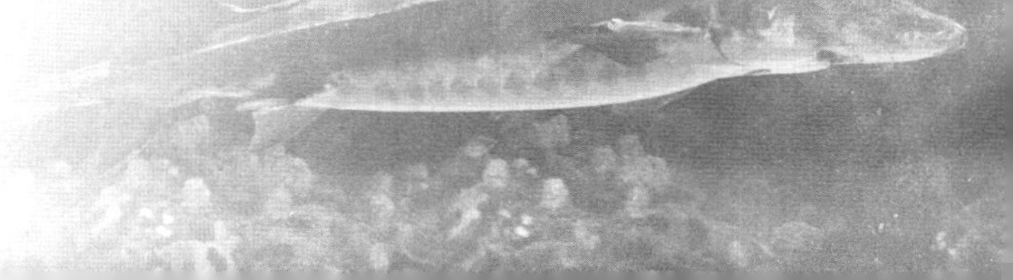

aquarium for a further 30 days acclimation to full freshwater.

Blood was sampled from the caudal vein of each fish at 9 days intervals, when the salinity reached 26.7 ‰, 21.3, 9.8, 2.5, and at thirty days thereafter at 0 ‰. While originally there were six fish, the freshwater samples are based on four fish only, due to the markers being lost on two fish after transfer to the aquarium tank. The same fish were used throughout the study. Blood was also sampled from the fish in the control group ($n = 6$) on the same days as from the acclimation group (at 0 ‰). Prior to the onset of the study, validation experiments had been conducted to confirm that repeated use of syringes on the same animal would not alter the serum biochemical composition being measured (i. e. stress effects) (Yang et al., 2005). In order to minimize the dietary influence on the metabolic status, fish were not fed on the day before blood sampling (Baker et al., 2005; Kiffer et al., 1998). The fish were removed rapidly and placed into a barrel where the individuals were anaesthetized with MS-222 (90 ppm). Blood was drawn via the caudal vessels using a heparin-coated needle and syringe (5 ml).

All biochemical parameters were measured using test kit preparations (SCKISUI, Sekisui Medical Co. Ltd, Kyoto, Japan). All samples were analyzed using an automated biochemical analyzer (OLYMPUS AU400). Serum biochemical parameters determined included: total bilirubin (TBIL), direct bilirubin (DBIL), indirect bilirubin (IBIL), total protein (TP), albumin (ALB), globulin (GLB), the albumin and globulin ratio (A/G), alanine aminotransferase (ALT), aspartate aminotransferase (AST), alkaline phosphatase (ALP), gamma glutamyl transpeptidase (GGT), total bile acid (TBA), cholinesterase (CHE), potassium (K^+), sodium (Na^+), chlorine (Cl^-), calcium (Ca^{2+}), phosphorus (P^{3+}), magnesium (Mg^{2+}), iron (Fe^{2+}), blood urea nitrogeon (BUN), inosine (CR), uric acid (UA), total cholesterol (CHO) and glucose (GLU).

Water quality was monitored daily. Temperature (T) and pH values were test using the HANNA convenient pH instrument (combined pH electrode, ATC with automatic temperature compensation, made in Italy). Oxygen (O_2) levels were measured by an infiltration membrane technique using HACH (made in America), NH_3/NH_4^+ and NO_2^- were measured by HACH Spectrophotometer (made in America). Water quality was maintained under the following conditions: T: 21-24 ℃, O_2: 7-10 mg/L, pH: 7.2-8.0, NH_3/NH_4^+: 0.01-0.05 mg/L, NO_2^-: 0-0.1 mg/L.

2.4 Data analyses

For all measurements, both treatment groups were analyzed for normality using Normality plots with tests. To examine whether there is any significant difference at each salinity level, species and treatment effects for each variable, parameters were assessed for the biochemical variables using a two-way ANOVA test. Duncan's test was also used to determine where differences occurred. Independent-Sample T Tests were used to determine if there were differences between the treatment group and the control group. Where the assumption of normality was violated, ANOVA on ranks (Friedman test) was used to confirm the results. The Mann-Whitney U test was used to determine whether there were differences between the treatment group and control group. Alpha (α) in all cases was 5 %($P < 0.05$).

All values are presented as mean ± SE. All statistics were performed using SPSS(ver. 11.0).

3 Results

Figure 1 presents the change in Na^+, Cl^- ion serum levels in the acclimation group, from sea water(26.7 ‰) to freshwater across the 57 days experimental period. The decline was both gradual and significant. During the subsequent 30 days period, when fish were maintained in fresh water, both the Na^+ and Cl^- values were observed to rise to that similar to fish in both in seawater(26.7 ‰ salinity) and in the control group. The other ions, Ca^{2+}, P^{3+}, Mg^{2+} serum Fe^{2+}, and ALT(Tables 1 and 2), showed a similar pattern of change to Na^+ and Cl^-. The only difference found was for the K^+ value, which showed no significant difference, despite a similar change being recorded. Also, TBA showed a decrease in concentration with declining salinity from 26.7 seawater to 2.5 slightly brackwater, but subsequently high levels were recorded in freshwater, although no significant difference was found.

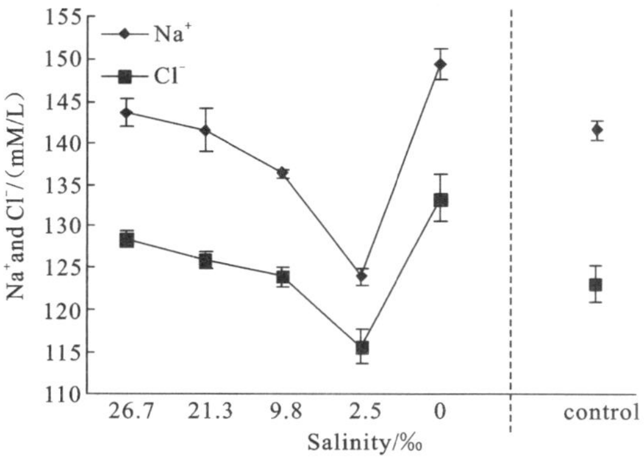

Fig. 1 The Na^+ and Cl^- ion serum levels of sturgeon *Acipenser sinensis* during the 27 day acclimation period from sea water(26.7 ‰) to slightly brackish water(2.5 ‰), and after a 30 day period in freshwater(0 ‰). The values represent mean ± SE. At each concentration $n = 6$ fish(the same each time), expect for 0 ‰ salinity when $n = 4$; control group $n = 6$ fish

The results presented in Tables 1 and 2 show no significant changes in TBIL, DBIL, IBIL TP levels during declining salinity. The only exception was TP, when compared with the control group. Also, an increase in AST levels was recorded at the salinity of 21.3, followed by a significant declining trend. After 1 month of fresh water acclimation, the AST value was lower than that recorded for the control fish. All salinity treatments showed no significant difference when compared to the control group. The change in ALB values followed a similar pattern to AST, although no significant difference was found. All salinity treatments(except for 2.5) showed no significant difference when compared to the control group.

Table 1 Serum biochemical parameters of *A. sinensis* at different salinity levels. Mean ± SE are presented. Data values with different letters indicate significant differences ($P < 0.05$). Two-way ANOVA test and Duncan's test were used for parameters K^+, Ca^{2+}, P^{3+}, Mg^{2+}, UA, TP, ALB, GLB, *A/G*, ALT, AST, ALP, CHE, Fe^{2+}. ANOVA on ranks (Friedmen tests) and two-way ANOVA tests were used for Na +, Cl^-, BUN, CR, TBIL, DBIL, IBIL, GGT, TBA, GLU. At each concentration, $n = 6$ fish (the same each time), except for 0 % salinity when $n = 4$; control group $n = 6$ fish

Parameter	Salinity/‰					
	26.73 (n = 6)	21.3 (n = 6)	9.8 (n = 6)	2.5 (n = 6)	0 (n = 4)	0 (Control) (n = 6)
K^+/(mM/L)		3.01 ± 0.15	2.69 ± 0.09	2.51 ± 0.08	2.71 ± 0.21	2.19 ± 0.16
Na^+/(mM/L)	143.47 ± 1.74^{ac}	141.44 ± 2.60^{ac}	136.27 ± 0.56^{ab}	123.98 ± 0.94^{b}	149.40 ± 1.86^{c}	141.30 ± 0.42
Cl^-/(mM/L)	128.35 ± 1.03^{ad}	125.91 ± 1.07^{ab}	124.00 ± 1.13^{b}	115.69 ± 2.07^{c}	133.31 ± 2.79^{d}	123.20 ± 1.26
Ca^{2+}/(mM/L)	1.88 ± 0.05^{a}	1.86 ± 0.05^{ab}	1.77 ± 0.03^{bc}	1.57 ± 0.01^{d}	1.73 ± 0.09^{c}	1.71 ± 0.02
P^{3+}/(mM/L)	2.93 ± 0.08^{a}	3.60 ± 0.23^{b}	2.84 ± 0.07^{a}	2.24 ± 0.03^{c}	2.83 ± 0.18^{a}	2.64 ± 0.07
Mg^{2+}/(mM/L)	0.99 ± 0.06^{ab}	1.08 ± 0.04^{b}	0.93 ± 0.03^{a}	0.81 ± 0.02^{c}	0.98 ± 0.09^{ab}	0.98 ± 0.01
BUN/(mM/L)	3.42 ± 0.40^{a}	2.77 ± 0.28^{a}	1.67 ± 0.48^{b}	1.75 ± 0.28^{b}	0.75 ± 0.21^{c}	0.42 ± 0.11
CR/(μM/L)	9.20 ± 1.13	6.57 ± 1.01	7.78 ± 0.89	4.12 ± 1.50	5.38 ± 1.48	4.53 ± 0.08
UA/(mM/L)	2.67 ± 1.40	0.80 ± 0.47	2.68 ± 1.14	1.13 ± 0.47	1.98 ± 0.62	1.75 ± 0.93
TBIL/(μM/L)	0.23 ± 0.03	0.17 ± 0.02	0.20 ± 0.03	0.18 ± 0.02	0.35 ± 0.13	0.33 ± 0.03
DBIL/(μM/L)	0.10 ± 0.02	0.08 ± 0.02	0.15 ± 0.02	0.12 ± 0.02	0.08 ± 0.03	0.22 ± 0.03
IBIL/(μM/L)	0.13 ± 0.04	0.08 ± 0.02	0.05 ± 0.03	0.07 ± 0.02	0.28 ± 0.11	0.12 ± 0.01
TP/(g/L)	16.12 ± 1.53	18.32 ± 1.04	17.63 ± 1.50	17.43 ± 0.98	17.88 ± 2.38	18.72 ± 1.03
ALB/(g/L)	4.53 ± 0.52	4.88 ± 0.41	4.47 ± 0.47	4.14 ± 0.35	3.95 ± 0.78	5.32 ± 0.41
GLB/(g/L)	11.58 ± 0.98^{a}	13.43 ± 0.65^{ab}	13.17 ± 1.05^{ab}	13.25 ± 0.64^{ab}	13.93 ± 1.62^{bc}	13.40 ± 0.67
A/G	0.39 ± 0.02^{a}	0.36 ± 0.02^{b}	0.34 ± 0.01^{c}	0.31 ± 0.01^{d}	0.28 ± 0.03^{e}	0.40 ± 0.02
ALT/(U/L)	56.70 ± 16.01^{a}	40.47 ± 5.48^{ab}	25.93 ± 3.43^{b}	23.63 ± 2.56^{b}	34.65 ± 4.65^{b}	15.55 ± 0.69
AST/(U/L)	142.33 ± 7.21^{a}	234.03 ± 33.89^{b}	188.40 ± 14.16^{c}	186.43 ± 18.35^{c}	152.68 ± 24.91^{ac}	163.22 ± 11.96
ALP/(U/L)	127.17 ± 5.65^{a}	989.83 ± 112.22^{b}	568.58 ± 68.31^{b}	450.67 ± 46.67^{c}	928.55 ± 122.42^{c}	508.48 ± 31.43
GGT/(U/L)	0.40 ± 0.16	4.87 ± 2.80	0.65 ± 0.07	0.20 ± 0.03	31.88 ± 16.51	2.32 ± 1.34
TBA/(μM/L)	0.73 ± 0.09	0.72 ± 0.11	0.54 ± 0.05	0.45 ± 0.03	1.10 ± 0.47	0.50 ± 0.08
CHE/(U/L)	3.83 ± 1.36	3.02 ± 0.66	2.18 ± 0.43	0.67 ± 0.19	4.48 ± 1.70	5.50 ± 1.08
Fe^{2+}/(mM/L)		2.21 ± 0.61^{a}	2.68 ± 0.44^{ab}	1.22 ± 0.30^{ac}	5.03 ± 0.60^{d}	4.50 ± 0.66
GLU/(mM/L)	1.82 ± 0.21^{a}	0.13 ± 0.03^{b}	0.93 ± 0.12^{c}	1.28 ± 0.09^{a}	0.63 ± 0.30^{c}	1.20 ± 0.10

Table 2 Showing the significant differences of serum biochemical parameters between the value for each salinity level and the control group. 'a' indicates significant differences ($P < 0.05$). NS indicates no significant difference. Independent-sample T Tests were used for parameters K^+, Ca^{2+}, P^{3+}, Mg^{2+}, UA, TP, ALB, GLB, A/G, ALT, AST, ALP, CHE, Fe^{2+}. Mann-Whitney U test were used for Na^+, Cl^-, BUN, CR, TBIL, DBIL, IBIL, GGT, TBA, GLU. At each concentration $n = 6$ fish (the same each time), except for 0 ‰ salinity when $n = 4$; control group $n = 6$ fish

Parameter	Salinity/‰				
	26.73($n = 6$)	21.3($n = 6$)	9.8($n = 6$)	2.5($n = 6$)	0($n = 4$)
K^+/(mM/L)	NS	a	a	NS	NS
Na^+/(mM/L)	NS	NS	a	a	NS
Cl^-/(mM/L)	a	NS	NS	a	a
Ca^{2+}/(mM/L)	a	a	NS	a	NS
P^{3+}/(mM/L)	a	a	a	a	NS
Mg^{2+}/(mM/L)	NS	NS	NS	a	NS
BUN/(mM/L)	a	a	a	a	NS
CR/(μM/L)	a	NS	NS	NS	NS
UA/(mM/L)	NS	NS	NS	NS	NS
TBIL/(μM/L)	NS	a	a	a	NS
DBIL/(μM/L)	a	a	NS	a	a
IBIL/(μM/L)	NS	NS	NS	NS	a
TP/(g/L)	NS	NS	NS	NS	NS
ALB/(g/L)	NS	NS	NS	a	NS
GLB/(g/L)	NS	NS	NS	NS	NS
A/G	NS	NS	a	a	a
ALT/(U/L)	a	a	a	a	a
AST/(U/L)	NS	NS	NS	NS	NS
ALP/(U/L)	a	a	NS	NS	a
GGT/(U/L)	NS	NS	NS	NS	NS
TBA/(μM/L)	NS	NS	NS	NS	NS
CHE/(U/L)	NS	a	a	a	NS
Fe^{2+}/(mM/L)	NS	a	a	a	NS
GLU/(mM/L)	NS	a	NS	NS	NS

Figure 2 shows that A/G levels fell significantly during acclimation to different salinity levels and during the 30 day period at 0 ‰ salinity. A significant difference was found for GLU with decreasing salinity. However, at a salinity of 21.3, a sharp decrease was recorded, followed by an immediate increase after 1 month in fresh water ($P < 0.05$). The recorded decrease was also statistically significant(Fig. 3). BUN, CR, UA and ALB values showed a similar pattern of change to *A/G*. The only differences were that CR, UA and ALB presented no statistical difference during acclimation to decreasing salinity, and UA showed no statistical difference when compared to the control group. In contrast, ALP, and GGT levels changed inversely when compared to GLU, but GGT showed no statistical significance.

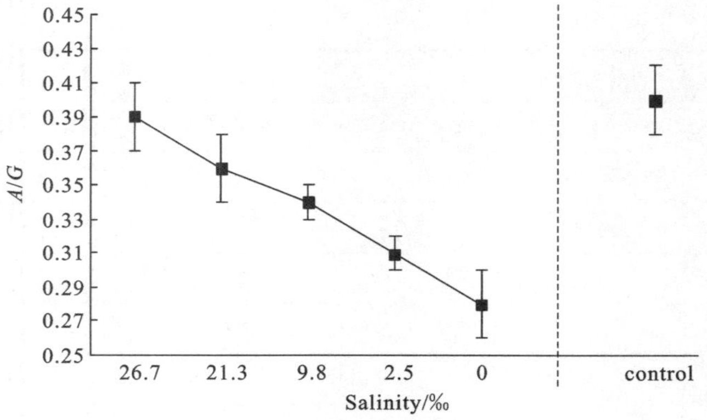

Fig. 2 The ratio of albumin and globulin (*A/G*) levels of sturgeon *Acipenser sinensis* during the 27 day acclimation period from sea water(26.7 ‰) to slightly brackish water (2.5 ‰), and after a 30 day period in freshwater(0 ‰). The values represent mean ± SE. At each concentration $n = 6$ fish(the same each time), expect for 0 ‰ salinity when $n = 4$; control group $n = 6$ fish

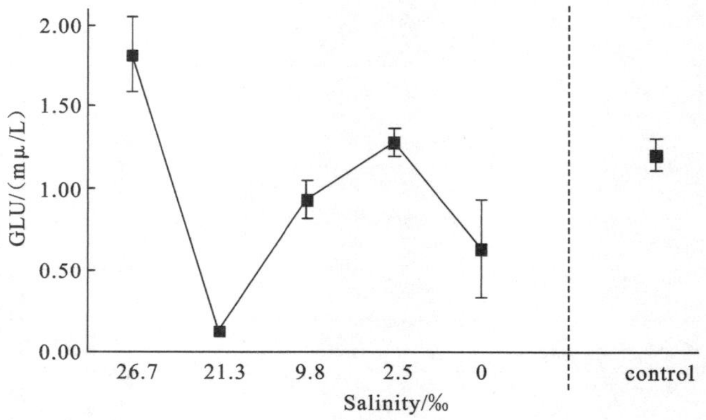

Fig. 3 Glucose(GLU) values in the serum of sturgeon *Acipenser sinensis* during the 27 day acclimation period from sea water(26.7 ‰) to slightly brackish water(2.5 ‰), and after a 30 day period in freshwater(0 ‰). The values represent mean ± SE. At each concentration $n = 6$ fish (the same each time), expect for 0 ‰ salinity when $n = 4$; control group $n = 6$ fish

There was a significant increase in GLB with decreasing salinity, with the final values being similar to that of the control group(Fig. 4). Thus, towards the end of the trial there was no significant difference for GLB at different salinity levels when compared with the control group(Table 2).

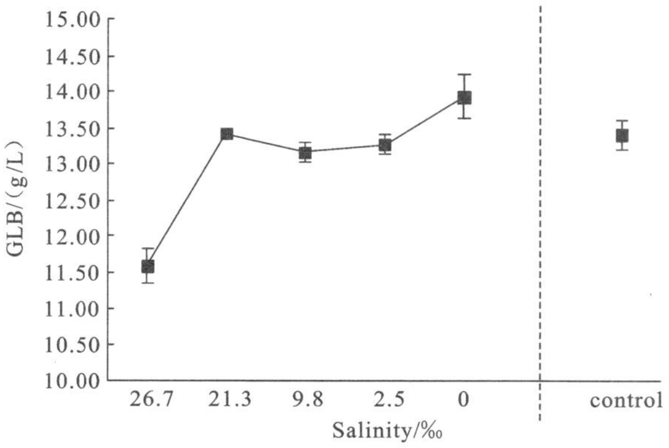

Fig. 4 Globulin (GLB) levels in the serum of sturgeon *Acipenser sinensis* during the 27 day acclimation period from sea water (26.7 ‰) to slightly brackish water (2.5 ‰), and after a 30 day period in freshwater (0 ‰). The values represent mean ± SE. At each concentration $n = 6$ fish (the same each time), expect for 0 ‰ salinity when $n = 4$; control group $n = 6$ fish

4 Discussion

In the present study, we found that ion concentrations may return to initial levels, when the sturgeon are moved from seawater to freshwater conditions. This finding indicates that, at least for the Chinese sturgeon, osmoregulatory processes fulfill their function, supporting existing studies of other species of sturgeon, including *A. naccarii* and *A. transmontanus* (Martínez-Álvarez et al., 2002; McEnroe et al., 1985). In a previous experiment on *A. fulvescens*, when salinity levels were below 25 ‰, the serum concentrations of Na^+, K^+, Cl^-, and Ca^{2+} ions were found to increase with increasing salinity (Lebreton et al., 1998). The different rhythms in acclimation recorded between this research and our study may be explained by differences in osmoregulatory ability among different species and even populations of sturgeon. Furthermore, in our study we found significant differences in Cl^-, Ca^{2+} and P^{3+} levels, but not in K^+, Na^+, Mg^{2+} levels, when comparing the seawater group (26.7 ‰) with the freshwater controls. This observation indicates that a difference exists in the ion levels between fish acclimated to seawater vs fish acclimated to freshwater. The study by McEnroe et al. (1985) on freshwater *Acipenser transmontanus* indicated that concentrations of Cl^- in the seawater group were greater than in the freshwater group. In contrast, the values of K^+ and Na^+ were found to be similar between both groups. In another study by Natochin et al. (1985), it was found that serum ion concentrations were similar between seawater and freshwater groups. This finding would account for the different serum ion concentrations that were recorded during fish acclimation in our study, both from seawater to freshwater, and during the prolonged period in freshwater.

The epithelium of the gill have a great permeability to the K^+, Na^+, Cl^-, Ca^{2+}, and excessive mono-ion was secreted through the gill in seawater teleosts, while divalent and trivalent ions are secreted in the form of urine (Lin, 1999). Hence, the kidney may play an important role in the regulation of P^{3+} and Mg^{2+} concentrations. In our study, P^{3+} and Mg^{2+} concentrations increased noticeably at a salinity of 21.3 ‰. This sudden rise may be explained by a major change in kidney function at this specific salinity level. In other words, the structure of the kidney may have been modified to adapt the alteration in salinity (Krayushkina et al., 1996).

CR, BUN, and UA comprised the main metabolic products that were excreted through the gills and kidney. The final products of protein in fish are CR, BUN, UA, NH_3 and TMAO etc. In freshwater teleosts, most nitrogenous waste (NH_3/NH_4^+) is discharged via the gill, while in marine teleosts, a large proportion of nitrogenous waste (TMAO) is discharged in the urine (Lin, 1999). In the present study, the decreasing trend of CR, BUN and UA with declining salinity in *A. sinensis* may indicate that, for this euryhaline species, when the chemical receptor in the gills sense an alteration in water salinity (Zhao et al., 2006), the type of nitrogenous waste and mode of excretion (i. e. via gills or in urine) may be adjusted in response to the switch from seawater to freshwater. As a result, there is an increase in the proportion of ammonia that is discharged from the gills. Thus, in the current study, the proportion of NH_3 increased with respect to total nitrogenous waste, while CR, BUN and UA decreased.

Our results indicated that, in *A. sinensis*, salinity acted as a stressor promoting the alteration of liver function. The transfer of the fish from seawater to brackish water caused a major change in the concentrations of TBA, ALT, *A/G*, GLB, AST, ALP, and GGT. Existing research has shown that the concentration of TP, ALB, GLB, *A/G*, TBA, ALT, AST, ALP, and GGT in plasma serve as indicators of liver function (Berneta et al., 2001). Changes in the levels of these parameters may explain the hypofunctioning of the liver cells under changing salinity levels. TP includes both ALB and GLB. GLB is related to immunity (Lin, 1999), the levels of which increase in the presence of infection or injury (Chen et al., 1987). In addition, in the present study, *A/G* levels continued to decrease after transferring the fish into the fresh water aquarium, which demonstrated that salinity variation and environment alteration act as stressors that boost body immunity. However, the effect of the duration and intensity of such stressors on the immunity of Chinese sturgeon requires further investigation, as it would indicate the species capacity to endure new environmental conditions and hence long-term survival.

There are a number of studies demonstrating that the reaction of fish to stress includes changes in hormone levels, blood cell parameters, enzyme activity, and other biochemical related processes (Martínez-Álvarez et al., 2002; Donaldson, 1981). Fan et al. (2000) noted a decline in erythrocyte function and a shortening of lifespan in *Sciaenops ocellatus* with decreasing salinity. In comparison, Huang et al. (2006) showed a decrease in erythrocyte counts and increase in MCV levels, however, for *Aclpenser schrenckii* a decline in the total function of erythrocytes did not occur during acclimation from freshwater to saline waters. In the current study, there was an increase in Fe^{2+} with decreasing salinity, which may indicate that erythrocyte count increased as a result of increasing Fe^{2+} serum levels when Chinese sturgeon occupy waters of lower salinity, however, this observation requires confirmation by further study.

It should be emphasized that TBIL, DBIL, CR, UA, and GLU levels decreased while TP, GLB, AST, ALP, GGT, P^{3+} and Mg^{2+} levels increased significantly at a salinity of 21.3 ‰. This record indicates that the physiology of Chinese sturgeon may be subject to specific regulation. It may also be inferred that a salinity of 20 % may represent the limit of physiological regulation in Chinese sturgeon. This observation supports existing research, in which a salinity of 20 % has been found to represents the limit of physiological regulation limit in *A. naccarii*, at which point poor growth and high mortality occurred (McKenzie et al., 1999; Cataldi et al., 1995).

Furthermore, Huang et al. (2006) noted that as environmental salinity increased, fish consumed more energy, while glucose and lipids provided the energy required for metabolism. Hence, when the available food source lacks sufficient energy, protein in the feed would be utilized as energy source (Lin, 1999). During this study, there was no alteration in the amount of food consumed by the sturgeon. Furthermore, TP levels remained stable, whereas there was a noticeable decline in GLU levels, both of which indicated that the osmoregulatory process required more energy to respond to these physiological changes.

When the fish remained in the freshwater aquarium for 30 days, GLU levels declined significantly to values lower than the controls, while ALT, ALP, Fe^{2+} levels increased significantly to levels higher than in the control. Sturgeon feeding levels were normal, indicating that the alteration in environmental conditions acted as a temporary stressor to regulate physiological responses while accelerating energy utilization.

In conclusion, when the sturgeon *A. sinensis*, is subjected to a decline in salinity to 2.5 ‰, a number of physiological responses of the fish occur. For example, serum ion levels and blood parameters related to metabolism and nutrition were affected. This observation indicated that osmotic stress counteracted by osmoregulation is involved in the acclimation of sturgeon to decreased salinities (Martínez-Álvarez et al., 2002). The ion values returned to normal for both seawater and freshwater conditions, after 1 month in salinity acclimation. This finding demonstrated that osmoregulatory processes may fulfill their function after a period of acclimation. However, there was a change in some blood parameters, related to metabolism and nutrition, which did not return to initial seawater and freshwater values. Hence, osmoregulatory processes in sturgeon may cause major alterations to the metabolic process, indicating that extreme change in ion regulation is energetically costly.

Acknowledgement

The authors are grateful to the Researcher Sun Xiulian (Virus Research Institute, Chinese Academy of Sciences) for providing strong assistance on the data analysis of this paper. This study was supported by the Essential Operation Fund of Freshwater Fisheries Research Center, Chinese Academy of Fishery Sciences.

References

Asadi F, Halajian A, Pourkabir M, et al. 2006a. Serum biochemical parameters of Huso huso. Comp. Clin. Pathol, 15: 245-248.

Asadi F, Masoudifard M, Vajhi A, et al. 2006b. Serum biochemical parameters of *Acipenser persicus*. Fish Physiol. Biochem, 32: 43-47.

Baker D W, Wood A M, Litvak M K, et al. 2005. Haematology of juvenile *Acipenser oxyrinchus* and *Acipenser brevirostrum* at rest and following forced activity. J. Fish Biol, 66: 208-221.

Berneta D, Schmidta H, Wahlia T, et al. 2001. Effluent from a sewage treatment works causes changes in serum chemistry of brown trout (*Salmo trutta* L.). Ecotoxicol. Environ. Saf, 48: 140-147.

Burggren W W, Randall D J. 1978. Oxygen uptake and transport during hypoxic exposure in the sturgeon *Acipenser transmmontanus*. Respir. Physiol, 34: 171-183.

Cataldi E, Ciccotti E, Dimarco P, et al. 1995. Acclimation trials of juvenile Italian sturgeon to different salinities: morpho-physiological descriptors. J. Fish Biol,47:609-618.

Chen D,Guan M. 1987. Some values of the blood biochemistry measured in Lipotes vexillifer. Acta Hydrobiol. Sinica,11:371-374.

Chen X. 2007. Biology and resources status of Acipenseriformes. Beijing:Ocean press.

Donaldson E M. 1981. The pituitary-interrenal axis as an indicator of stress in fish//Pickering A D. Stress and fish. New York:Academic Press:11-41.

Fan R,Jiang M,Ru S,et al. 2000. Preliminary study on the changes ultrastructure of the blood sell of Sciaenops ocellatus under different osmotic pressure. Mar. Sci,24:48-53.

Huang X,Zhang L,Zhuang P,et al. 2006. Hematological parameters of Amur sturgeon,*Acipenser schrencki*,during different salinity domestication. Mar. Fish,28:177-184. [in Chinese].

Kiffer J D,Tufts B L. 1998. Effects of blood deprivation on white muscle energy reserves in rainbow trout(*Oncorhynchus mykiss*): the relationships with body size and temperature. Fish Physiol. Biochem,19:239-245.

Krayushkina L S. 1998. Characteristics of osmotic and ionic regulation in marine diadromous sturgeons *Acipenser brevirostrum* and *A. oxyrhinchus*. J. Ichthyol,38:684-692.

Krayushkina L S,Panov A A,Gerasimov A A,et al. 1996. Changes in sodium,calcium and magnesium ion concentrations in sturgeon(*Huso huso*) urine and in kidney morphology. J. Comp. Physiol. B, Biochem. Syst. Environ. Physiol, 165: 527-533.

Kynard B. 1997. Life history,latitudinal patterns,and status of the shortnose sturgeon,*Acipenser brevirostrum*. Environ. Biol. Fishes,48:319-334.

Lebreton G T O,Beamish F W H. 1998. The influence of salinity on ionic concentrations and osmolality of blood serum in lake sturgeon,Acipenser fluvescens. Environ. Biol. Fishes,52:477-482.

Lin H. 1999. The physiology of fishes. Guangzhou:Guangdong higher education press.

Martínez-Álvarez R M,Hidalgo M C,Domezain A,et al. 2002. Physiological changes of sturgeon *Acipenser naccarii* caused by increasing environmental salinity. J. Exp. Biol,205:3699-3706.

Maxime V,Nonote G,Peyraud C,et al. 1995. Circulatory and respiratory effects of an hypoxic stress in Siberian sturgeon. Respir. Physiol,100:203-212.

McEnroe M,Cech J J. 1985. Osmoregulation in juvenile and adult white sturgeon,*Acipenser transmontanus*. Environ. Biol. Fishes,14:23-30.

McKenzie D J, Piraccini N, Galli C, et al. 1997. Oxygen consumption and ventilator reflex responses are influenced by dietary lipids in sturgeon. Fish Physiol. Biochem,16:365-379.

McKenzie D J,Cataldi E,Marco P D,et al. 1999. Some aspects of osmotic and ionic regulation in Adriatic sturgeon *Acipenser naccarii*. II:morpho-physiological adjustments to hyperosmotic environments. J. Appl. Ichthyol,25:61-66.

Natochin Y V,Lukianenko V I,Kirsanov V I,et al. 1985. Features of osmotic and ionic regulations in Russian sturgeon (*Acipenser güdenstäti* Brandt). Comp. Biochem. Physiol. A,Comp. Physiol,80:297-302.

Wei Qiwei,Chen X,Yang D,et al. 2005. Variations in spawning stock structure of *Acipenser sinensis* within 24 years since damming of Gezhouba Dam. J. Fish. Sci. Chin,12:452-457.

Yang Y,Yao F,Yu W,et al. 2005. Changes in serum and muscle biochemical parameters of *Pelteobagrus fulvidraco* by starvation in winter. J. Yangtze Univers. (NATSci. Edit) Agri. Sci. V 4:42-44.

Zhao F,Zhuang P,Zhang L,et al. 2006. Effects of different saltwater acclimation on growth and feeding of Amur sturgeon *Acipenser schrenckii*. J. Fish. Sci. Chin. ,13:945-950.

Zheng W D,Li D. 2007. Blood biochemical parameters of juvenile Chinese sturgeon,*Acipenser sinensis*. J. Huazhong Agri. Univers. ,26:95-97.

17β-雌二醇对中华鲟(*Acipenser sinensis* Gray 1834)幼鱼血细胞和血生化参数的影响

张晓雁[1]　杜　浩[2,3]　张艳珍[1]　刘志刚[2,3]
柴　毅[4]　乔心美[2,3]　危起伟[2,3,*]

(1. 北京信沃达海洋科技有限公司(北京海洋馆)　北京　100081
2. 农业部淡水生物多样性保护重点实验室 中国水产科学研究院长江水产研究所　湖北武汉　434000
3. 中国水产科学研究院　北京　100039　　4. 长江大学生命学院　湖北荆州　434025)

摘　要　通过体侧肌肉注射17β-雌二醇,监测注射及停止注射28天中华鲟血细胞和血液生化参数值,分析不同剂量(0.5 mg/kg 和 0.1 mg/kg)外源雌激素对中华鲟的雌激素效应及对生理过程的影响,旨在为促进中华鲟性成熟技术提供借鉴。注射组血清雌二醇与外源 E2 呈剂量效应关系。睾酮(T)变化趋势与 E2 相反。与对照组相比,注射租的血清生化指标与血细胞参数值变化趋势各异。结论认为,外源雌激素对中华鲟可产生雌激素效应,但对健康也产生影响,呈剂量效应关系,可以 0.1 mg/ kg 的注射剂量作为进一步试验的参考值。

关键词　中华鲟　17β-雌二醇　血细胞参数　血液生化参数

Hematological and Biochemical Responses of Juvenile Chinese Sturgeon (*Acipenser sinensis* Gray 1834) to Exogenous 17β-Estradiol

Zhang Xiaoyan[1]　Du Hao[2,3]　Zhang Yanzhen[1]　Liu Zhigang[2,3]
Chai Yi[4]　Qiao Xinmei[2,3]　Wei Qiwei[2,3,*]

(1. Beijing Aquarium, Beijing 100081, China;
2. Key Laboratory of Freshwater Biodiversity Conservation and Utilization, Ministry of Agriculture, Yangtze River Fisheries Research Institute, Chinese Academy of Fisheries Sciences, Wuhan 434000, China;
3. Chinese Academy of Fishery Sciences, Beijing 100039, China;
4. College of Animal Science, Yangtze University, Jingzhou, 434025, China)

Abstract　One of the handicaps to the success of the large-scale controlled reproduction of cultured Chinese sturgeon (*Acipenser sinensis*) is that broodstocks were mostly not well developed due to the lacking of certain stressors to trigger the hypothalamus-pituitary-gonad axis in the captive environment. Supplying with exogenous hormone is one of the possible ways to promote the gonad maturation. The present study aims at learning the physiological response of the Chinese sturgeon to the exogenous 17β-estradiol (one of the primary female sex

* 通讯作者(Corresponding author), E-mail: weiqw@ yfi. ac. cn

hormone), and estimating its possible effects to the gonad development. Serum hormone testosterone (T) and 17β-estradiol (E2), biochemical parameters, and hematological parameters were tested for 10 weeks (injected with E2 for the first 42 days then cultured for a further 28 days) with 4-year-old captive Chinese sturgeons with 3 groups: high dose group (HDG, 0.5 mg/kg, $n=3$), low dose group (LDG, 0.1 mg/kg, $n=3$), control (injected with the 5 ml saline/time/sturgeon, $n=2$). The results showed that serum E2 concentrations increased during injection stage of E2 in both HDG and LDG group, and then decreased after the end of injections, when serum testosterone showed the opposite trend to E2. This trend was mirrored by serum triglyceride concentrations, Other five serum biochemical parameters (total protein, albumin, globulin, total cholesterol, serum magnesium) showed ascent during the70 days period, while blood glucose decreased, albumin to globulin ratio decreased rapidly after increasing in early days of injection. The predominant hematological finding was a severe anemia. The blood parameters in HDG changed more significant than that in the LDG. Depending on the serum hormone and the increasing biochemical parameters examined, the potential estrogenic effect of exogenous E2 on Chinese sturgeon might be produced. However the decreased value for the hematological parameters indicated that the health status was somewhat declined after the injection of E2. From the present findings, it is followed that the physiological responses of Chinese sturgeon to E2 supplementation is predominant, in a dose-dependent relationship. The results also showed lower E2 concentration of 0.1 mg/kg may be sufficient to constitute the baseline dosage in further studies.

Key words *Acipenser sinensis* 17β-estradiol Hematological Blood parameters

1 Introduction

Due to the rapid decline in the natural reproduction of Chinese sturgeon, *Acipenser sinensis*, population enhancement through releases of the second filial generation fingerlings is a vital measure for the conservation of the species. Although a cultured broodstock reserve has already been developed, most females (the oldest close to 17 years of age) remain in stage II gonad development. Thus, the low sexual maturity rate of females remains a major difficulty in large-scale controlled reproduction of sturgeons for restocking (Wei et al., 2013). As the culture environment is apparently not fully sufficient for sturgeon to proceed with a normal functioning of the endocrine system, the initiation of a rapid gonad development towards maturation that is controlled by reproductive hormones of the brain, pituitary and gonads in farmed sturgeon, cannot be achieved reliably (Webb et al., 2011). It has been shown that the application of exogenous estrogen can induce vitellogenesis and promote gonadal development (Zhuang et al., 2010; Verslycke et al., 2002), however, it can also disturb the normal hormone metabolism, leading to endocrine and other metabolic disorders as well as organ malformations and anaemia. Consequently, mortality may also occur in a dose-dependent manner (Zhang et al., 2012; Ma et al., 2004).

In order to evaluate the effect of different dosages of exogenous estrogen on the physiology of Chinese sturgeon and to provide a reference for the technology of induced sexual maturity of farmed specimens, the properties of blood cells, serum biochemical parameters and serum E2 and T levels were examined during and after the injection of E2.

2 Materials and methods

2.1 Materials

The research was performed at the Beijing Aquarium from May to July in 2012. The cultured sturgeons were produced in 2008(3-year-old F1-2008). Their baseline fork length at the initiation of this study was (102.81 ± 4.41) cm, weight (9.41 ± 1.58) kg, and gonad development at stage II (gonadal tissue was collected through surgical operations; tissue sections (Chen et al., 2004) were later prepared to determine the different maturity stages of the gonad). For the convenience of later research, the sturgeon were numbered according to their morphological characteristics. Fish were maintained in circular tanks (diameter 7.0 m, water depth 0.75 m; water exchange in 45 min; temperature 21.2-23.7 ℃; dissolved oxygen 7.0-7.8 mg/L; pH 7.6-7.8). Bottom waste was removed regularly and the sand filtration tank was backwashed every week. Fresh baitfish with moderate vitamin supplementation provided once per day (1.2 % of body weight) included frozen capelin, *Mallotus villosus*, *Loligo chinensis*, *Penaeus vannamei* (at ratios of 50 %, 30 %, 20 % of the total diet, respectively).

2.2 Experimental design

Test sturgeon were divided into three groups: control ($n = 2$); low dose (LDG) E2 (0.1 mg/kg, $n = 3$); and high dose (HDG) E2 (0.5 mg/kg, $n = 3$). The E2 (250 mg) was dissolved in 25 ml of saline to obtain injection concentrations of 10 mg/ml (the injection volume for each fish was calculated based on weight). The control group was injected with 5 ml saline/time/sturgeon (placebo). Duration of the experiment was 70 days. The injections (lateral muscle) were performed at a fixed time every 7 days, and after 42 days of injections the sturgeon were kept for another 28 days. Blood samples were collected every 7 days, and every 14 days after the period of injections.

2.3 Sampling and measurements

To reduce the stress response, gentle handling is necessary while also catching the fish as quickly as possible (within 5 min). Feeding was stopped one day before handling and blood sampling. Blood samples (5 ml) were taken from the caudal vasculature with a syringe, and later processed for different types of analysis. Samples for the erythrocyte parameters were heparin treated, while other samples were centrifuged at 4 470 g for 20 min to obtain serum (stored at −4 ℃). Test methods of the various blood parameters were:

Red blood cells (RBC): blood samples were diluted 200 times using 0.05 % NaCl solution as the diluent, and the erythrocyte count was done using a Neubauer's hemocytometer under light microscopy. Hemoglobin (Hb) was measured using the Sahlis method (Chen et al., 2007). Hematocrit values (Ht) were obtained using Winchester's manual procedure (Ye, 1991).

Indices also calculated were: mean corpuscular volume (MCV), mean corpuscular hemoglobin (MCH), and mean corpuscular hemoglobin concentration (MCHC). The total number of leukocytes was also counted and differential white blood cell counts were performed.

Serum biochemical parameters included total protein (TP), albumin (ALB), globulin (GLB), albumin:globulin ratio(A/G), total cholesterol(CHO), blood glucose(GLU), triglyceride(TG), and magnesium(Mg), all of which were measured with an Automatic Biochemical Analyzer(OLYMPUS AU400, Japan; Yingke Xinchuang Technology Co., Ltd., China)(Chen at el., 2007). Hormone concentrations(E2 and T) were measured with the electrochemiluminescence immunoassay(ADVIA Centaur CP, Siemens Healthcare Diagnostics Inc., Germany)(Zhang et al., 2011a). The detection limit for technical E2 was 10 pg/ml, and for technical T was 10 ng/dl. Samples were run in duplicate, and variation coefficients of TP, ALB, Mg, CHO, TG, GLU, E2, and T were not more than 3, 3, 8, 5, 5, 5, 3 and 3 %, respectively.

2.4 Statistical analysis

Data were analyzed using SPSS v13.0 software(SPSS Inc., Chicago, IL) and represented as means ± SD. Differences in the blood parameters between the various injection time points and among the three different groups at various sampling times were evaluated using one-way analysis of variance (ANOVA). Multiple comparisons were performed using the Student-Neuman-Keuls(SNK) method, and correlations between E2 and T levels were assessed using Pearson's correlation analysis. Differences were considered to be significant at $P < 0.05$.

3 Results

3.1 Serum hormone(E2 and T) alterations

The serum E2 levels(Fig. 1a) increased significantly in E2-treated groups after the injections in a dose-dependent manner. However, E2 concentrations quickly decreased after the injection period(7-day intervals) was terminated(on day 42). This applies particularly for the high dose group, whereas E2 in the low dose group began to decrease already before the end of the repeated injections. The differences in E2 concentrations between the high and low dose groups at the time of the last two injections were significant($P < 0.05$). Concentrations of T(Fig. 1b) had a different trend from the E2, as this was significantly decreased in E2-treated groups after the first injection, and these were significantly different again between the two groups after the second injection($P < 0.05$). These values returned to the normal concentrations(baseline) after 28 days(at the end of E2-injections; $P > 0.05$). Furthermore, E2 concentrations were significantly negatively correlated with concentrations of T in the high dose group($r = 0.429$, $P < 0.05$). In the control, E2 and T revealed no significant changes during the 70 days.

3.2 Hematological and biochemical findings

Six serum parameters revealed a significant increase during the experiment compared to controls (Table 1, Fig. 2a-d). TG(Fig. 2a) showed the same trend as the E2 serum. TP(Fig. 2b) in E2-treated fish increased significantly from day 14, after the first injection compared to controls. The concentrations increased continually in the high dose group compared to little change in the low dose group after the injections were finished. Furthermore, TP increased more in the high dose group than in the low dose group during the experiment. ALB, GLB, Mg^{2+} and CHO concentrations showed similar changes to TP. In the E2-treated group, GLU showed the opposite trend(Fig. 2c). Figure 2

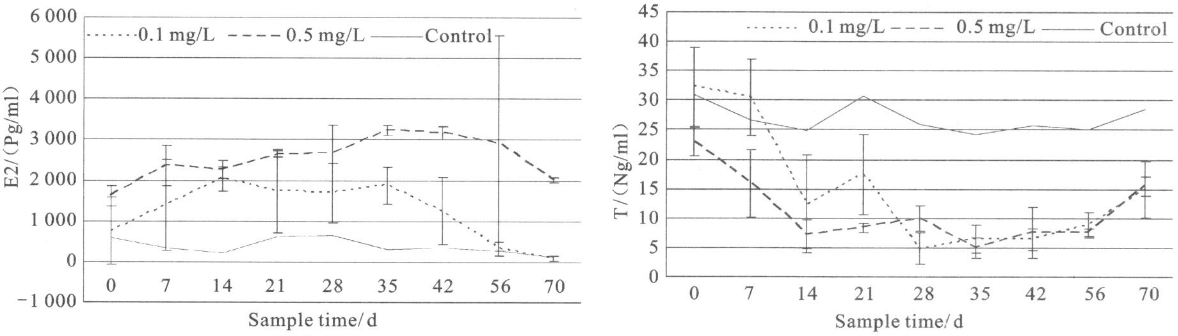

(a)　　　　　　　　(b)

Fig. 1: Variations in the serum 17β-estradiol (E2, 1a) and testosterone (T, 1b) concentrations in the three groups of Chinese sturgeon after E2 injections (low dose (0.1 mg/kg) vs. high dose (0.5 mg/kg), control group (injected with 5 ml saline/time/sturgeon)) at different time points. Injections ended on day 42.

(d) provides an overview of the *A/G* ratio. The ratio in all three groups increased rapidly during the early stage of the study, fell rapidly in the E2-treated fish, with a more significant reduction in the high dose group. However, the control group leveled off.

Table 1　Comparison of the hematological and biochemical parameters of different groups of Chinese sturgeon after 17β-estradiol injections (low dose (0.1 mg/kg) vs. high dose (0.1 mg/kg)) at different time points. Data are means + SD (*n* = 2 control, only 2 values and only the means can be reported; *n* = 3 low dose; *n* = 3 high dose). Different letters denote statistically significant differences among control, low, and high doses. Injections ended on day 42.

Sample time/d	Parameter	Control	Low dose (0.1 mg/kg)	High dose (0.5 mg/kg)	Parameter	Control	Low dose (0.1 mg/kg)	High dose (0.5 mg/kg)
0	ALB/(g/L)	10.10	9.18 ±0.43	8.32 ±2.53	CHO/(mmoL/L)	2.20	1.88 ±0.30	2.47 ±0.39
7		7.63	10.07 ±2.85	11.58 ±2.79		1.95	2.68 ±0.01	4.11 ±0.51[a]
14		12.98	15.55 ±0.88	23.28 ±2.51[a]		2.91[a]	4.59 ±0.74[b]	6.69 ±0.28[c]
21		9.83[a]	20.38 ±3.18[b]	33.02 ±2.14[c]		2.83[a]	4.62 ±1.18[b]	7.22 ±0.12[c]
28		11.53[a]	26.20 ±4.54[b]	37.85 ±2.17[c]		4.11[a]	6.28 ±1.39[ab]	7.18 ±0.47[b]
35		11.15[a]	28.35 ±5.64[b]	39.53 ±1.89[c]		2.35[a]	5.57 ±1.88	8.64 ±1.03
42		10.98[a]	30.65 ±6.19	40.10 ±2.52		2.09[a]	5.95 ±1.86	8.65 ±1.11
56		9.53[a]	29.63 ±8.97	40.55 ±4.35		2.75[a]	5.64 ±2.36[ab]	9.45 ±1.51[b]
70		10.40[a]	29.62 ±8.35[b]	42.43 ±2.14[c]		2.88[a]	6.57 ±2.01[b]	13.01 ±0.52[c]
0	GLB/(g/L)	24.98	21.17 ±1.54	21.85 ±6.75	Mg^{2+}/(umoL/L)	1.79	1.40 ±0.38	1.73 ±0.30
7		11.80	15.53 ±3.53	20.03 ±5.61		0.90[a]	1.21 ±0.17[ab]	1.60 ±0.28[bc]
14		17.48	19.13 ±0.88	28.30 ±6.11		1.29	1.52 ±0.04	2.12 ±0.24[a]
21		13.30	22.17 ±4.71	55.92 ±9.56[a]		0.90[a]	1.76 ±0.26[b]	3.17 ±0.35[c]
28		16.33	32.18 ±9.50	63.27 ±8.83[a]		1.12[a]	2.07 ±0.33	2.38 ±0.06
35		15.73[a]	37.20 ±11.41[b]	71.50 ±8.97[c]		0.86[a]	1.90 ±0.31	2.27 ±0.08
42		14.38	42.50 ±14.62	75.42 ±13.23[a]		0.84[a]	1.91 ±0.24	2.13 ±0.02

continue

Sample time/d	Parameter	Control	Low dose (0.1 mg/kg)	High dose (0.5 mg/kg)	Parameter	Control	Low dose (0.1 mg/kg)	High dose (0.5 mg/kg)
56		14.30[a]	41.60 ±18.59[ab]	81.63 ±26.10[b]		1.01[a]	2.15 ±0.35	2.46 ±0.35
70		15.65	40.92 ±15.39	90.25 ±15.76[a]		1.12[a]	2.06 ±0.23	2.21 ±0.00
0	Hb/(g/L)	58	54 ±8	57 ±5	MCH/pg	137	131 ±30	145 ±18
7		63	55 ±7	52 ±6		154	121 ±11	162 ±30
14		51	51 ±4	43 ±6		115	117 ±17	107 ±16
21		58[a]	52 ±6[ab]	43 ±3[b]		136	118 ±5	105 ±12
28		64	48 ±8	45 ±8		137	121 ±34	126 ±24
35		58	51 ±8	42 ±5		129	139 ±23	117 ±38
42								
56		60	50 ±13	39 ±7		124	110 ±11	95 ±16
70		59[a]	48 ±7[ab]	37 ±6[b]		150	115 ±13	106 ±24
0	MCV/fl	723	690 ±182	677 ±107	MCHC/(g/L)	723	690 ±182	677 ±107
7		679	553 ±94	722 ±101		679	553 ±94	722 ±101
14		666[a]	545 ±28	573 ±21		666[a]	545 ±28	573 ±21
21		670	573 ±48	533 ±66		670	573 ±48	533 ±66
28		651	595 ±221	580 ±39		651	595 ±221	580 ±39
35		629	635 ±129	562 ±168		629	635 ±129	562 ±168
42								
56		563	486 ±73	463 ±86		563	486 ±73	463 ±86
70		722[a]	534 ±55	511 ±48		722[a]	534 ±55	511 ±48
0	WBC/(10^9/L)	15.90	12.93 ±5.46	17.73 ±2.89	Lym/%	41	39 ±5	44 ±5
7		10.20	10.20 ±3.86	21.40 ±10.67		40	49 ±8	51 ±4
14		19.60	15.13 ±5.14	17.67 ±1.29		42	45 ±2	46 ±11
21		21.40	11.87 ±4.06	19.93 ±5.94		43	43 ±6	49 ±7
28		29.50	19.53 ±2.87	20.07 ±4.94		47	47 ±5	42 ±7
35		14.30	11.13 ±2.32	10.87 ±2.58		42	51 ±4	41 ±6
42						43	54 ±3	38 ±5
56		23.30	16.00 ±3.47	10.73 ±4.32		53	46 ±4	44 ±7
70		26.90	21.33 ±1.22	11.00 ±6.97		39	54 ±4	36 ±6
0	Neut/%	50	25 ±2	48 ±5	Eos/%	9	11 ±5	8 ±3
7		50	29 ±5	42 ±2		10	7 ±4	6 ±4
14		43	29 ±4	49 ±9		15	9 ±5	5 ±4
21		44	33 ±6	41 ±5		13	13 ±10	10 ±2
28		45	37 ±6	38 ±6		9	7 ±4	20 ±2
35		50	43 ±6	45 ±6		9	12 ±6	14 ±11
42		46	44 ±6	45 ±4		12	6 ±5	17 ±8
56		40	48 ±8	47 ±7		8	11 ±5	9 ±8
70		46	58 ±5	51 ±3		15	7 ±3	13 ±6

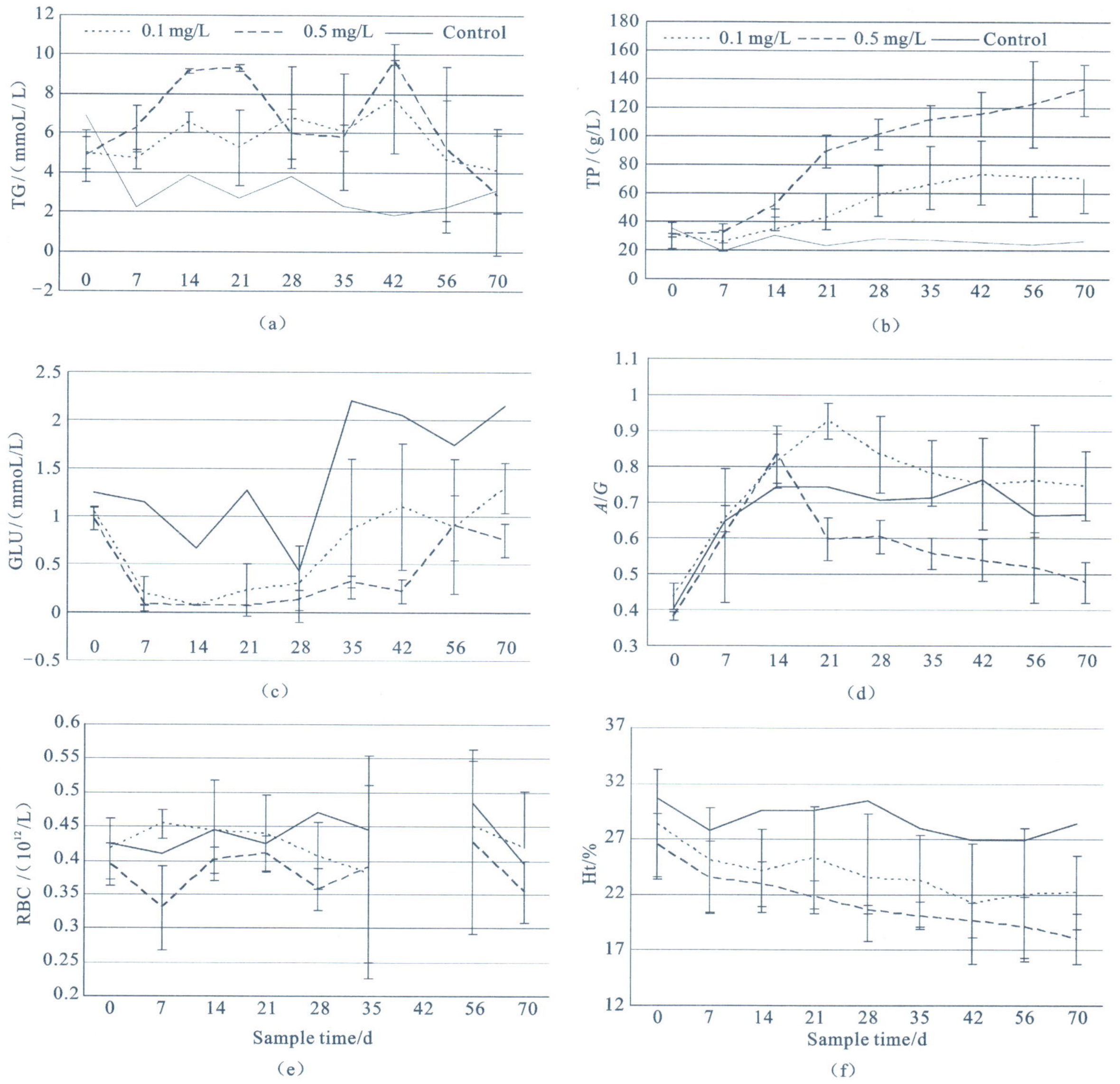

Fig. 2: Changes in serum triglyceride (TG, a), total protein (TP, b), blood glucose (GLU, c), red blood sell (RBC, e) and hemoglobin (Ht, f) concentrations and albumin to globulin ratio (the *A/G* ratio, d) in three groups of Chinese sturgeon after 17β-estradiol injections (low dose (0.1 mg/kg) vs. high dose (0.5 mg/kg), control group (injected with 5 ml saline/time/sturgeon)) at different time points. Injections ended on day 42.

The predominant haematological finding in the high dose group was a substantial decrease in the RBC values (Fig. 2e); the differences were significant compared to controls on day 28 after the first injection, thereby indicating a severe anaemia. No significant changes were found in the low dose group. Hb revealed a similar change as in the RBC values. With regard to the Ht (Fig. 2f), the decrease was in the treated groups, especially in the high dose group. There was a significant difference between the E2-treated groups and the controls at the end of the experiment ($P < 0.05$). Only the high dose fish showed a high elevation of MCV after the first injection, with a significant decrease thereafter. The low dose group, however, showed a slight decline during the experiment. MCH and the MCHC in E2-treated fish did not reveal any significant changes throughout the test period. WBC and the DLC did not differ in any of the three groups throughout the study.

Most hematological and biochemical parameter concentrations did not return to the initial values after 28 days, which was the end of the E2-injections.

4 Discussion

The present study indicates that the exogenous estrogen application can 'disturb' the normal metabolism of estrogen in fish, and that there is a dose-dependent relationship between the serum E2 concentration and accumulation of 17β-estradiol. The underlying mechanism that led to the elevation of the serum E2 is purportedly a consequence of the increase in activity of aromatase (shown to be the key enzyme of E2 synthesis from T in fish), which is promoted by exogenous estrogen. As it follows, fish showed the enhancement of E2 and a decrease in T. According to Zhang et al. (2011b), the high concentration of E2 will inhibit the T synthesis in animals, which might also be to some extent responsible for the opposite change from E2 in the present study.

The data in this study showed that exogenous E2 promoted a predominant increase in the TP, ALB, GLB, CHO, Mg^{2+} and TG concentrations. When the gonad develops rapidly, the E2 has been shown to increase. The concentrations required to induce the significance elevated the level of liver vitellogenin (a larger molecule protein including lipid and glucose can combine the metal ion such as Fe^{2+} and Cu^{2+} into developing oocytes). In addition, the vitellogenin materials are from ALB, CHO and TG; others such as phospholipid and carotene are transferred by GLB, which will contribute to the enhancement of the above parameters during rapid gonadal development. As a result, the enhancement of the above parameters could be used as indicators of assessing gonad development (Zhang et al., 2007; Srivastava et al., 1994). At least for the initiation of the next maturation cycle, an increase in Mg^{2+}, ALB, CHO and TG concentrations have been demonstrated to be indicators in wild sturgeon while in captivity. Moreover, the increase in the GLB concentration can indicate the adolescence of captive females (Zhang, 2012). The present investigation revealed a significant increase in the TP, ALB, GLB, CHO, and Mg^{2+} concentrations with the accumulated volumes of 17β-estradiol in E2-treated fish, which might be due to the estrogenic effect in a dose-dependent manner.

Thus far, many studies have revealed that glucose or glycogen synthesis from amino acids or TG occurred through gluconeogenesis, mainly in the hepatocytes (Lin, 2004). A poor gluconeogenesis level due to the application of exogenous estrogen (the hormone responsible for vitellogenesis) was considered as responsible for the decrease in the lower plasma glucose concentration (Washburn et al., 1993).

The present data indicated that the general health status was severely affected in E2-treated fish. E2 led to a severe anaemia, as shown by a significant reduction of RBC, Hb Ht, especially in the high dose group. The E2 that is beyond the physiological level can cause the enhancement of serum CHO and TG, which means that the mixed pool of plasma fatty acids relied by the mature erythrocytes metabolizing is altered. This effect is responsible for the decreased membrane fluidity and the increased osmotic fragility, further causing hematoclasis (Wang et al., 2001; Brown et al., 1993). In order to compensate the lower erythrocytes, an increase in immature stages of erythrocytes (reticulocytes) occurred. In the present hematological investigations there was a significant decrease in RBC, but an increase in MCV (in fish, the size of reticulocytes is larger than that of erythrocytes (Yuan et al., 2011; Liu et al., 2006)) after the first E2-injection in the high dose group. As the kidney and spleen are the main hematogenic organs (Ma et al., 2004; Schwaiger et al., 2000), exogenous estrogen can influence their functions and may cause histopathological alterations. In the present data, MCV continually decreased more after the first E2 enhancement than that of the RBC, indicating that the proportion of reticulocytes decreased.

The WBC and differential counts did not differ significantly between the E2-treated groups and the controls, thus there was no significant difference between the WBC parameters induced by the currently applied dosages.

It should be noted that although serum E2 and T returned to normal values after the 28-day injection period, most hematological and biochemical parameters did not return to their initial concentrations (especially in the high dose group), which revealed that the residual influence on the physiology of the E2-treated animals had not been removed completely.

5 Conclusions

Concerning the E2-treatment, the mode of the application and the dosages used in the present study might induce an estrogenic effect, but the general health status of *Acipenser sinensis* was severely affected at the same time. Furthermore, the residual effect was significant after the treatment ended, which is an obviously dose-dependent response. As a result, the relatively lower effects mediated by the low dose of 0.1 mg/kg would be useful as the baseline dose in future studies.

Acknowledgements

This study was supported by the National Natural Science Foundation of China (31172413) and Fund for Agro-Scientific Research in the Public Interest (NO. 201203086).

References

Brown S A, Hutchinson R, Morrisett J D, et al. 1993. Plasma lipid, lipoprotein cholesterol, and apoprotein distributions in selected US communities. The Atherosclerosis Risk in Communities (ARIC) Study. Arterioscler. Thromb. Vasc. Biol, 13: 1139-1158.

Chen X H, Wei Q W, Yang D Q, et al. 2004. Histological studies on gonadal origin differentiation of cultured *Acipenser sinensis*. Acta Zool. Sinica, 28: 633-639.

Chen Y Q, Yu S Y. 2007. Some biochemical and physiological indices in blood of *Chrysolophus pictus*, *Alectoris chukar* and *Phasianus colchicus*. Acta Zool. Sinica, 53: 674-681.

Liu H B, Hua Y P, Qu Q Z, et al. 2006. Microstructure and ultrastructure of peripheral blood cells of Amur sturgeon *Acipenser schrencki* Brandt. Acta Hydrobiol. Sinica, 30: 214-220.

Lin H R. 2004. Fish physiology. Guangzhou: Guangdong Higher Education Press: 19.

Ma T W, Wang Z J, Chen J F, et al. 2004. Toxic effects of 17-α-ethinylestradiol to the kidney of rare minnow (Gobiocypris rarus). Acta Sci. Circumst, 24: 487-491.

Schwaiger J, Spieser O H, Bauer C, et al. 2000. Chronic toxicity of nonylphenol and ethinylestradiol: haematological and histopathological effects in juvenile common carp (*Cyprinus carpio*). Aquat. Toxicol., 51: 69-78.

Srivastava S J, Srivastava S K. 1994. Seasonal changes in liver and serum proteins, serum calcium, inorganic phosphate and

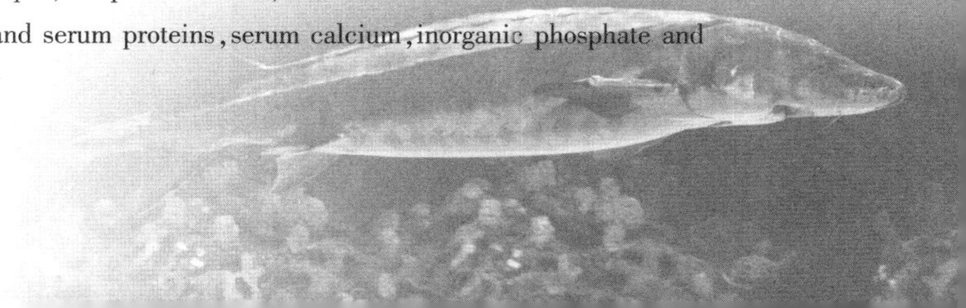

magnesium levels in relation to vitellogenesis in a freshwater catfish, *Heteropneustes fossilis* (Bloch) . Ann. Endocrinol, 55: 197-202.

Verslycke T, Vandenbergh G F, Versonnen B, et al. 2002. Induction of vitellogenesis in 17α-ethinylestradiol-exposed rainbow trout (*Oncorhynchus mykiss*) : a method comparison. Comp. Biochem. Physiol. C Toxicol. Pharmacol, 132: 483-492.

Wang X Y, Zhang M L, He Y P, et al. 2001. Effects of estrogen of mature females on member lipid fluidity of blood sell and microcirculation. J. Chin. Microcirc, 5: 232-233.

Washburn B S, Krantz J S, Avery E H, et al. 1993. Effects of estrogen on gluconeogenesis and related parameters in male rainbow trout. Am. J. Physiol. Regul. Integr. Comp. Physiol, 264: 720-725.

Webb M A, Doroshov S I. 2011. Importance environmental endocrinology in fisheries management and aquaculture of sturgeons. Gen. Comp. Endocrinol, 170: 313-321.

Wei Qiwei, Li L X, Du H, et al. 2013. Research on technology for controlled propagation of cultured Chinese sturgeon (*Acipenser sinensis*) . J. Fish. Sci. Chin, 20: 1-11.

Ye Y W. 1991. National guide to clinical laboratorial procedures. Nanjing: Publishing House of Southeast University: 8.

Yuan S B, Zhu A Y, Jiang L H, et al. 2011. Observations on the developments of blood cells in Nibea japonica. J. Fish. Chin, 31: 1374-1381.

Zhang L Z, Zhuang P, Zhang T, et al. 2007. Haematological biochemical norm of different age groups of cultured Amur sturgeon (Acipenser schrenckii) . J. Fish. Chin, 31: 159-164.

Zhuang P, Ding J W, Hou J L, et a;. 2010. Introduction and determination of vitellogenin in *Acipenser baerii* exposed to 17β-estradial. Mar. Fish, 32: 251-256.

Zhang T, Tian M P, Zhuang L Z, et al. 2011a. Estrogenic effect of estradiol-17β on juvenile Siberian sturgeon *Acipenser baerii*. Asian J. Ecotoxicol, 6: 415-421.

Zhang X, Li J, Liu Y S, et al. 2012. Toxic effects on adult male zebrafish (*Danio rerio*) following exposure to tris-(2, 3-dibromopropyl) isocyanurate and 17β-estradiol. Acta Sci. Circumst, 32: 450-456.

Zhang X Y. 2012. Blood biochemistry and its applications in Chinese sturgeon (*Acipenser sinensis*) . Wuhan: Institute of Hydrobiology, Chinese Academy of Sciences.

Zhang Y, Sun D J, Qu Q Z, et al. 2011b. Effect of temperature and estradiol-17β on gonad development and serum sex steroid in Amur sturgeon (*Acipenser schrenckii*) . J. Northeast Agrc. Univ, 42: 118-124.

养殖中华鲟的产后康复*

张晓雁[1,2,**]　杜　浩[1]　危起伟[1,***]　张艳珍[2]　王彦鹏[2]
蔡经江[2]　刘志刚[1]　乔新美[1]　熊　伟[1]

(1.农业部淡水生物多样性保护重点实验室 中国水产科学研究院长江水产研究所　湖北武汉　430223
2.北京信沃达海洋科技有限公司(北京海洋馆)　北京　100081)

摘　要　观测了初次繁殖的养殖中华鲟(*Acipenser sinensis*)亲本产后的摄食行为、生长、血清中甲状腺指标、血液理化指标,分析产后亲本11个月的康复效果。使用多春鱼(*Mallotus villosus*)可诱导产后亲本恢复摄食及转食混合鲜饵。随摄入营养水平提高,雌性异常行为7个月后恢复正常。与同龄参照组相比,亲本体质量和体长的增长率均处于较高水平。亲本多项血清中甲状腺指标和血液理化指标在恢复培养期明显升高,多接近参照组平均水平,表明产后亲本生理情况基本恢复。研究表明,加强产后康复培养,可以提高初次繁殖亲本,尤其是雌性的健康恢复和生长效果,对养殖中华鲟资源的重复利用和再次成熟后生殖力的提高至关重要。

关键词　中华鲟　产后　摄食行为　生长　血清中甲状腺指标　血液生理指标　血清生化指标

中华鲟(*Acipenser sinensis*)是国家一级保护动物。由于人类活动的干扰,自然繁殖群体数量急剧减少,自然种群灭绝的危险提高[1]。开展全人工繁殖是实现人工保种的有效途径[2]。由于中华鲟体型大,性成熟晚,性周期时间长,养殖过程风险大、成本高,养殖达到初次性成熟的子代个体不仅数量有限,相比于长江野生雌性最小性成熟个体,其体质量和怀卵量亦明显减小[2,3]。另外,就野生产后个体而言,经历长距离繁殖洄游期间不摄食,繁殖后体质虚弱,易引起死亡,通过人工干预法可克服恢复摄食的障碍,获得康复及性腺再次发育[4,5]。对于养殖中华鲟亲本,与野生相比,没有产卵洄游前的育肥阶段,性腺发育期间多停食,产后恢复困难[2,6]。因此,进行养殖亲本产后康复和生长研究,是促进产后亲本成活率和再次成熟后生殖力提高的关键,为开展养殖中华鲟资源的重复利用提供支撑。同时,通过产后亲鱼培育效果分析,为产前亲鱼培育技术提供借鉴。

本研究通过对养殖产后亲本的摄食行为进行观察、测量生长指标及监测血清中甲状腺指标和血液理化指标,通过与同龄性腺未成熟的养殖个体(参照组)比较,以确定养殖亲本恢复效果,为建立养殖亲本产后恢复培养技术提供指导。本研究对中华鲟物种保护具有现实意义。

1　材料与方法

1.1　养殖设施和养殖时间

本文产后恢复培养从2012年11月至次年9月止,共计11个月。在临时养殖池(养殖池

*基金项目:公益性行业(农业)科技专项(201203086);国家自然科学基金项目(31172413);973计划课题(2015CB150702)。

**第一作者,女,1973年出生,山东栖霞人,高级工程师,博士,主要从事濒危鱼类生理生态学和水族馆生态学方向的研究,E-mail:zn7@sina.com

***通信作者(Corresponding author):危起伟,男,1960年出生,江西南昌人,研究员,博导,主要研究方向为珍稀鱼类保护,E-mail:weiqw@yfi.ac.cn

的池体规格 32.0 m×12.0 m×3.8 m，水温 19.0～19.5 ℃，溶氧 6.0～8.4 mg/L，流速 0.2～0.5 m/s）培养前 2 个月，于 12 月 28 日转入驯养池（驯养池的池体规格 29.0 m×11.0 m×4.4 m，展示观察窗规格 20.0 m×3.0 m，水温 21.0～22.5 ℃，溶氧 7.0～7.8 mg/L，循环水过滤系统的循环率为 50 min/次，养殖密度 2.2 kg/m^3）继续产后恢复培养 9 个月。

1.2 养殖亲本和饲养管理

雌、雄产后亲本[2]（编号 F51，F38）分别为长江水产研究所 2001 年和 1998 年秋季通过野生中华鲟人工繁殖的养殖个体。于 2012 年 10 月 28～30 日实施人工催产。繁殖前、后即刻测得的体长分别为 180 cm，180 cm 和 186 cm，186 cm；体质量分别为 57 kg，51 kg 和 64 kg，58 kg。

在同一养殖条件下，接近性成熟年龄的未产个体（12 龄养殖个体），血清 Mg^{2+}，TP，GLB，ALP 和 Ht 值（本文检测的血液生理、生化指标）略低于处于性腺发育间隔期的野生成熟个体[7]。本文为分析产后亲本的恢复情况，在驯养池内随机选取健康的雌雄鱼各 4 尾（同为长江水产研究所人工繁殖个体，分别与雌雄亲本同龄，处于性腺慢速发育阶段）作为参照组，12 月体长分别为（206±6）cm 和（189±4）cm，体质量为（89±9）kg 和（72±4）kg。

产后亲本采用直接投洒冰鲜多春鱼（*Mallotus villosus*）的方式诱导开口。在转入驯养池后，由潜水员水下采用多春鱼携带方式驯化投喂混合饵料，逐渐增加混合饵料量，直至完全摄食。每次投喂量以自主摄食量为主，隔日喂食。同期，参照组采用混合饵料由潜水员水下投喂，平均投饵率为体质量的 0.91 %。将一定比例的冰鲜多春鱼、蓝点马鲛（*Scomberomorus niphonius*）、中国枪乌贼（*Loligo chinensis*）、南美白对虾（*Penaeus vanname*）绞碎，添加适量维生素和螺旋藻，搅拌均匀后黏合成混合鲜饵，多春鱼和混合饵料的营养成分见表 1。

表 1 产后亲本饵料的营养成分

Tab. 1 The composition of feed of post-spawned cultured *Acipenser sinensis*

饵料 Feed	水分/% Moisture	粗蛋白/% Crude protein	粗脂肪/% Crude fat	粗灰分/% Ash	总磷/% Phosphorus	钙/% Calcium
多春鱼 *Mallotus villosus*	75.20	14.39	7.68	2.55	0.37	1.36
混合料 Self-designed diet	64.01	24.44	5.73	2.49	0.31	0.66

1.3 摄食行为与生长和生理观察

摄食行为观察项目包括：潜水员潜入水底喂食前后产后亲本的游泳区域、游泳水层、游泳平稳性，摄食积极性，吞咽饵料行为过程，摄食量等。各项目的计算方法如下。

游泳区域：亲本在观察时段内经常游动的区域。区域划分方法为，将驯养池按照长度平均分成 6 个区域，约 5 m/区，顺序记为 1～6，其中 3～4 为中华鲟的日常水下投食区；从池底向池水表面垂直分为 4 个水层，1.1 m/层，从水体底部向表面顺序记为底层、下层、中层、上层（接近水体表面记为表层）。在池内外设置区域和水层划分记号。

游泳平稳性：通过观察亲本游动时躯体左右摇摆程度及上浮和下潜的平稳性确定。平稳性好-游动时身体平稳，可自如下潜或上浮；平稳性较好-游动时身体左右摆动，但不明显，可自如下潜或上浮；平稳性较差-游动时身体左右摆动较明显，下潜或上浮时不仅尾部摆动幅度明显加大，身体左右摆动较明显；平稳性差-游动时身体左右摆动明显，下潜或上浮时尾部摆动幅度大，且下潜或上浮困难。

摄食积极性：通过观察潜水员投递饵料时亲本口的伸展程度表示。强：口完全伸展，伸展长度≥10 cm。较强：口伸展长度约为5～10 cm；较弱：口伸展长度≤5 cm；弱：口不主动伸展。

吞咽饵料过程：观察描述行为表现和每次喂食时亲本吞吐和吞咽饵料的平均用时。

摄食量：记录每次所摄食种类和摄食量。

为降低捕捞对产后亲本的影响，从转入驯养池开始，每3个月捕捞测定生长参数和血液参数。尾静脉采血5 ml，血细胞参数的测定用肝素钠抗凝，其余参数测定用3 000 r/min离心20 min后取血清，-4 ℃保存备用。各项血液参数的测定方法如下。

红细胞计数（red blood sell，RBC）：以浓度为0.5％ NaCl溶液为稀释液将血样稀释200倍后，用Neubarner氏计数板在显微镜下计数。

血红蛋白量（hemoglobin，Hb）：采用比色法测定。

红细胞比积（hematocritvalue，Ht）：采用温氏法测定。

实验测定的19项血清指标包括：血清促甲状腺激素（serum thyroid-stimulating hormone，TSH）、血清总T3（serum T3，TT3）、血清总T4（serum T4，TT4）、血清游离T3（Serum free T3，FT3）、血清游离T4（serum free T4，FT4）、钙（calcium，Ca^{2+}）、无机磷（phosphorus，P^{3+}）、镁（magnesium，Mg^{2+}）、丙氨酸氨基转移酶（glutamate-pyruvate，ALT）、天冬氨酸氨基转移酶（aspartate aminotransferase，AST）、碱性磷酸酶（alkaline phosphatase，ALP）、胆碱酯酶（cholinesterase，CHE）、总蛋白（total protein，TP）、白蛋白（albumin，ALB）、球蛋白（globulin，GLB）、甘油三酯（triglycerides，TG）、胆固醇（total cholestero，CHO）、血糖（glucose，GLU）。使用英科新创（厦门）科技有限公司生产的试剂盒在全自动生化分析仪（OLYMPUS AU400）上测定生化指标，使用西门子生产的试剂盒在全自动化学发光免疫分析仪（ADVIA Centaur CP）上测定激素指标。

1.4 数据处理与计算公式

体长生长率（growth rate of length，*GRL*）

$$GRL=(l_2-l_1)/l_1$$

体重生长率（growth rate of weight，*GRW*）

$$GRW=(W_2-W_1)/W_1$$

式中：l_1，l_2为单位时间内始、末体长，cm；W_1，W_2为单位时间内始、末体重，kg。

参照组数据统计分析用SPSS 13.0软件完成。用单因素方差分析（ANOVA）参照组血液指标差异，以$P<0.05$为显著水平。参照组结果为平均值±标准差。

2 结果

2.1 摄食行为和摄食量

在人工催产繁殖后，采用产前培育期的主要饵料多春鱼诱导开口，产后亲本2周后摄食，每日仅食少量多春鱼，摄食过程中反复吞吐饵料，并从鳃孔溢出大量饵料碎渣。在转入驯养池后（12月28日），潜水员通过多春鱼携带混合饵料方式伴随游动，驯化喂食（携带量逐渐增加），20 d后产后亲本可直接摄食混合料，摄食量逐步提高，摄食过程中吞吐饵料及从鳃孔溢出饵料碎渣行为逐渐减少。雌性亲本从4月开始，摄食期间降至池体下层进入喂食区域寻食，摄食积极性和摄食量明显提高，游泳平稳性逐渐转好，于5月可直接吞咽饵料。雄性摄食行为恢复快于雌性，于4月可直接吞咽饵料（表2）。雌性恢复培养初期摄食期间行为明显异常，主要表现为随吞吐饵料时鳃孔溢出大量气泡，身体明显下沉，然后冲出水面吸气调节身体平衡。随着 4 月后摄食量持续提高，此行为明显减轻至消失。培养后期雌性摄食量增高幅度明显大于雄性亲本。从转入驯养池至培育期末，摄食量占体质量比例为雌性 0.52 % ~ 2.04 %，雄性0.39% ~ 1.60 %。

表 2　产后亲本在驯养池的摄食行为

Tab. 2　The feeding behaviour of post-spawned cultured *Acipenser sinensis* during recovering in the aquarium

月份 Month	摄食量/体质量/% Feeding quantity		游泳区域和泳层 Swimming area and depth		游泳平稳性 Swimming balance		摄食积极性 Feeding activity	
	雌 Female	雄 Male	雌 Female	雄 Male	雌 Female	雄 Male	雌 Female	雄 Male
12	0.40	0.39	不固定	3 ~ 4，下层	+	+ +	+	+ + + +
1	0.91	0.87	不固定	3 ~ 4，下层	+	+ + + +	+ + +	+ + + +
2	1.05	0.94	1 ~ 6，下层 ~ 中层	3 ~ 4，下层	+ +	+ + + +	+ + +	+ + + +
4	1.40	1.28	3 ~ 4，下层	3 ~ 4，下层	+ + +	+ + + +	+ + + +	+ + + +
5	2.04	1.60	3 ~ 4，下层	3 ~ 4，下层	+ + + ~ + + + +	+ + + +	+ + + +	+ + + +

注：+ + + + 表示好（强）；+ + + 表示较好（较强）；+ + 表示较差（较弱）；+ 表示差（弱）。

Note：+ + + + means the best degree；+ + + means the better degree；+ + means the worse degree；+ means the worst degree.

2.2　生长效果

雌雄亲本的体长和体质量产后培育期间变化趋势相似（图 1），至培育期末，体长分别为 189 cm 和 192 cm，体质量为 55.5 kg 和 63.5 kg。从转入驯养池到培育期末，体长增长率分别为 5.00 %和 3.23 %，体质量增长率分别为 44.16 %和 23.30 %。雌、雄参照组在培育期体长和体质量的增长率分别为 2.91 %，2.65 %和 15.73 %，5.56 %，明显低于产后雌雄亲本。

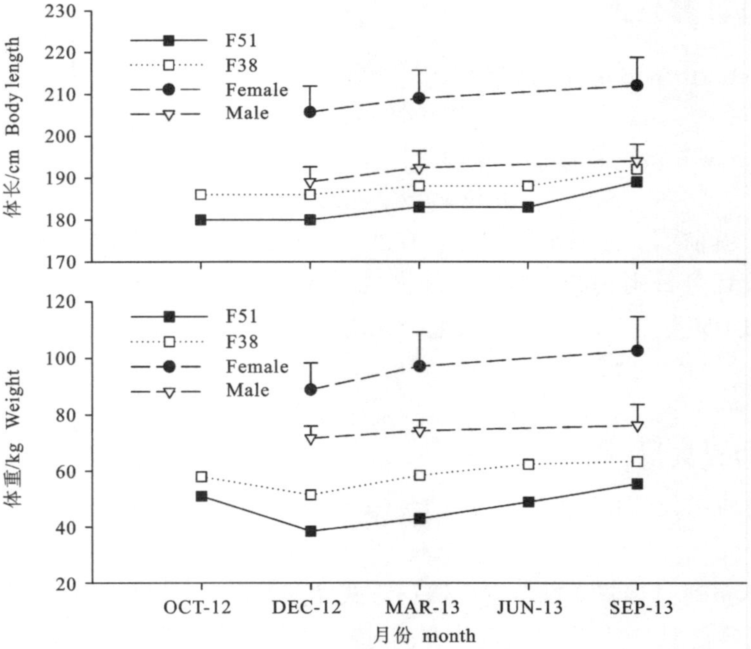

图 1　产后培育期间亲本体长和体质量的变化

Fig. 1　Changes of body length and body weight in post – spawned cultured *Acipenser sinensis* during the recovering

2.3 血清中甲状腺指标和血液理、化指标的变化

雌性亲本多项血液指标培育初期低于参照组平均水平，经产后培育，15 项指标可恢复。其中，6 项指标 Ca^{2+}，P^{3+}，AST，CHE，GLU，RBC 在较短时间（产后 5 个月左右）可恢复；9 项指标 TT4，FT4，ALP，TP，ALB，GLB，CHO，TG，Hb 产后较长时间（产后 8 ~ 11 个月）恢复至参照组平均水平。4 项指标培育期末未恢复，TT3，FT3，ALT 高于参照组平均值，TT3，FT3 近 3 倍于参照组平均水平，ALT 约 1.7 倍于参照组平均水平，Ht 低于参照组平均值。TSH，Mg、*A/G* 培育期间无明显变化，接近参照组平均水平（表 3）。雄性亲本产后培育期间血液指标恢复与雌性亲本有所不同（表 4）。12 项指标可恢复，Ca^{2+}，AST，GLB，RBC 4 项指标在较短时间可恢复，FT4，TT4，TP，ALB，*A/G*，CHO，Ht，Hb 8 项指标产后较长时间恢复，TT3，FT3 变化趋势与雌性亲本较相似，约 2 倍于参照组平均值，ALP 低于参照组平均值。TSH，Mg^{2+}，P^3，ALT，CHE，GLU，TG 7 项指标培育初期接近参照组平均水平，以后变化不明显。与雌性亲本相比，雄性亲本培育初期多项指标稳定，GLB 值升高速度快于雌性，至培育期末，多数指标值恢复。这表明至培育期末，产后亲本生理情况基本恢复，雄性亲本在初期恢复效果优于雌性亲本。

表 3 雌性亲本产后培育期间血液指标与参照组（$n=4$）比较

Tab.3 Comparison of the blood indices between the post-spawned female and the referenced female（$n=4$） during recovery

血液指标 Blood index	产后雌性 The post-spawned female				雌性参照组 The referenced female		
	2013.09	2012.12	2013.03	2013.06	2013.09	2012.12	2013.03
血清促甲状腺激素 TSH/（μIU/ml）	0.02	0.02	0.02	0.01	0.01 ±0	0.02 ±0	0.01 ±0
血清总 T3 TT3/（ng/ml）	0.02	0.02	0.64	1.26	0.33 ±0.04	0.33 ±0.11	0.47 ±0.08
血清总 T4 TT4/（μg/dl）	0.01	0.02	0.13	0.36	0.48 ±0.22	0.38 ±0.10	0.23 ±0.13
血清游离 T3 FT3/（pg/ml）	0.01	0.19	5.02	6.39	2.56 ±0.15	2.89 ±0.26	2.67 ±0.44
血清游离 T4 FT4/（ng/dl）	0.02	0.04	0.20	0.25	0.16 ±0.12	0.15 ±0.10	0.24 ±0.08
钙 Ca^{2+}/（mmol/L）	0.86	1.68	1.71	1.84	1.93 ±0.07	1.91 ±0.06	1.83 ±0.07
无机磷 P^{3+}/（mmol/L）	3.28	2.13	2.19	2.88	2.65 ±0.20	2.32 ±0.12	2.40 ±0.20
镁 Mg^{2+}/（μmol/L）	0.98	0.97	0.89	0.88	0.90 ±0.08	0.89 ±0.07	0.81 ±0.08
丙氨酸氨基转移酶 ALT/（U/L）	1.0	1.5	3.5	6.0	2.4 ±0.8	2.5 ±0.6	3.8 ±2.6
天冬氨酸氨基转移酶 AST/（U/L）	110.5	61.5	75.5	81.0	75.6 ±17.7	50.8 ±11.8	65.3 ±17.8
碱性磷酸酶 ALP/（U/L）	23.0	31.4	27.4	73.5	109.9 ±51.9	72.4 ±13.7	56.0 ±7.0
胆碱酯酶 CHE/（mmol/L）	69.0	157.5	187.5	124.0	193.3 ±25.0	141.3 ±12.4	133.8 ±0.1
总蛋白定量 TP/（g/L）	2.50	20.10	33.65	33.90	39.59 ±1.85	40.11 ±2.84	37.75 ±4.37
白蛋白定量 ALB/（g/L）	0.75	6.30	11.50	12.30	13.18 ±2.24	12.55 ±1.69	12.55 ±1.86
球蛋白 GLB/（g/L）	1.75	13.80	22.15	21.60	27.04 ±0.91	27.56 ±2.87	24.40 ±3.24
白球比 *A/G*	0.48	0.46	0.52	0.57	0.51 ±0.11	0.46 ±0.08	0.52 ±0.09
总胆固醇 CHO/（mmol/L）	0.26	2.10	1.34	2.85	3.96 ±0.26	3.39 ±0.54	3.11 ±0.55
甘油三酯 TG/（mmol/L）	0.14	2.10	0.74	1.77	1.83 ±0.40	1.77 ±0.68	1.75 ±0.65
血糖 GLU/（mmol/L）	0.58	1.36	1.34	2.10	1.30 ±0.20	1.83 ±0.34	2.31 ±0.89
红细胞计数 RBC/（10^{12}/L）	0.18	0.59	0.46	0.59	0.55 ±0.12	0.59 ±0.12	0.52 ±0.09
血红蛋白量 Hb/（g/L）	34	55	74	74	80 ±7	85 ±5	83 ±13
红细胞比积 Ht	12.2	21.1	26.9	21.9	31.2 ±5.3	30.8 ±3.4	29.1 ±3.3

表 4　雄性亲本产后培育期间血液指标与参照组($n=4$)比较

Tab. 4　Comparison of the blood indices between the post-spawned male and the referenced male($n=4$) during recovery

血液指标 Blood index	产后雌性 The post-spawned male				雌性参照组 The referenced male		
	2012.12	2013.03	2013.06	2013.09	2012.12	2013.03	2013.09
血清促甲状腺激素 TSH/(μIU/ml)	0.02	0.02	0.02	0.01	0.01 ±0	0.02 ±0	0.01 ±0.01
血清总 T3 TT3/(ng/ml)	0.02	0.01	0.26	0.52	0.33 ±0.21	0.32 ±0.22	0.22 ±0.14
血清总 T4 TT4/(μg/dl)	0.02	0.77	0.21	0.33	0.70 ±0.16	0.33 ±0.21	0.35 ±0.24
血清游离 T3 FT3/(pg/ml)	0.19	1.29	2.46	2.84	2.14 ±0.96	1.94 ±0.62	1.38 ±0.58
血清游离 T4 FT4/(ng/dl)	0.04	0.82	0.40	0.17	0.31 ±0.10	0.15 ±0.17	0.16 ±0.11
钙 Ca^{2+}/(mmol/L)	1.22	1.70	1.80	1.75	1.90 ±0.14	1.86 ±0.10	1.81 ±0.07
无机磷 P^{3+}/(mmol/L)	2.16	2.37	2.67	2.34	2.59 ±0.42	2.23 ±0.21	2.33 ±0.02
镁 Mg^{2+}/(μmol/L)	0.97	0.98	1.06	0.99	0.92 ±0.07	0.91 ±0.04	0.92 ±0.06
丙氨酸氨基转移酶 ALT/(U/L)	2.5	2.5	2.0	4.0	2.3 ±1.0	1.3 ±0.5	3.3 ±1.7
天冬氨酸氨基转移酶 AST/(U/L)	133.0	56.0	51.5	103.0	82.5 ±18.0	58.5 ±17.5	83.8 ±23.7
碱性磷酸酶 ALP/(U/L)	23.9	21.3	23.2	25.6	71.2 ±36.3	57.8 ±23.7	50.5 ±17.4
胆碱酯酶 CHE/(mmol/L)	230.0	93.5	80.0	182.0	171.1 ±48.9	124.0 ±33.7	116.3 ±36.7
总蛋白定量 TP/(g/L)	8.45	29.00	38.10	38.70	43.90 ±5.83	41.38 ±4.31	38.45 ±3.20
白蛋白定量 ALB/(g/L)	4.00	9.15	18.80	13.50	18.93 ±0.69	17.89 ±0.74	17.68 ±1.36
球蛋白 GLB/(g/L)	4.45	19.85	19.30	25.20	25.00 ±5.22	23.49 ±4.48	21.43 ±3.20
白球比 *A/G*	0.90	0.46	0.97	0.54	0.78 ±0.14	0.78 ±0.16	0.86 ±0.12
总胆固醇 CHO/(mmol/L)	0.85	1.86	0.93	1.55	2.66 ±1.29	2.14 ±0.80	2.21 ±0.95
甘油三酯 TG/(mmol/L)	1.52	1.20	0.77	1.04	2.30 ±0.45	1.88 ±0.87	1.91 ±1.12
血糖 GLU/(mmol/L)	1.49	1.35	1.37	1.81	1.40 ±0.44	1.81 ±0.44	1.84 ±0.58
红细胞计数 RBC/(10^{12}/L)	0.17	0.52	0.56	0.41	0.56 ±0.10	0.63 ±0.12	0.51 ±0.10
血红蛋白量 Hb/(g/L)	20	40	57	62	75 ±10	75 ±4	78 ±8
红细胞比积 Ht	6.1	27.2	22.2	26.7	28.8 ±3.9	30.3 ±3.2	30.0 ±2.9

3　讨论

3.1　人工条件下可实现养殖中华鲟产后康复和继续生长

本文产后养殖亲本为长江水产研究所 2012 年初次全人工繁殖的个体,产后身体虚弱。通过本次康复培养实验,逐渐恢复摄食,甲状腺激素分泌水平提高,血液理化指标值趋向于同龄未产组平均水平,生长优于未产组,雌性异常行为恢复正常。这说明产后养殖亲本可以获得康复并继续生长。这也使得中华鲟养殖亲本再次成熟后生殖力的提高成为可能,对养殖中华鲟资源的重复利用具有重要意义。

3.2　养殖中华鲟产后的摄食行为

在鱼类繁殖后,恢复摄食的效果和摄入的营养水平是影响亲本成活率及再次繁殖周期长短的主要因素[8]。在天然水域中,产后中华鲟食物可获得性受到环境限制,存活率低,再次繁殖的间隔时间是 2 年甚至更长时间[9];而蓄养产后野生中华鲟可摄食营养丰富的食物,存活率高，产后 7～12 个月行为恢复，再次发育的间隔时间多为 2～3 年[5]。在本研究中，产后亲本摄食欲差，特别是雌性身体虚弱，提供营养丰富的适口性饲料并促进其快速摄食是康复成功的关键。雌性亲本产后初期摄食期间泳层与游泳区域不固定，不能直接游至下层寻食，且游动平稳性差，通过潜水员伴随游动投递食物方式，提高了食物的可获得性，降低摄食时的体能消耗。随体力逐渐加强，摄食积极性增强，摄食期间可降低泳层主动寻食，摄食量也逐步提高。中华鲟对食物有较长时间记忆[5]，摄取食物后，主要靠口腔触觉和味觉判断适口性，最终吞入或摒弃食物，对部分食物的选择性可以被逐步驯化[4,10]。在本文中，先采用产前主要食物多春鱼诱导产后亲本开口，然后混合以多春鱼为原料的自制鲜饵继续驯化，利于亲本快速选食[11]。雄性亲本产前即摄食混合饵料，因此适应混合料的时间明显短于雌性。相较于野生产后中华鲟通过灌食促进恢复摄食而言，养殖中华鲟较容易恢复摄食，但采用多春鱼驯化摄食混合饵料也需要 20 d[4]，与野生中华鲟灌食到自主摄食的时间相似，这可能与中华鲟消化有关的酶类及消化有关的神经内分泌系统需要一段时间平衡有关[8]。另外，野生中华鲟主动摄食后泳层即下降，雌性亲本则远长于野生，可能受其虚弱体质影响所致[4,5]。混合饵料的营养均衡性优于多春鱼，利于促进产后亲本快速康复。

行为是动物对外界环境和内部环境变化的外在反应。本文雌性亲本产后初期摄食行为明显异常。分析认为，一方面，养殖个体类似于天然水域野生个体，在性腺发育过程中停食或少食，依靠体内的脂肪完成性腺的最后成熟[9]，但是由于缺少野生中华鲟繁殖洄游前的育肥阶段，性腺发育过程又要消耗大量营养物质（特别是雌性），致使产后体内营养物质积累不足，易导致体质虚弱；另一方面，大型雌性动物一般在人工繁殖过程中受到的损伤和应激大于雄性，这也容易导致产后胃肠神经紊乱和虚弱，致厌食甚至死亡等[12]。中华鲟靠气鳔保持浮力，气鳔开口于食道后段，健康个体可以将吞入的空气通过鳔管送入鳔内，也可以通过气鳔周围肌肉的收缩经鳔管把空气排出以调节浮力[13]。雌性亲本产后体弱，可能胃肠神经紊乱，吞吐饵料时食道和气鳔周围肌肉不能协调运动，摄食时浮力控制出现异常。随摄食逐渐恢复，摄入营养量提高，雌性亲本产后 7 个月体力明显增强，行为恢复正常，与蓄养野生产后中华鲟行为恢复正常的时间相似[5]。这说明，在人工养殖条件下，营养丰富的适口饵料和促进摄食的方法可以提高养殖亲本（特别是雌性）的产后恢复效果，这也是本文研究获得成功的关键。本文混合饵料仅是初步尝试，深入探讨亲本在性腺进入快速发育阶段前的营养强化方式，研究产后恢复培养时具有诱食效果的强化饵料，可以成为中华鲟亲鱼培育技术新的研究方向。

3.3　初次繁殖的养殖中华鲟产后生长特点

危起伟等[2]研究了初次性成熟的养殖中华鲟生物学情况，发现雌性年龄小于记录到的最小野生雌性中华鲟性成熟年龄，性成熟体长和体质量值远小于长江野生雌性最小成熟个体值，认为这不利于中华鲟的保护和规模化培育。本文亲本出现产后体长继续生长的特点，与天然水域产后亲本可持续生长的特点相同[9]。养殖亲本生长速度均优于未进入性腺快速发育的同龄未产组，这可能是长期饥饿后摄入丰富的食物时出现的补偿生长现象[13]。另外，水深、空阔、水质清新的大型养殖空间有利于大体型成体中华鲟的生长[14]。这说明，在人工条件下，通过加强产后培养，可以提高亲本的生长率，这使得初次繁殖亲本个体小的情况得以改善成为可能，对于提高再次成熟后的生殖力具有现实意义。

3.4　产后亲本的血清甲状腺指标、血液理化指标与摄食、健康情况及生长性能的关系

鱼类的甲状腺素主要是 T3 和 T4（T3 是由 T4 在外周组织中，主要是肝脏，脱碘转化而来

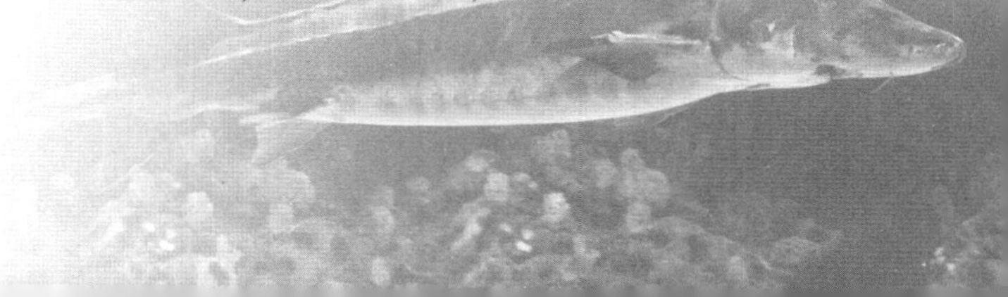

的)、对代谢活动、生长、行为等有重要作用,T3 的生物活性比 T4 强[15]。饥饿会影响下丘脑-垂体-甲状腺轴,使甲状腺对 TSH 刺激的敏感性降低,T4 向 T3 的转化率下降[16]。在本文培养期间,产后亲本 TSH 持平于参照组,而食物摄取量提高后甲状腺激素水平明显升高,表明食物摄取量对甲状腺激素的分泌影响较大。这与对黑鲷(*Sparus macroehhalus*)、虹鳟(*Oncorhynchus mykiss*)等研究结果较为一致[17]。食物的营养成分(特别是蛋白质)是影响甲状腺激素分泌的重要因子[18],本文产后培养采用的主要饲料中蛋白质含量较高,可能也更有利于促进甲状腺激素的产生。雄性亲本产后初期 TT3,FT3,TT4 水平和体质量增长率大于雌性,而 6 月以后,雌性 TT3,FT3 水平和体质量增长率明显高于雄性。另外,相比于参照组,雌、雄产后亲本实验末期 TT3,FT3 和雌性体长、体质量及雄性体长均处于较高水平,说明了甲状腺激素与生长关系密切。这在其他鱼的生长研究中也有类似报道[17,18]。

RBC 与 Ht 培养期间快速增高,Hb 升高速度滞后于红细胞增生的速度。鱼类红细胞由脾脏等造血器官产生,主要有携带和运输氧气及排除二氧化碳的功能。随摄食逐渐恢复,红细胞快速增生,说明红细胞生成过程受营养水平影响大,RBC 值增高显示机体代谢活动增强。

产后亲本 Ca^{2+},Mg^{2+},P^{3+} 三月恢复至参照组平均水平。Ca^{2+}、P^{3+} 在骨骼形成和维持酸碱平衡等代谢中起重要作用,Mg^{2+} 除了参与骨盐形成外,还是很多酶的激活剂。三种离子含量快速恢复,表明随着恢复摄食,产后亲本可通过自身的调节快速维持内环境的稳定。较雄性亲本相比,雌性的 Ca^{2+} 及钙磷比值在 12 月时均明显低于参照组水平,这可能受到卵巢发育过程中卵黄蛋白原合成的影响所致[19]。

ALT 主要存在于肝细胞胞浆中,肝细胞受损,细胞通透性增强,血清 ALT 活性上升。AST 也存在于肝脏,也可反映肝脏的生理状态。CHE 主要由肝脏合成,故能灵敏地评价肝细胞的合成能力,与 ALT,AST 可作为指示肝脏功能的酶学指标[20,21]。本文产后亲本 3 月时 ALT,AST,CHE 接近参照组平均水平,说明肝脏的代谢活动已接近健康的未产个体。培育末期,雌性产后亲本 ALT 增高至参照组 1.7 倍的水平,这主要与其后期快速生长,甲状腺功能亢进有关[22]。ALP 主要来源于肝脏、肾脏、成骨细胞等脏器和组织,是反映成骨细胞活性、骨生成状况和钙、磷代谢及肾脏功能的重要生化指标,甲状腺激素影响 ALP 活性[22,23]。雌性亲本培育末期 ALP 快速升高,与甲状腺激素分泌水平和继续加速生长趋势一致。另外,ALP 值接近参照组平均水平也说明肾脏功能逐渐恢复。雄性 ALP 在恢复培养期间较稳定,至培育末期略有提高,仍然低于参照组平均水平。这与其体长增长 6 月后才快于参照组较为一致。

GLU 是机体组织生化活动所需能量的来源,CHO,TG 是脂肪的代谢产物,与 TP 同受营养水平的影响显著[24,25]。产后亲本 TP,ALB,GLB 值随摄食量提高持续增高,表明受营养水平影响明显,与对施氏鲟(*Acipenser schrenckii*)[26]的研究结果较为一致。鱼的生长主要是依靠蛋白质在体内构成组织和器官,摄入蛋白质总量提高后,机体蛋白质代谢增强,合成速率提高,因此 TP,ALB,GLB 与体质量表现出一致的增长趋势,这也说明 TP,ALB,GLB 能较好地反映机体的营养水平。雌性亲本 TP,ALB,GLB 值 12 月远低于参照组,TG,CHO,GLU 也有类似表现,而雄性相对差别较小,表明雌性性腺发育和繁殖过程消耗的营养和能量物质多,易造成产后体内营养积累不足,导致体质虚弱。这也提示,加强产后营养护理是促进雌性亲本康复的关键。

综上所述,本次养殖亲本产后康复培养初次获得成功,尽管亲本获得较好的康复和生长效果,然而由于本研究主要依据 2012 年全人工繁殖的产后亲鱼恢复培养实验数据得来,受研究样本量限制,尚存在偶然因素影响,产后初期数据完整性也存在不足,还需继续增加实验样本,积累多批次恢复培养数据。

参考文献

[1] Wang J H, Wei Q W, Zou Y C. Conservation strategies for the Chinese sturgeon, *Acipenser sinensis*, an overview on 30 years of practices and future needs. Journal of Applied Ichthyology, 2011, 27(2): 176-180

[2] 危起伟,李罗新,杜浩,等. 中华鲟全人工繁殖技术研究. 中国水产科学,2013,20(1):1-11.

[3] Webb M A, Doroshov S I. Importance of environmental endocrinology in fisheries management and aquaculture of sturgeons [J]. General and Comparative Endocrinology, 2011, 170(2): 313-321.

[4] 张晓雁, 刘鉴毅,危起伟,等. 不同喂食方法对野生产后中华鲟的摄食促进. 动物学杂志,2007,42(2): 142-146.

[5] Wei Qiwei, Zhang Xiaoyan, Zhang Xianfeng, et al. Acclimating and maintaining Chinese sturgeon Acipenser sinensis in a large public aquarium environment. Journal of Applied Ichthyology, 2011, 27(2): 533-540.

[6] 郭柏福,常剑波,肖慧,等. 中华鲟初次全人工繁殖的特性研究. 水生生物学报,2011, 35(6):940-945.

[7] 张晓雁. 中华鲟血液生化学的研究与应用. 武汉:中国科学院水生生物研究所,2012.

[8] 柳凌,张洁明,郭峰,等. 日本鳗鲡人工催产后亲鱼恢复培养与再催产效果. 中国水产科学,2004,11(1):54-58.

[9] 陈细华. 鲟形目鱼类生物学与资源现状. 北京:海洋出版社,2007: 96-99.

[10] 庄平,章龙珍,罗刚,等. 长江口中华鲟幼鱼感觉器官在摄食行为中的作用. 水生生物学报,2008,32(4): 475-481.

[11] Kasumyan A Q. Sturgeon food searching behaviour evoked by chemical stimuli: a reliable sensory mechanism. Journal of Applied Ichthyology, 2002, 18(4/6): 685-690.

[12] 于彦强. 母猪产后不食症的综合防治方法. 山东畜牧兽医,2012,33(22):34-35.

[13] 谢从新. 鱼类学. 北京:中国农业出版社,2010:304.

[14] 杨德国,危起伟,王凯,等. 淡水养殖中华鲟成鱼的三种方式及效果比较. 南方水产,2006,2(2):1-5.

[15] 林浩然. 鱼类生理学. 广州:广东高等教育出版社,2004:222.

[16] Pedro N D, Delgado M J, Gancedo B, et al. Changes in glucose, glycogen, thyroid activity and hypothalamic catecholamines in tench by starvation and refeeding. Journal of Comparative Physiology B, 2003, 173(6): 475-481.

[17] 邓利,张为民,林浩然. 饥饿对黑鲷血清生长激素、甲状腺激素以及白肌和肝脏脂肪、蛋白质含量的影响. 动物学研究,2003,24(2):94-98.

[18] 王桂芹,周洪琪,陈建明,等. 饲料蛋白对翘嘴红鲌生长和内分泌激素的影响. 水生生物学报,2008,32(4): 543-550.

[19] Matsche M A, Rosemary K M, Brundage H M. Hematology and plasma chemistry of wild shortnose sturgeon Acipenser brevirostrum from Delawarw River, USA. Journal of Applied Ichthyology, 2013, 29(1): 6-14.

[20] 仲其军,张淑芳. 有机化学检验技术. 武汉:华中科技大学出版社,2012:196-200.

[21] 王转丁,杨晶晶,李留安,等. 不同强度游泳应激对小鼠血清 ALT, AST 和 ALP 活性的影响. 黑龙江畜牧兽医, 2013(11):164-166.

[22] 苏会璇,钟宇华,高碧华. 甲状腺功能亢进症者肝功能的检测及其临床意义. 广西医科大学学报,2002,22(1): 91-92.

[23] 王石莹,闫素梅. 碱性磷酸酶在动物骨骼代谢中的研究进展. 饲料博览,2009(4):14-16.

[24] 强俊,杨弘,王辉,等. 饲料蛋白水平对低温应激下吉富罗非鱼血清生化指标和 HSP70 mRNA 表达的影响. 水生生物学报,2013,37(3):434-443.

[25] de Pedro N, Guijarro A I, Lpez-patio M A, et al. Daily and seasonal variations in hematological and blood biochemical parameters in the tench, *Tinva tinca* Linnaeus, 1758. Aquaculture Research, 2005, 36(12): 1185-1196.

[26] 章龙珍,庄平,张涛,等. 人工养殖不同年龄施氏鲟的血液生化指标. 水产学报,2007,31(2):159-164.

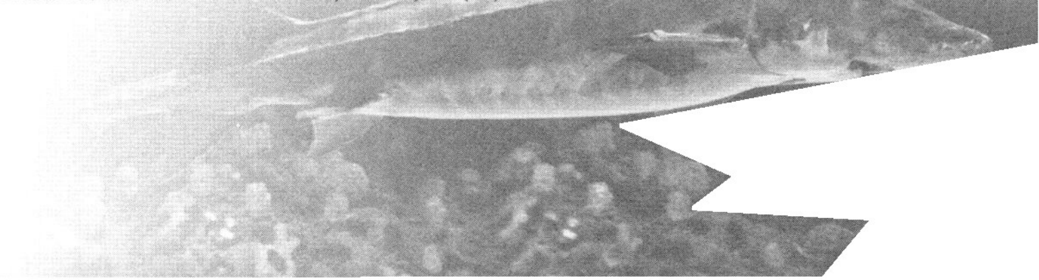

The Recovery of Post-Spawned Cultured Chinese Sturgeon(*Acipenser sinensis*)

Zhang Xiaoyan[1,2] Du Hao[1] Wei Qiwei[1]
Zhang Yanzhen[2] Wang Yanpeng[2] Cai Jingjiang[2]
Liu Zhigang[1] Qiao Xinmei[1] Xiong Wei[1]

(1. Key Laboratory of Freshwater Biodiversity Conservation, Ministry of Agriculture of China, Yangtze River Fisheries Research Institute Chinese Academy of Fishery Sciences, Wuhan 430223, China;
2. Beijing Xinwoda Marine Technology Co. Ltd(Beijing Aquarium), Beijing 100081, China)

Abstract In this study we recorded the feeding behavior, the growth indicators and the blood indices for 11 months after the spawning of cultured Chinese sturgeon. The growth indicators and the blood samples of the 2 referenced groups were also analyzed, and these 2 groups were in the slow gonad development stage and the same age as the female and male post-spawned individuals. We tested 22 blood indices. We found that the re-feeding of the post-spawned sturgeons can be induced with *Mallotus villosus*, the preferred feed of the two sturgeons before the controlled reproduction. Then the sturgeons started to feed on self-designed diet composed of *Mallotus villosus* , *Scomberomorus niphonius*, *Loligo chinensis and Penaeus vanname*. The feeding of the post-spawned female was initially abnormal due to the weakness, and then the feeding quantity increased to 2.04 % of the body weight(to 1.60 % of body weight in the post-spawned male at the same time) and the feeding turned back to normal 7 months post-spawning. As the feeding quantity increased over 9 months of recovery, the post-spawned cultured sturgeons showed the higher growth rates of the body length(5.00 % and 3.23 % for the female and the male respectively) and the body weight(44.16 % and 23.30 % for the female and the male respectively) than those of the two referenced groups. The concentrations of serum T3(TT3), serum T4(TT4), serum free T3(FT3), serum free T4(FT4) and many other blood indices of the two post-spawned sturgeons were lower than those of the two referenced groups in the early recovery, then most of the indices increased nearly to the average levels in the referenced groups, and the levels of TT3 and FT3 were much higher than the average values in the two referenced groups. These implied that the two post-spawned sturgeons were recovered. Our data suggested that refined culture technique could effectively improve the recovery and elevate the growth rate of the post-spawned cultured sturgeon(especially for the female). Our study is also crucial for the reuse of cultured Chinese sturgeon resources and for improving the fecundity after re-maturation.

Key words *Acipenser sinensis* After-spawning Feeding behavior Growth Serum thyroid indicators Blood cell indices Serum biochemical indices

中华鲟精子在扫描透射电子显微镜下的超微结构和形态

危起伟[1,3,4]　厉　萍[1,3]　M. Psenicka[2]　S. M. Hadi Alavi[2]
沈　丽[1,4]　刘　健[1,4]　J. Peknicova[5]　O. Linhart[2,*]

(1. 农业部淡水鱼类种质资源与生物技术重点实验室 中国水产科学研究院长江水产研究所　湖北荆州　434023
2. 南波西米亚大学水生生物与鱼类养殖研究所　Vodnany, Czech Republic　38925
3. 华中农业大学水产学院　湖北武汉　430070
4. 中国水产科学研究院淡水鱼类研究中心　江苏无锡　214081
5. 捷克科学院分子遗传研究所　Prague, Czech Republic)

摘　要　中华鲟是我国长江中的一种溯河洄游性濒危鱼类。在本文中,我们采用低温冷冻技术透过扫描透射电子显微镜,研究了中华鲟精细胞的超微结构和形态。结果显示,精子由不同顶体和细胞核区组成的细长头部、精子中段和尾部鞭毛组成。精子的头部加中段、鞭毛和全长的平均长度分别为 4.48 μm,33.3 μm 和 37.8 μm。精子核区位于比后部窄的细长梯形顶体的尾部。此外,在顶体前段可观察到颗粒物质和肌丝,内管 3～5 根,精子中段沿其纵向轴线左右对称。至目前的研究结果显示,与其他种类鲟鱼精子相比,中华鲟与白鲟的亲缘关系更近。

关键词　中华鲟　扫描电镜　透射电镜　精子　顶体　鱼类生物学

Ultrastructure and Morphology of Spermatozoa in Chinese Sturgeon (*Acipenser sinensis* Gray 1835) Using Scanning and Transmission Electron Microscopy

Wei Qiwei[1,3,4]　Li Ping[1,3]　M. Psenicka[2]　S. M. Hadi Alavi[2]
Shen Li[1,4]　Liu Jian[1,4]　J. Peknicova[5]　O. Linhart[2,*]

(1. Key Laboratory of Freshwater Fish Germplasm Resources and Biotechnology, Ministry of Agriculture, Yangtze River Fisheries Research Institute, Chinese Academy of Fisheries Sciences, Jingzhou, China;
2. University of South Bohemia, Research Institute of Fish Culture and Hydrobiology, 38925 Vodnany, Czech Republic;
3. Huazhong Agricultural University, Fishery College, Wuhan, China;
4. Freshwater Fisheries Research Center, Chinese Academy of Fisheries Science, Wuxi, China;
5. Institute of Molecular Genetics, Academy of Sciences of the Czech Republic, Prague, Czech Republic)

Abstract　The Chinese sturgeon (*Acipenser sinensis* Gray 1835) is an endangered anadromous sturgeon inhabiting the Yangtze River in China. In this study, the ultrastructure and morphology of spermatozoa was studied using transmission and scanning electron microscopy with a cryo-holder. The spermatozoon consisted of an elongated head with a distinct acrosome and nucleus region, a midpiece and a flagellum. The mean length of the head and midpiece, the flagellum and total length of spermatozoon were 4.48, 33.3 and 37.8 μm, respectively.

* 通讯作者(Corresponding author), E-mail: linhart@vurh.jcu.cz

The nucleus was an elongated trapezoid shape with anterior(acrosome) end narrower than the posterior. Granular material and an actin filament were observed within the anterior acrosome. Three to five endonuclear canals were present. The midpiece was eudipleural along its longitudinal axis. Compared to other sturgeon species, the data from the present study suggest a more recent evolutionary linkage between Chinese sturgeon and white sturgeon (*Acipenser transmontanus* Richardson 1836).

Keywords *Acipenser sinensis* SEM TEM Spermatozoon Acrosome Fish biology

1 Introduction

Study of the ultrastructure and morphology of fish spermatozoa provides information for understanding their possible taxonomic and evolutionary relationships at family[1,2], subfamily and species[3] levels, as well as for optimizing artificial reproduction, prevention of polyspermy problems and development of cryopreservation techniques[4-6]. Several different types of spermatozoa have been distinguished in fish. Nevertheless, little research has been carried out to describe sturgeon diversity as revealed by spermatozoa structure.

The order *Acipenseriformes* consists of two families comprising 27 species[7], but ultrastructure and morphology of spermatozoa have been described only for Russian sturgeon (*Acipenser gueldenstaedti* Brandt& Ratzeburg 1833)[4], stellate sturgeon (*A. stellatus* Pallas 1771)[8], white sturgeon (*A. transmontanus* Richardson 1836)[9], Atlantic sturgeon (*A. oxyrhynchus* Mitchill 1815)[10], shortnose sturgeon (*A. brevirostrum* Lesueur 1818)[11], lake sturgeon (*A. fulvescens* Rafinesque 1817)[12], pallid sturgeon (*Scaphirhynchus albus* Forbes & Richardson 1905)[13], Siberian sturgeon (*A. baerii* Brandt 1869)[14] and paddlefish (*Polyodon spathula* Walbaum 1792)[15].

Chinese sturgeon (*Acipenser sinensis* Gray 1835) is an endangered anadromous species[16] belonging to the family *Acipenseridae*[17] and found only in China at Yangtse River[18]. As with other sturgeon species, its stock has declined dramatically due to overfishing, loss of natural habitat for reproduction and interference by other human activities[16,18-20]. Since the 1990s, several attempts have been made to conserve the species by artificial culture and restocking with fingerlings[21,22].

The main objective of this study was to investigate the morphology and ultrastructure of Chinese sturgeon spermatozoa by scanning and transmission electron microscopy.

2 Materials and methods

Electron microscopy was carried out in China (Laboratory of Electron Microscopy, Institute of Hydrobiology, Chinese Academy of Sciences) and the Czech Republic (Laboratory of Electron Microscopy, Institute of Parasitology, Academy of Sciences of the Czech Republic in Ceske Budejovice).

2.1 Sperm sampling

Three mature male Chinese sturgeons were captured during their spawning migration (October-December) in the Yangtze River. To induce spermiation, the males were injected intramuscularly with 3.5 μg/kg luteinizing hormone releasing hormone analogue (LHRHa). Sperm was collected by catheter 6-7 h after hormone injection from the urogenital papilla, stored in sealed plastic bags with oxygen, placed on ice, and transported to the laboratory for controlling sperm motility and collection of samples. Non-active sperm samples were used to study the morphology and ultrastructure of spermatozoa. The percentage of motile spermatozoa and total period of motility were in range of 20 %-100 % and 4-8 min.

2.2 Electron microscopy

For SEM study, sperm samples were fixed for 3 h in freshly prepared 2.5 % glutaraldehyde in 0.1 M phosphate buffer (pH 7.4) at 0-4 ℃, rinsed in the same buffer and immersed for 2 h in 1.0 % osmium tetroxide in 0.1 M phosphate buffer (pH 7.4) at 20 ℃. After rinsing in the buffer, sperm were dehydrated in a graded ethanol series and critical-point dried. Samples were examined and photographed using a scanning electron microscope (SEM) (Hitachi-S570, Tokyo, Japan and JSM 6300, JEOL Ltd., Akishima, Tokyo, Japan) for morphological parameters and a JSM 7401 S (JEOL Ltd., Akishima, Tokyo, Japan) in cryo-regime for fine details.

For TEM study, sperm was centrifuged at 200 × *g* for 30 min and the supernatant discarded. It was prepared with a two-step fixation following the method described above. After several rinses the sample was dehydrated in a graded ethanol series and embedded in Epon812. Hardened blocks were sectioned at 50-70 nm and sections mounted on copper grids. These were post-stained with 2 % uranyl acetate in 50 % ethanol, followed by staining with Reynolds' lead citrate. Finally, the grids were examined and photographed using transmission electron microscopy (TEM) (JEM-1230, Tokyo, Japan and JEOL 1010, JEOL Ltd., Tokyo, Japan).

2.3 Measurements and data analysis

Measurements of spermatozoa parameters were taken using CorelDRAW 12™ software with the reported scale from our electron microscopy. The measurement parameters, and methods used, are in Psenicka et al.[14], and include: acrosome length and width; anterior and posterior head width; midpiece length and width; head length (nucleus + acrosome length) and total length (total head length + midpiece length + flagellum). Other parameters, including diameter of the endonuclear canal and microtubules and length of flagellum, were also measured. Average values of the measured parameters and standard deviation (S. D.) were calculated from 12 measurements of 3 fish.

3 Results

3.1 General morphology

The Chinese sturgeon spermatozoon was an elongate, slightly conical cylindrical shape with radial symmetry (Fig. 1). It consisted of a head, midpiece and flagellum. The head was elongated with an umbrella-like acrosome, a common structural feature of sturgeon spermatozoa, and an elongate nuclear region with the posterior end wider than the anterior. The Chinese sturgeon spermatozoon possessed a small midpiece, filler-like in shape, and a single flagellum surrounded by a plasma membrane. Morphological parameter measurements are given in Table 1.

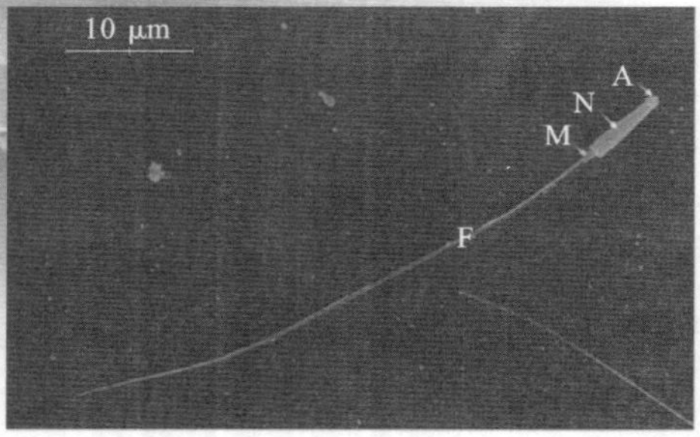

(a)

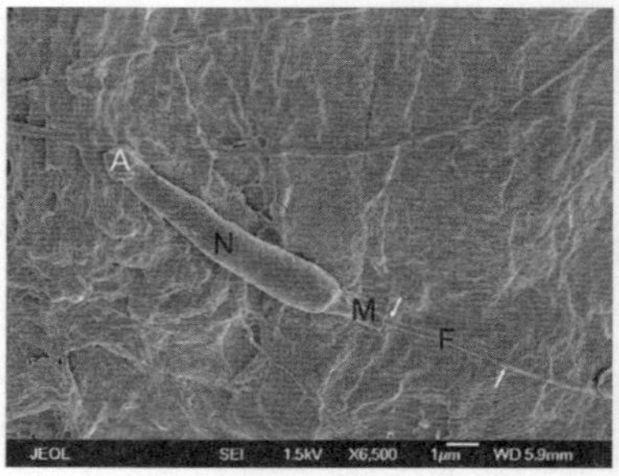

(b)

Fig. 1 (a) SEM[JSM 6300] showing sperm cell morphology(scale bar 10 μm);(b) SEM[JSM 7401 S] showing sperm cell morphology, arrows point to the origin of the fins along the flagellum(scale bar 1 μm). Acrosome(A), nucleus(N), midpiece(M), flagellum(F).

Table 1 Measurements of Chinese sturgeon sperm components from SEM and TEM

Characteristic	Dimension/μm	Characteristic	Dimension/μm
Acrosome length	0.54 ±0.15	Endonuclear canal diameter	0.08 ±0.03
Acrosome width	0.68 ±0.06	Posterolateral projections	0.37 ±0.04
Anterior head width	0.59 ±0.05	Flagellum diameter	0.32 ±0.01
Posterior head width	1.84 ±0.45	Lateral extensions of flagellum	0.34 ±0.11
Total head length	3.27 ±0.20	Microtubule diameter	0.01 ±0.00
Midpiece length	2.17 ±0.36	Flagellum length	33.26 ±2.74
Midpiece width	1.57 ±0.27	Total length	38.70 ±0.37

Dimensions are the mean ± S. D.

3.2 Head

The head contained an acrosome and a nucleus(Fig. 1 and Fig. 2a-f). In longitudinal section, the acrosome covered the anterior end of the spermatozoon. There was a granular material in the anterior end of the acrosome and probably an actin filament reaching out from its center(Fig. 2c). The acrosomal vesicle bounded by the acrosomal membrane was located in the outermost portion. Ten posterolateral projections(PLP) arise from the posterior edge of the acrosome(Fig. 2a, b and f). These PLPs were 0.37 μm in length(Table 1), extending along the anterior sides of the nucleus. Between the nucleus and the acrosomal vesicle was the subacrosome region(Fig. 2b and e). Fig. 2d shows the surface of the anterior nucleus, under the acrosome, after removing of the acrosome.

The nucleus, extending from the rear of the granular material to the midpiece, consisted of electron-dense homogeneous chromatin surrounded by a nuclear membrane. It was an elongate trapezoid in longitudinal section and circular in cross section. Three intertwining endonuclear canals, which measured 0.082 μm in diameter, traversed the nucleus from the acrosomal end to the basal nuclear fossa region(Fig. 2b and e-g). The implantation fossa, an organelle connecting the axial portion of nucleus with the midpiece, was located at the posterior end of the nucleus(Fig. 3a and e).

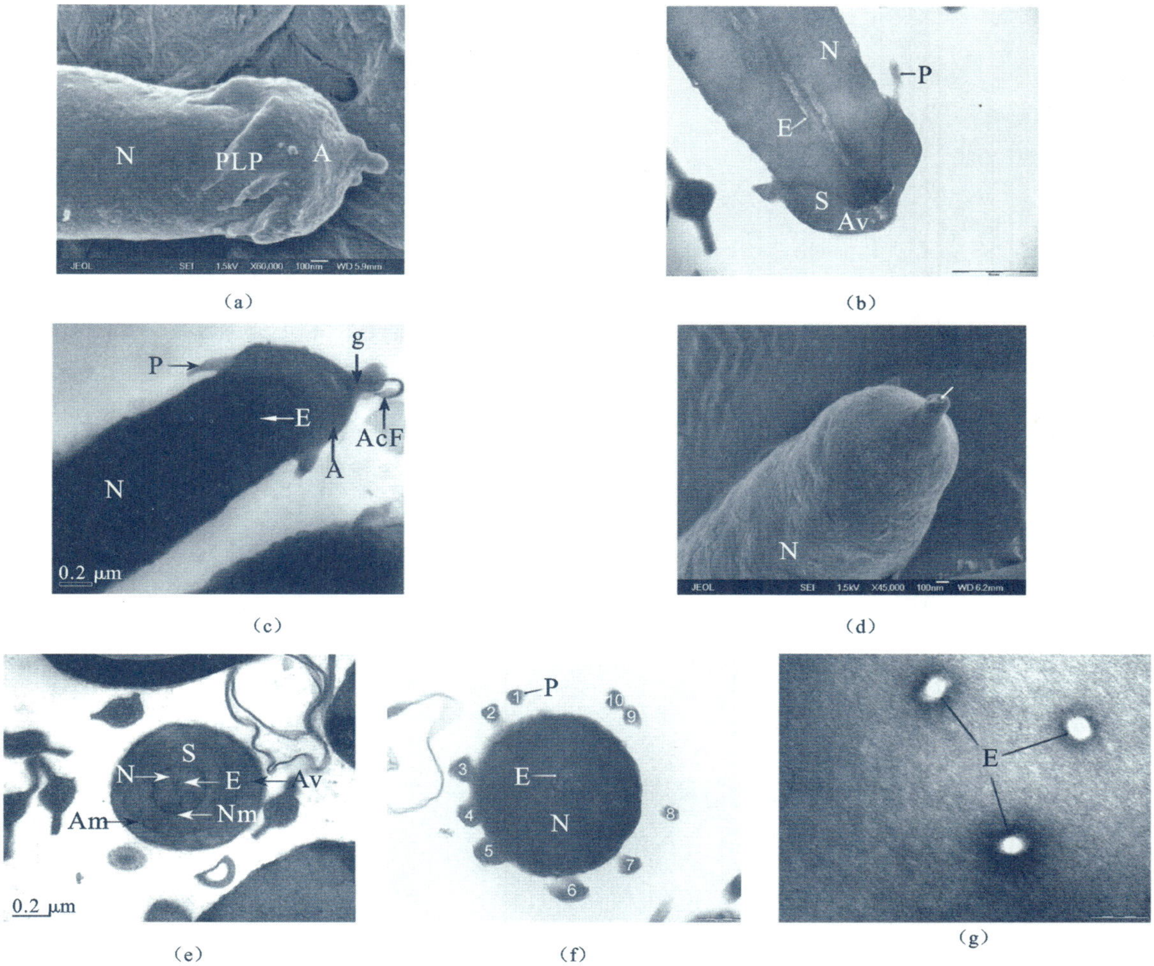

Fig. 2. (a) SEM[JSM 7401 S] of the acrosome showing posterolateral projections(PLP)(scale bar 100 nm);(b) TEM [JEOL 1010] sagittal section between the nucleus and acrosome, scale bar 500 nm; (c) TEM [JEM-1230] sagittal section between the nucleus and acrosome showing the acrosome vesicle(Av) and the subacrosome(S) with the posterolateral projections(PLP) and the endonuclear canals(ENC) traversed in the nucleus(N), actin filament(AcF), granular material(g)(scale bar 200 nm);(d) SEM[JSM 7401 S] of the anterior surface of nucleus(N) after abruption of the acrosome, scale bar 100 nm. The arrow shows an opening of an endonuclear canal; (e) TEM [JEM-1230] transverse section between the nucleus and the acrosome showing acrosome membrane(Am), acrosome vesicle(Av), subacrosome(S), nuclear membrane(Nm), nucleus(N), and endonuclear canals(E)(scale bar 200 nm); (f) TEM [JEOL 1010] transverse section between the nucleus and the acrosome showing posterolateral projections(PLP) 1-10 (scale bar 200 nm); (g) TEM [JEOL 1010] transverse section of the nucleus showing the endonuclear canals traversed in the nucleus(scale bar 100 nm).

3.3 Midpiece

The midpiece was small, filler-like in longitudinal section and cuneiform in transverse section (Fig. 3), and contained mitochondria and the centriolar apparatus. Around the flagellum there was a cytoplasmic channel, extracellular space between the midpiece and the flagellum, formed by an invagination at the plasmalemma(Fig. 3a). Three to eight ovoid mitochondria in one slice that deliver

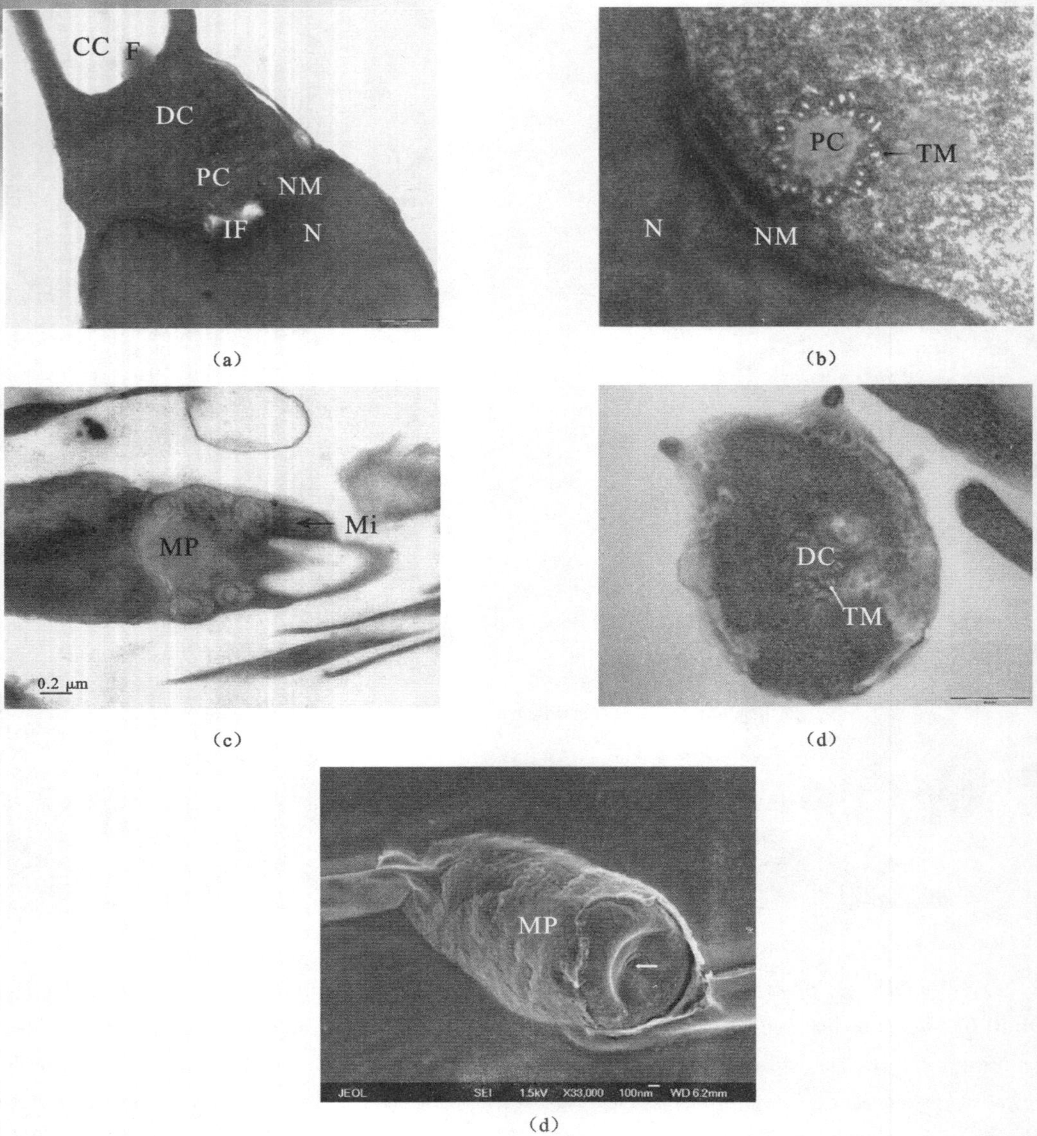

Fig. 3. (a) TEM [JEOL 1010] sagittal section of the midpiece showing the distal(DC) and the proximal centriole (PC), the implantation fossa(IF), the nucleus(N) surrounded by the nuclear membrane(NM) and the flagellum(F) separated by the cytoplasmic channel (CC) (scale bar 500 nm); (b) TEM [JEM-1230] showing the proximal centriole(PC) with triplets of microtubules(TM) (scale bar 100 nm); (c) TEM showing the midpiece(MP) with numerous mitochondria(Mi) irregularly dispersed in the cytoplasm(scale bar 200 nm); (d) TEM [JEOL 1010] transverse section at midpiece showing distal centriole (DC) with nine triplets of microtubules (TM) (scale bar 200 nm); (e) SEM[JSM 7401 S] of the midpiece after abruption of nucleus using. showing the proximal centriole (arrow) under the implantation fossa region(scale bar 100 nm).

the ATP required for flagellar beatings were found in the midpiece. Ridges of the mitochondria were well developed(Fig. 3c). The centriolar apparatus consisted of proximal and distal centrioles. The ovoid proximal centriole((196.46 ± 22.22) nm × (198.76 ± 20.70) nm) was located immediately posterior to the implantation fossa. The proximal centriole was connected to the implantation fossa by a structure resembling a fibrous body. The distal centriole was observed near the flagellar base. The axoneme was located immediately posterior to the distal centriole. Both orthogonally centrioles were formed of nine peripheral triplets of microtubules arranged in a cylindrical pattem(Fig. 3). The distal centriole was connected with the anterior of the proximal centriole, and the axoneme was immediately posterior to the distal centrioles. The diameter of the cytoplasmic channel, an invagination of the plasma membrane in the midpiece separating the cytoplasm of the midpiece almost completely from the flagellum, measured (479.18 ± 38.16) nm and was located between the cytoplasmic sheath and the flagellum.

3.4 Flagellum

The flagellum contained a cytoplasmicmembrane and an axoneme comprising 9 + 2 microtubules (Fig. 4). The internal diameter of the microtubules measured 9 nm. Dynein arms are not visible in the micrograph. The mean width of the peripheral and the central microtubules were (37.42 ± 8.07) and (50.26 ± 12.22) nm, respectively. The flagellum unthreaded in the cytoplasmic canal and gradually developed into two independent lateral extensions of cytoplasmic membrane (Figs. 1b and 4). These fins were observed parallel to the plane of the two central microtubules of the flagellum. The first and second fins originated approximately 0.65 and 4.01 μm post midpiece, respectively and extended along the flagellum terminating 5.58 and 7.97 μm from the end of flagellum, respectively. The average length and diameter of flagellum were (33.26 ± 2.74) and (0.33 ± 0.01) μm, respectively.

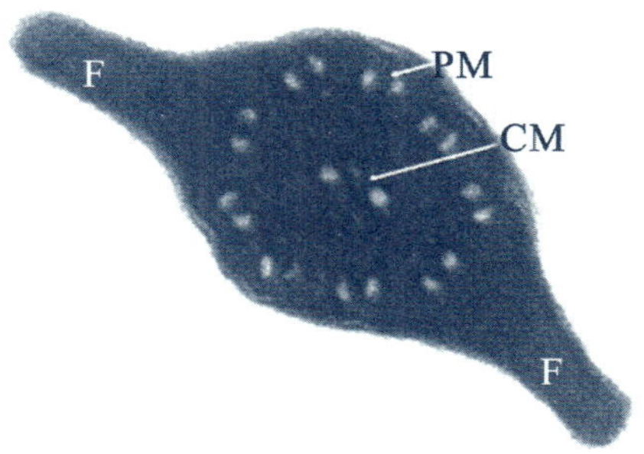

Fig. 4 TEM [JEOL 1010] transverse section of the flagellum showing the peripheral doublets of microtubules (PM) and central doublets of microtubules (CM). The fins (F) make the flagellum more effective (scale bar 50 nm).

4 Discussion

The spermatozoa of sturgeons are primitive in organization, with cells having almost radial symmetry [14,23]. Sperm is discharged directly into the water and sperm motility is long in terms of sperm activation, 4-8 min[24]. The Chinese sturgeon spermatozoon is similar in shape to those of most other sturgeon species (Table 2). It is very close to shortnose and lake sturgeon in terms of head shape and to Atlantic sturgeon in terms of head size (Table 2). The present study further revealed differences among sturgeons in terms of spermatozoon morphometry. Morphological parameters of spermatozoon in previous studies of sturgeons are summarized in Table 2. The observed differences suggest species-specific characteristics among sturgeons similar to those reported for teleosts[3], which could be useful for taxonomic purposes[1,13,25]. For example, DiLauro et al.[12] suggested that white and lake sturgeons are similar phylogenetically and systematically based on observed similarities in structure of spermatozoon. However, in the case of the Chinese sturgeon spermatozoa, we could not conclude which species it is closer to evolutionarily (Table 2).

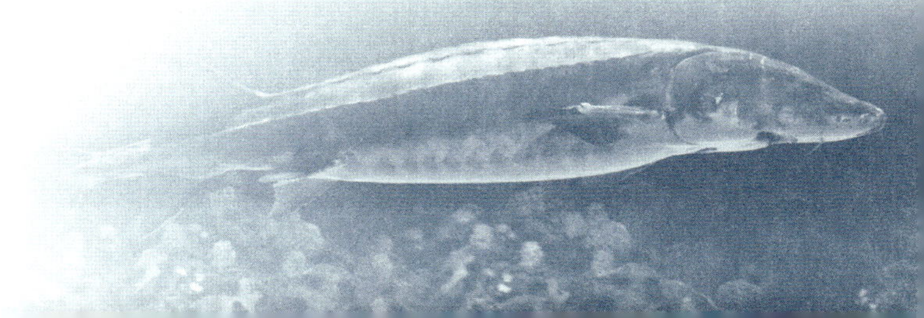

Table 2 Comparison of sperm cell morphology of stellate[8], white[9], Atlantic[10], shortnose[11], lake[12], pallid[13], Siberian[14] and Chinese sturgeon (present study)

Species	AL	AW	NL	ANW	PNW
Stellate sturgeon	0.97	1.22	6.66	0.98	1.49
White sturgeon	1.31	1.34	9.21	1.25	1.44
Atlantic sturgeon	0.83(0.11)	1.00(0.07)	3.15(0.36)	0.92(0.06)	0.55(0.08)
Shortnose sturgeon	0.78(0.08)	0.91(0.06)	6.99(0.83)	0.75(0.11)	1.21(0.12)
Lake sturgeon	0.73(0.14)	0.81(0.07)	5.69(0.43)	0.68(0.07)	1.04(0.08)
Pallid sturgeon	1.07(0.10)	0.82(0.06)	3.78(0.33)	0.68(0.04)	0.89(0.06)
Siberian sturgeon	0.95(0.17)	0.93(0.12)	5.87	0.87(0.13)	1.14(0.18)
Chinese sturgeon	0.54(0.15)	0.68(0.06)		0.59(0.05)	1.835(0.45)

Species	ML	MW	AHML	FL	TL	n
Stellate sturgeon	3.43	1.38	11.05	40-70	51.05-81.05	1
White sturgeon	2.13	1.08	11.82	30-40	41.82-51.82	1
Atlantic sturgeon	1.37(0.16)	0.51(0.07)	5.66(0.37)	37.08	42.74	12
Shortnose sturgeon	1.91(0.35)	0.81(0.09)	9.71(0.73)	36.7	46.41	15
Lake sturgeon	2.68(0.43)	0.70(0.08)	9.10(0.53)	47.53	56.63	14
Pallid sturgeon	1.23(0.16)	0.67(0.08)	6.07(0.48)	37.16	43.23	16
Siberian sturgeon	1.09(0.43)	0.81(A), 0.57(P)	7.01(0.83)	44.75(4.93)	51.76	8
Chinese sturgeon	2.17(0.36)	1.57(0.27)	5.44	33.26(2.74)	38.7	3

Data are the means and S. D. AL, acrosome length, AW, acrosome width; NL, nuclear length; ANW, anterior nucleus width; PNW, posterior nucleus width; ML, midpiece width; AHML, length of acrosome, head and midpiece; FL, flagellum length; TL, total length.

The longest flagellum and total length have been reported in stellate sturgeon[8], but the shortest flagellum and total length were observed in the Chinese sturgeon in the present study. In addition, the present study confirmed the presence of an acrosome in spermatozoa of sturgeon, a group of ancient animals called "living fossils". This is similar to elasmobranch and teleost fishes with internal fertilization, but in contrast to other fishes, such as other teleosts with external fertilization[26,1,10,3], indicating the evolutionary development of fish reproductive systems[3].

4.1 Acrosome

A common characteristic of higher vertebrate spermcell acrosomes is that they possess an acrosome vesicle[27], as, for example, in tree frogs[28], birds[29] and mammals[30]. These structures are not found in the spermatozoa of all sturgeons. Some species possess only an acrosome vesicle, a subacrosome and posterolateral projections. This is the case for white sturgeon[9,31], stellate sturgeon[8], Atlantic sturgeon[10] and Siberian sturgeon[14]. The acrosome, possibly functioning to aid in egg penetration during fertilization, is considered to be derived from the Golgi apparatus and is commonly cap-shaped[32]. The present study showed some variations in length and width of the acrosome among sturgeon species (Table 2). This may reflect an inter-specific diversity.

The physio-biochemical function of the acrosome in fertilization of sturgeon eggs is unclear.

Granular material and probably actin filament, which describe Gary et al. [33] in white sturgeon, are noted within the anterior end of the acrosome of spermatozoon in Chinese sturgeon as well as in white sturgeon[9]. However, the acrosome of lake, shortnose and Atlantic sturgeon exhibits granular, but not filamentous, material in this region[12]. Fine granular material is evenly dispersed within the anterior end of the spermatozoon of lamprey acrosome, with distinct clumps present against the posterior wall[34]. It is known that such structure corresponds to acrosomal function, so the granular material and actin filament may adapt the spermatozoon to fertilization mode. Further studies are needed to understand its role during sperm penetration into the egg and egg activation, or egg activation alone.

The sturgeon spermatozoon head usually terminates in posterolateral projections (PLP) that are radially distributed lobes[10-14] and present study. As with the acrosome, there are no data to explain PLP function in sperm physiology. DiLauro et al. [12] speculated they might function as barbs to assist effective penetration into the egg. The literature shows variations in the number and size of PLPs. The present study showed the presence of 10 PLPs in Chinese sturgeon, which differs from the pallid, which have 8 PLPs[13]. The length of PLPs in lake, Atlantic, shortnose, pallid and Siberian sturgeon has been reported to be 0.324, 0.233, 0.246, 0.760 and 0.950 μm, respectively. The length of PLPs in the Chinese sturgeon (0.369 μm) is thus close to that of the lake sturgeon.

4.2 Nucleus

The nucleus morphology and the presence or absence of nuclear vesicles are key parameters in terms of teleost taxonomy[3,35]. Nuclear vesicles have been observed in the spermatozoon of teleost fish, such as the yellow catfish (*Pseudobagrus fulvidraco* Richardson 1846)[36] and tench (*Tinca tinca* Linnaeus 1758)[37]. It has been suggested that theymay decrease by stress of mechanical injury, and thus the impact on the germ plasma was reduced when the spermatozoon swam incessantly during fertilization[30]. Interestingly, no nuclear vesicles are reported in sturgeons, which contain only electron-dense homogeneous chromatin, confirming that sturgeons are more primitive than teleosts.

One or more endonuclear canals have been reported in teleosts, with the exception of garfish[3,38]. In the case of sturgeons and paddlefish, three endonuclear canals are reported in white[39], stellate[8], shortnose[11], and Siberian sturgeon[14], and in paddlefish[15], but two canals in Atlantic sturgeon[10] spermatozoa. Spermatozoa of Chinese sturgeon have three to five endonuclear canals that traverse the entire nucleus in a helical arrangement. Further studies are needed to determine whether the number of canals is an intra- or inter-specific characteristic and why the arrangement of canals is helical. Cherr and Clark[9] presumed the endonuclear canals assisted in transferring a centriole into the egg after sperm-egg fusion. In general, it seems that the endonuclear canals traversing the nucleus from the nuclear fossa to the acrosome act on the eggs through the granular material of the acrosome. The spermatozoa of Atlantic sturgeon is the smallest among the species studied and its chromatin and nucleic acids are the most densely packed, suggesting a more advanced evolutionary arrangement[11].

Similar to other sturgeon species[8-12,14,15], the implantation fossa is present in the Chinese sturgeon spermatozoon and is located at the posterior end of the nucleus. Mattei and Mattei[40] described this structure as an enlargement of the endonuclear canal and DiLauro et al. [12] suggested it connects the endonuclear canals with the centrioles of the midpiece. DiLauro et al. [10] observed a structure in Atlantic, lake and shortnose sturgeon that is similar to the fibrous body of the garfish spermatozoon[38] and proposed that the structure, which has nearly the same diameter as the endonuclear canals, might serve as a channel for the transport of genetic material between the centriole and the nuclear fossa.

4.3 Midpiece

In previous studies, the midpiece of sturgeon spermatozoa was considered to be an elongate and slightly conical cylinder with nearly radial symmetry[10]. The present study showed that the midpiece of Chinese sturgeon spermatozoa is eudipleural (bilaterally symmetrical) about the longitudinal axis (Figs. 3d and 4).

There is a great diversity in the shape and the structure of the spermatozoa midpiece between fish species. The symmetry of the midpiece, the arrangement of the centriole, the number of mitochondria, the distribution of vacuoles and the presence of vacuoles are all regarded as important systematic characteristics for teleosts[3,35]. Normally, there are a few vacuoles with membranes between mitochondria and nucleus in most teleost fish, such as *Sebastiscus marmoratus* Cuvier 1829, Nile tilapia (*Tilapia nilotica* Linnaeus 1758) and common carp (*Cyprinus carpio* Linnaeus 1758)[29]. No similar structures were observed in the midpiece of Chinese sturgeon in the present study, or in white[9], lake[12] or stellate[8] sturgeon. Vacuoles containing lipid droplets have been reported in Atlantic[10], shortnose[11] and Siberian[14] sturgeon.

In all sturgeon species, both proximal and distal centrioles have been observed in the midpiece, with variation in size but with similar function[8-15]. The proximal centriole is apparently attached to the nucleus by several thin strands in the center of the nuclear fossa, called the fibrous body[10,14], which connects the nuclear fossa with the proximal centriole. The distal centriole is located adjacent to the proximal centriole; posterior to the distal centriole is the axoneme. The arrangement of the centriolar complex in Chinese sturgeon spermatozoon was T-shaped as in other sturgeons.

The number of mitochondria may differ among sturgeon species[11] as in teleosts[39]. The number of sperm mitochondria in other sturgeon is reported to be more than one[11]. For example, three to six mitochondria have been observed in the Siberian sturgeon, whilst six mitochondria were present in the periphery of the midpiece of the Chinese sturgeon (Fig. 3a). However, to determine developmental changes of structure and behavior of spermatozoa in *Acipenseriformes*, there is a need to study and compare the mitochondrial number, their physiological function during spermatozoa motility and ATP generation, and the effects of storage procedures, both short and long term (cryopreservation) among species.

4.4 Flagellum

In the mid stage of spermatozoa ontogeny, the distal centriole outgrows the axoneme, extending to form the flagellum. In the Chinese sturgeon, the flagellum is separated from the midpiece by the cytoplasmic channel as in other fish spermatozoa[1,3,40] with the characteristic 9 + 2 structure of microtubules. The mean length of the Chinese sturgeon flagellum is 33.26 μm (Table 1) and is the shortest flagellum of those sturgeons examined, although close to that of white sturgeon (30-40 μm)[9].

In both teleosts and sturgeons[1,3,14] the flagellum is separated from the midpiece by the cytoplasmic channel as in other fish spermatozoa. The axoneme originates from the posterior end of the distal centriole in the classical 9 + 2 arrangement of microtubules, giving rise to the flagellum in all described sturgeon spermatozoa. The present study and those of Ginsburg[8], Cherr and Clark[9], DiLauro et al.[10,11] and Psenicka et al.[14] revealed that the sturgeon spermatozoon has a flagellum with a paddle- or fin-like shape, which is adapted for locomotion during external fertilization. The same structure has been also observed in some teleosts[36,41,42]. In the Siberian sturgeon, it has been shown that the fins were flat, indicating no rotation, so that the first and second fins started 0.7 and 5.3 μm post midpiece, respectively, and extended along the flagellum. The first and second fins terminated 3.4 and 5.1 μm, respectively, from the end of Siberian sturgeon spermatozoon flagellum[14]. Compared to the Siberian sturgeon, both first and second fins of Chinese sturgeon arose closer to the midpiece (0.65 and 4.01 μm post midpiece, respectively) and extended farther along the end of flagellum (terminating 5.58 and 7.97 μm from the end of flagellum, respectively). As with the Siberian sturgeon, the observed fins were flat, not in a spiral configuration.

5 Conclusion

(1) The present study shows the diversity of spermatozoa ultrastructure between sturgeon species, similar to that observed and reviewed in teleost fishes[1,3], due to adaptation to various modes of fertilization according to the physio-chemical, environmental and related parameters, which regulate fertilization mechanism in fishes.

(2) Comparison of spermatozoa between Chinese and other sturgeons in Table 2 suggests four characteristics as parameters that will be considered to study phylogenetic relationships among species: (i) the shape and size of the sperm head with its organelles, (ii) the presence of granular material and the actin filament, (iii) the number of endonuclear canals and (iv) the absence of fibrous body and vacuole in the midpiece. Taken together, based on ultrastructural similarities, the Chinese sturgeon may be more closely related to white sturgeon than to other sturgeons.

(3) Compared to teleost fishes, data on physiological functions of sturgeon spermatozoa organelles are rare and need further study especially to understand the roles of the acrosome, PLP, and fin structure of the flagellum.

Acknowledgments

This study was supported by the Fund of Commonwealth Study of China (No. 2000DIB50177), Key Program of Foundational Work of China (No. 2002DEA1004), the Fund of Conservation in Ecology & Environment of Three Gorges Project (No. SX97-25/HB), the Key Project of Natural Science Foundation of China (No. 30490231), by USB RIFCH, Czech Republic no. MSM6007665809, GACR, Czech Republic no. 524/06/0817, Kontakt ME 742, Czech Republic and Biodiversity LC06073, Czech Republic. We thank the research team for excellent technical assistance.

References

[1] Jamieson B G M. Fish evolution and systematics: evidence from spermatozoa. Cambridge, UK: Cambridge University Press, 1991

[2] Mattei X. Spermatozoon ultrastructure and its systematic implications in fishes. Can J Zool, 1991, 69: 3038-3055

[3] Lahnsteiner F, Patzner R A. Sperm morphology and ultrastructure in fish//Alavi S M H, Cosson J, Coward K, et al. Fish spermatology. Oxford: Alpha Science Ltd, 2007: 1-61

[4] Ginzburg A S. Fertilization in fishes and the problem of polyspermy. Moscow: Nauka Press, 1968

[5] Billard R. Ultrastructure of trout spermatozoa: changes after dilution and deep freezing. Cell Tissue Res, 1983, 228: 205-218

[6] Billard R, Cosson J, Perchec G, et al. Biology of sperm and artificial reproduction in carp. Aquaculture, 1995, 129 (1-4): 95-112

[7] Bemis W E, Birstein V J, Walkman J R. Sturgeon biodiversity and conservation: an introduction. Environ Biol Fish, 1997, 48(1-4): 13-14

[8] Ginsburg A S. Fine structure of the spermatozoan and acrosome reaction in *Acipenser stellatus*//Beijaev D K. Problemy eksperimentalnioj biology. Moscow: Nauka, 1977: 246-256

[9] Cherr G N, Clark W H. An acrosome reaction in sperm from the white sturgeon *Acipenser transmontanus*. J Exp Zool, 1984, 232: 129-139

[10] DiLauro M N, Kaboord W, Walsh R A. Sperm-cell ultrastructure of North American sturgeons. I. The Atlantic sturgeon (*Acipenser oxyrhynchus*). Can J Zool, 1998, 76: 1822-1836

[11] DiLauro M N, Kaboord W S, Walsh R A. Sperm-cell ultrastructure of North American sturgeons. II. The shortnose sturgeon(*Acipenser brevirostrum*, Lesueur, 1818). Can J Zool, 1999, 77: 321-330

[12] DiLauro M N, Kaboord W S, Walsh R A. Sperm-cell ultrastructure of North American sturgeon. III. The Atlantic sturgeon(*Acipenser oxyrhynchus*). Can J Zool, 2000, 78: 438-447

[13] DiLauro M N, Walsh R A, Peiffer M. Sperm-cell ultrastructure of North American sturgeons. IV. The pallid sturgeon (*Scaphirhynchus albus* Forbes and Richardson, 1905). Can J Zool, 2001, 79: 802-808

[14] Psenicka M, Alavi S M H, Rodina M, et al. Morphology and ultrastructure of Siberian sturgeon, *Acipenser baerii*, spermatozoa using scanning and transmission electron microscopy. Biol Cell, 2007, 99: 103-115

[15] Zarnescu O. Ultrastructural study of spermatozoa of the paddlefish, *Polyodon spathula*. Zygot, 2005, 13: 241-247

[16] Birstein V J, Bemis W E, Waldman J R. The threatened status of acipenseriform species: a summary. Environ Biol Fish, 1997, 48: 427-435

[17] Bemis W E, Findeis E K, Grande L. An overview of Acipenseriformes. Environ Biol Fish, 1997, 48: 25-71

[18] Wei Qiwei, Ke F E, Zhang J, et al. Rueqiong Z, et al. Biology, fisheries, and conservation of sturgeons and paddlefish, in China. Environ Biol Fish, 1997, 48: 241-255

[19] Billard R, Lecointre G. Biology and conservation of sturgeon and paddlefish. Rev Fish Biol Fish, 2001, 10(4): 355-392

[20] Wei Qiwei. Reproductive behavioral ecology of Chinese sturgeon(*Acipenser sinensis* gray) with its stock assessment. Doctoral dissertation. Institute of Hydrobiology, Chinese Academy of Sciences, 2003

[21] Kynard B, Wei Qiwei, Ke F. Use of ultrasonic telemetry to locate the spawning area of Chinese sturgeons. Chin Sci Bull, 1995, 40: 668-671

[22] Wei Qiwei, Yang D, Ke F, et al. Technique of ultrasonic telemetry for Chinese sturgeon, Acipenser sinensis, in Yangtze River. J Fish China, 1998, 22(3): 211-221

[23] Baccetti B. Evolutionary trends in sperm structure. Comp Biochem Physiol A, 1986, 85: 29-36

[24] Linhart O, Kudo S, Billard R, et al. Morphology composition and fertilization of carp eggs: a review. Aquaculture, 1995, 129: 75-93

[25] Alavi S M H, Cosson J, Karami M, et al. Spermatozoa motility in the Persian sturgeon, *Acipenser persicus*: effects of pH, dilution rate, ions and osmolality. Reproduction, 2004, 128: 819-828

[26] Wolenski T S, Hart N H. Scanning electron microscope studies of sperm incorporation into the zebrafish (*B. rachydanio*) egg. J Exp Zool, 1987, 243: 259-273

[27] Jamieson B G M. Evolution of tetrapod spermatozoa with particular reference to amniotes//Jamieson B G M, Ausio J, Justine J. Advances in spermatozool phylogeny and taxonomy. Paris: Museéum National of Histoire Naturelle, 1995: 343-358

[28] Lin D J, You Y L, Zhong X R. The structure of the spermatozoon of *Hyla chinensis* and its bearing on phylogeny. Zool Res, 1999, 20: 161-167

[29] Lin D J, You Y L A. Study on the changes during the maturation of the late spermatid and the structure of the spermatozoon of *Sebastiscus marmoratus*. Zool Res, 1998, 19: 359-366

[30] Chen D Y. Impregnation biology-impregnation mechanism and procreation engineering. Beijing: Science Press, 2000: 1-52, 105-138

[31] Cherr G N, Clark W H. Gamete interaction in the white sturgeon *Acipenser transmontanus*: a morphological and physiological review. Environ Biol Fishes, 1985, 14: 11-22

[32] Nicander L. Comparative studies on the fine structure of vertebrate spermatozoa//Baccetti B. Comparative spermatology. Rome: Academia Nazionale Dei Lincei, 1980: 47

[33] Gary N, Cherr C J R, Wallis C J R. An acrosome Reaction in Sperm the White Sturgeon, *Acipenser transmontanus*. J Exp Zool, 1984, 232: 129-139

[34] Stanley H P. The fine structure of spermatozoa in the lamprey *Lampetra planeri*. J Ultrastruct Res, 1967, 19: 84-89

[35] Liu X Z, Yang W X. Recent progresses of study on spermatozoon structure of teleost and its application in reproductive evolution studies. Dong Hai Mar Sci, 2002, 20: 32-37

[36] You Y L, Lin D J. The ultrastructure of the spermatozoon of the teleost, *Pseudobagrus fulvidraco*. Acta Biol Exp Sin, 1996, 29(3): 235-245

[37] Psenicka M, Rodina M, Nebesarova J, et al. Ultrastructure of spermatozoa of tench *Tinca tinca* observation with scanning and transmission electron microscopy. Theriogenology, 2006, 64: 1355-1363

[38] Afzelius B A. Fine structure of the garfish spermatozoon. J Ultrastruct Res, 1978, 64: 309-314

[39] Linhart O, Slechta V, Slavik A. Fish sperm composition and biochemistry. Bull Inst Zool Acad Sinica, 1991, 16: 285-311

[40] Mattei C, Mattei X. Spermiogenesis and spermatozoa of the elopomorpha (Teleost fish)//Afzelius B A. The functional anatomy of the spermatozoon. Oxford: Pergammon Press, 1974: 211-221

[41] Mattei X, Siau Y, Seret B. Etude ulrastructurale du spermatozoide du coelacanthe: *Latimeria chalumnae*. J Ultrastruct Mol Struct Res, 1988, 101: 243-251

[42] You Y L, Lin D J. The release of vesicles from the nucleus of *Tilopia nilotica* spermatid during spermiogenesis. Acta Zool Sinica, 1998, 44(3): 257-263

几种不同浓度的离子及单糖对中华鲟精子活力的影响*

刘鉴毅[1,2,3,**]　甘　芳[1,2,3]　危起伟[1,2,3,***]　杜　浩[1,2,3]　朱永久[1,2,3]

(1. 中国水产科学研究院长江水产研究所　湖北荆州　434000

2. 农业部淡水鱼类种质资源与生物技术重点开放实验室　湖北荆州　434000

3. 中国水产科学研究院淡水渔业研究中心　江苏无锡　214081)

摘　要　本文研究在不同浓度 Na^+，K^+，Ca^{2+}，Mg^{2+}，Cu^{2+} 及葡萄糖、果糖溶液中，中华鲟(*Acipenser sinensis* Gray)精子活动情况。结果表明：Na^+ 浓度为 2 ‰时，中华鲟精子寿命时间(life time，LT)最长，为 347 s；而 Na^+ 浓度为 1 ‰时，中华鲟精子剧烈运动时间(acute movement time，AT)最长，为 98.67 s；精子激活率在 Na^+ 浓度小于等于 2 ‰时，为 100 %，随着 Na^+ 浓度的增加中华鲟精子激活率明显下降。与 Na^+ 溶液不同，在 K^+ 浓度为 0.005 ‰时中华鲟精子活力最佳，其 AT 和 LT 最长，分别为 80 s 和 174 s；而精子激活率随 K^+ 浓度增加而增强，在浓度为 0.005 ‰时激活率最强为 60 %，然后快速下降。在 Ca^{2+}，Mg^{2+}，Cu^{2+} 三种溶液中，随着三种溶液浓度的增加，中华鲟精子剧烈运动时间和寿命逐渐缩短，精子激活率逐步下降。与 Ca^{2+}，Mg^{2+}，Cu^{2+} 三种溶液完全不同的是，随着葡萄糖和果糖两种单糖浓度的增加，中华鲟精子 AT 及精子 LT 逐渐延长。结果说明适量浓度的 Na^+，K^+ 可以延长中华鲟精子 AT 及其 LT，激活精子活力；Ca^{2+}，Mg^{2+}，Cu^{2+} 三种溶液即使在较低浓度下都表现出对中华鲟精子 AT、精子 LT 及其激活率明显的抑制作用；而葡萄糖和果糖则能有效地延长中华鲟精子 AT 和 LT。

关键词　中华鲟　剧烈运动时间　精子寿命　激活率　精子活力　单糖

中华鲟(*Acipenser sinensis* Gray)为中国特有的一级保护动物，是一种典型的江海洄游性鱼类，主要分布于长江、东海、黄海等水域[1]。在自然条件下，长江上游江段产卵孵化的中华鲟幼鲟，随水流而进入海洋肥育，待性成熟后返回产卵场繁殖[1,2]。葛洲坝水利工程的修建阻隔了中华鲟产卵洄游的通道，对其种群资源造成影响，国家决定采取人工放流的方法对中华鲟种群进行增殖和保护。1983 年首次获得葛洲坝下中华鲟人工繁殖成功[3]。

在中华鲟人工繁殖过程中雄鱼不足的问题时常发生，因此对其精子生物学的研究非常必要，并且可以为进一步进行精子超低温保存提供参考依据。人工授精是鱼类繁殖过程的一项重要技术，获取具有受精能力的精子是人工授精的重要步骤。精子活力不仅是评价精液质量的重要指标，也是精子保存、人工授精等其他相关研究的基础。许多研究表明，精子活力的抑制与激活和盐度、温度、pH、渗透压、离子、激素及季节变化等因素有关，这是鱼类精子生理特性与环境相适应的结果[4]。研究不同化学物质对中华鲟精子活力的影响，对提高精子受精率，增强人工增殖能力，恢复物种种群具有重大意义。本实验是针对中华鲟精子在不同化学物质的不同浓度下对其精子活力进行试验和分析，旨在为今后科学研究者提供借鉴和参考。

---

* 基金项目：国家公益研究专项(2000DIB50177)；国家基础性工作重点专项项目(2002DEA1004)；国务院三峡办重点项目(SX97-25/ HB)；国家自然科学基金重大项目(30490231)。

** 第一作者，男，1965 年出生，江西吉安人，硕士，副研究员，从事珍稀水生动物保护生物学和繁殖生理研究。

*** 通讯作者：(Corresponding author)，E-mail：weiqw @ yfi. ac. cn

1 材料与方法

1.1 试验材料

1.1.1 卵子、精子的获得

中华鲟亲本是2004~2005年10~11月份从宜昌葛洲坝下通过渔民捕捞获取的。雌性亲本由中华鲟专用运输车运到荆州中华鲟增殖放流基地进行人工催产获得的卵子;雄性亲本则在宜昌鲟鱼研究所进行人工催产获得精液,低温保存运输至荆州中华鲟增殖放流基地,放入2℃冰箱充氧保存,每4 h换气一次。

1.1.2 试验溶液的配制

精子激活剂的配置:药品均购于市场上的分析纯,用去离子水稀释。分3批进行试验,每个药品浓度设3个平行组。在0.5‰~5‰ NaCl配制9个梯度;在0.0025‰~0.0156‰ KCl配置5个梯度;在1‰~4‰ $CaCl_2$ 配置4个梯度;在1‰~7‰ $MgCl_2$ 配制7个梯度;$CuCl_2$ 在0.005‰~0.05‰配置4个梯度;在0.05%~0.80%葡糖糖配制7个梯度;果糖设置了0.1%,0.3%,0.6%三个浓度。

1.2 试验方法

用尖头针蘸精液涂在载玻片上,用胶头吸管吸0.5 ml不同药品溶液将精液涂均匀,立即在100倍的显微镜下观察精子的运动情况,每个实验重复做3次。参照文献[5,6]的方法及文献[7]关于精子活力的分级标准,稍加修改。用剧烈运动时间(AT)、总运动时间(精子寿命,LT)和激活率等指标来评价精子活力。剧烈运动时间是指从精液与激活液混合开始,到约70%运动精子转入缓慢运动为止的时间;总运动时间是指精液从精液与激活剂混合开始,到视野中约90%的精子停止活动为止的时间;激活率以显微镜下同一视野中的活动精子百分比表示;分别观察并记录每种运动形式的时间。

1.3 统计分析

实验数据用SPSS 11.0统计软件进行统计分析。

2 结果和分析

2.1 NaCl溶液对中华鲟精子活力的影响

中华鲟精子随着NaCl溶液浓度不断升高,精子寿命开始不断上升。当NaCl浓度为2‰时,精子寿命最长,为347 s。而后随着NaCl浓度的增加,精子寿命急速下降。精子激活率在NaCl浓度为2‰之前一直保持为100%。后随着NaCl浓度的继续升高,激活率快速下降。在浓度为0.5%以后精子100%死亡,其精子激活率为0。精子AT和LT变化规律相似,但精子剧烈运动变化较缓,在NaCl浓度为小于1‰时精子AT呈上升趋势;NaCl浓度为1‰时时间最长,为98.67 s;NaCl浓度大于1‰以后精子AT逐渐下降,3‰以后精子几乎没有明显地剧烈运动现象。说明在适量的NaCl存在的情况下,能够促进精子活力,当其含量超过一定程度会对精子的活动有抑制作用(图1)。对数据进行统计分析表明,NaCl浓度为2‰时中华鲟精子LT与其他浓度之间差异极显著($p<0.01$)。AT的变化规律与LT相似,但AT变化规律不明显,各浓度下的AT之间差异不显著($p>0.05$)。NaCl浓度为2‰时与小于2‰的NaCl浓度之间差异不显著($p>0.05$)与大于 2 ‰的 NaCl 浓度之间差异显著($p<0.05$)。

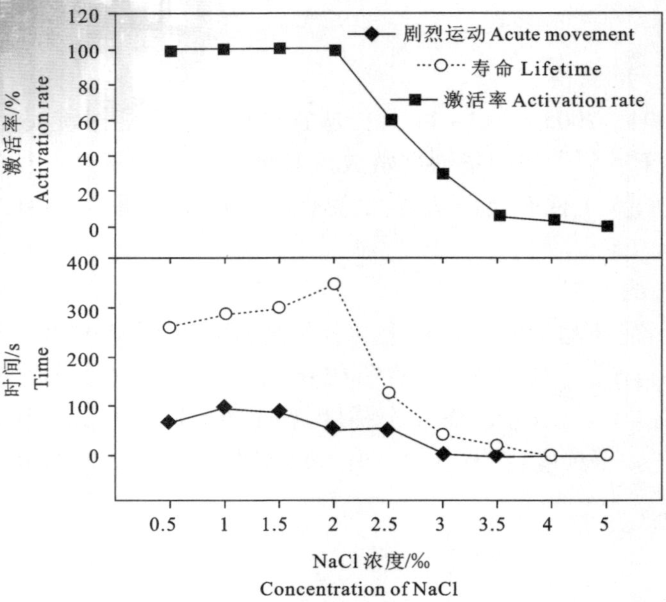

图 1　不同浓度的 NaCl 溶液对中华鲟精子寿命、剧烈运动及激活率的影响

Fig. 1　The effect of NaCl solution under different concentration on sperm longevity, acute movement and activation rate of *Acipenser sinensis*

2.2　KCl 溶液对中华鲟精子活力的影响

与 NaCl 溶液不同,KCl 溶液对中华鲟精子活力的影响变化规律较为复杂。从图 2 中明显看出,精子剧烈运动变化规律与精子寿命的变化相近,随着 KCl 溶液浓度的增加,精子 LT 和 AT 变化首先呈逐渐上升的状态,在其浓度为 0.005 ‰时,精子 LT 最长,为 174 s;AT 也最长,为 80 s;当 KCl 溶液浓度超过 0.005 ‰后中华鲟精子 AT 和 LT 急速下降,在 KCl 溶液浓度为 0.01 ‰时,精子 LT 和 AT 最短,分别为 108 s 和 39 s;随着 KCl 溶液浓度的逐步增加,精子剧烈运动趋势与精子寿命又呈缓慢上升的趋势。而 KCl 溶液对中华鲟精子激活率的影响较为简单,开始随着 KCl 溶液浓度的增加,精子激活率明显上升,当浓度为 0.005 ‰时,对精子的激活作用最强,激活率为 60 %;随着 KCl 溶液浓度的进一步增加,激活率逐渐下降。由此可见在溶液中加入适量的 KCl 有利于延长鱼类精子 AT 和 LT,提高精子激活率。KCl 溶液浓度为 0.005 ‰时的中华鲟精子 AT 与其他浓度之间差异极显著($p<0.01$)。KCl 溶液浓度为 0.005 ‰时的 LT 与 KCl 浓度为 0.015 6 ‰之间差异不显著($p>0.05$),与其他浓度之间差异极显著($p<0.01$)。

2.3　$CaCl_2$ 溶液对中华鲟精子活力的影响

在精液滴入 $CaCl_2$ 溶液,中华鲟精子 AT 和 LT 的变化规律相似。随着 $CaCl_2$ 浓度的逐渐增加,中华鲟精子 AT 和 LT 都呈明显下降趋势;当其浓度大于 4 ‰以后,精子大多数只是在微微颤动，极少数缓慢游动。而精子激活率与 AT 和 LT 变化规律稍有不同，随着 $CaCl_2$ 浓度的增加，精子激活率几乎呈逐步下降趋势，但在 $CaCl_2$ 浓度为 2 ‰ 时出现突变，精子激活率突然上升，90 %的精子被激活。从中我们可以得出，较低浓度的 $CaCl_2$ 溶液对中华鲟精子剧烈运动及寿命、激活率都会产生较强的抑制作用，影响精子活力（图 3）。

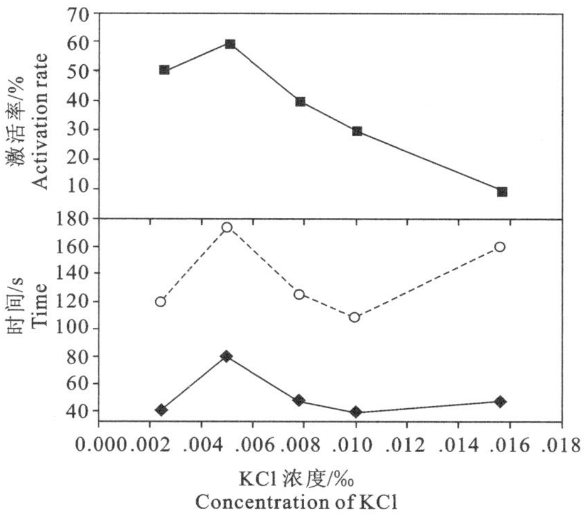

图 2　不同浓度的 KCl 溶液对中华鲟精子精子寿命、剧烈运动及激活率的影响（图例同图 1）

Fig. 2　The effect of KCl solution under different concentration on sperm longevity, acute movement and activation rate of *Acipenser sinensis*

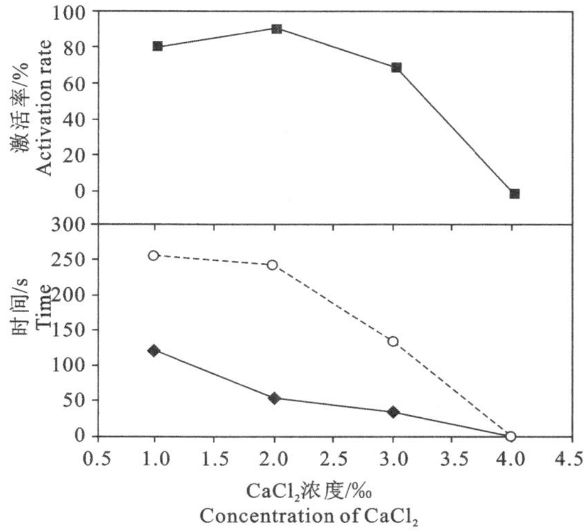

图 3　不同浓度的 $CaCl_2$ 溶液对中华鲟精子精子寿命、剧烈运动及激活率的影响（图例同图 1）

Fig. 3　The effect of $CaCl_2$ solution under different concentration on sperm longevity, acute movement and activation rate of *Acipenser sinensis*

2.4　$MgCl_2$ 溶液对中华鲟精子活力的影响

$MgCl_2$ 溶液浓度从 1 ‰变化到 7 ‰时，中华鲟精子 AT 变化规律与精子激活率的变化规律相似。随着 $MgCl_2$ 浓度逐渐增加，其精子剧烈运动逐步变缓慢，精子激活率也逐渐下降；当 $MgCl_2$ 溶液浓度增加到 7 ‰以后，其精子活动较缓慢，无涡动现象。随着 $MgCl_2$ 浓度的逐步增加，中华鲟精子 LT 表现出下降趋势；但在变化过程中，在 $MgCl_2$ 溶液浓度为 3 ‰和 6 ‰时，出现了两个突变点，在这两个浓度下精子寿命比其他浓度下的精子寿命更长。结果显示：随着 $MgCl_2$ 浓度的增加，中华鲟精子活力及其 $MgCl_2$ 溶液对精子激活作用较差，明显产生了对精子的抑制作用（图 4）。

2.5　$CuCl_2$ 溶液对中华鲟精子活力的影响

精液中滴入 $CuCl_2$ 溶液后，随着 $CuCl_2$ 溶液浓度的增加，中华鲟精子 AT 和 LT 缩短。当 $CuCl_2$溶液浓度小于 0.05 ‰，精子寿命为 0。从中可以得出：重金属离子对中华鲟精子活力及激活率都具有很强的抑制作用，即使在很低的浓度下，精子也能够很快死亡（图 5）。

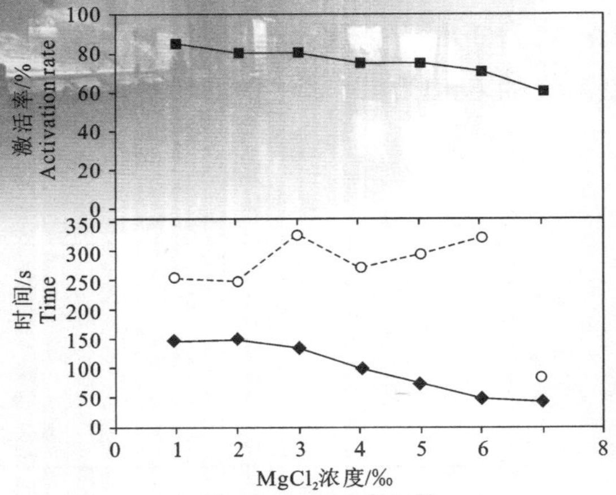

图4 不同浓度的 $MgCl_2$ 溶液对中华鲟精子精子寿命、剧烈运动及激活率的影响(图例同图1)

Fig. 4 The effect of $MgCl_2$ solution under different concentration on sperm longevity, acute movement and activation rate of *Acipenser sinensis*

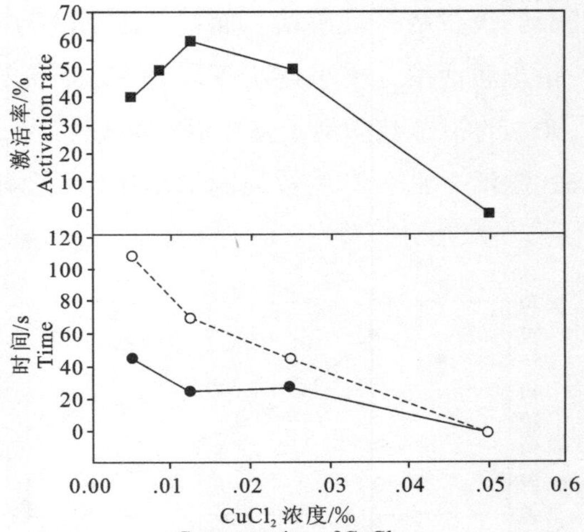

图5 不同浓度的 $CuCl_2$ 溶液对中华鲟精子寿命、剧烈运动及激活率的影响(图例同图1)

Fig. 5 The effect of $CuCl_2$ solution under different concentration on sperm longevity, acute movement and activation rate of *Acipenser sinensis*

2.6 葡萄糖、果糖对中华鲟精子活力的影响

葡萄糖与果糖对中华鲟精子剧烈运动时间和寿命的影响相似,随着葡萄糖、果糖溶液浓度的逐渐增加,中华鲟精子 AT,LT 延长;而精子 AT 随着葡萄糖、果糖溶液浓度的增加,延长趋势较缓。这种现象可能是由于葡萄糖和果糖在精子运动过程能够为精子提供能量,以弥补在活动过程中消耗的能量,使其死亡时间延长(图6,图7)。

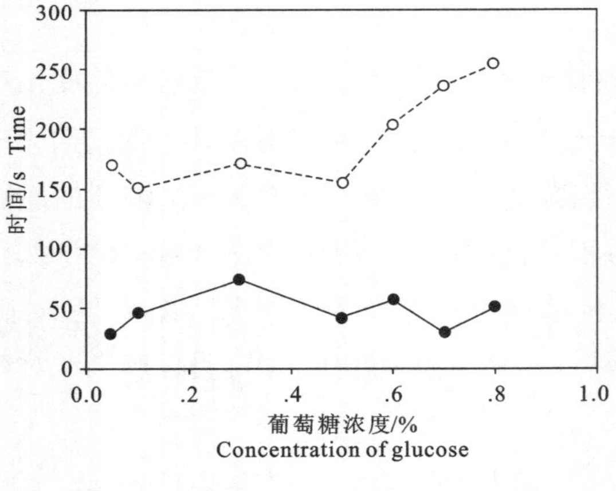

图6 不同浓度的葡萄糖溶液对中华鲟精子精子寿命及剧烈运动的影响(图例同图1)

Fig. 6 The effect of glucose solution under different concentration on sperm longevity, acute movement and activation rate of *Acipenser sinensis*

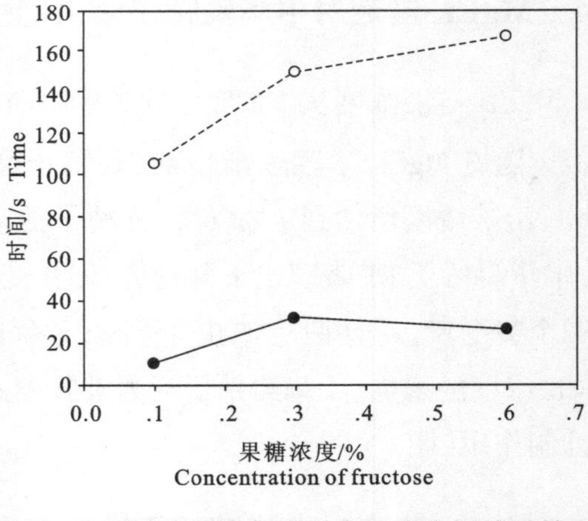

图7 不同浓度的果糖溶液对中华鲟精子精子寿命及剧烈运动的影响(图例同图1)

Fig. 7 The effect of fructose solution under different concentration on sperm longevity, acute movement and activation rate of *Acipenser sinensis*

3 讨论

3.1 NaCl,KCl 溶液对精子活力的影响

鱼类精子在精巢中是不活动的,但是却存在着极强的潜在活动能力[8]。由于精巢中存在着诸多对鱼类精子产生抑制作用的因素,减缓精子运动强度,使得鱼类精子的成熟和储存得到很好的保障。这也正是很多雄性鱼在整个繁殖期内可以一直产出活力良好精子的主要原因。多种鱼类精浆中含有浓度较高的 Na^+,K^+,它们是造成鱼类精巢内环境高渗透压的主要离子[9]。潘德博等研究广东鲂,发现在 Na^+浓度为3 ‰时精子快速运动时间最长[10];江世贵等研究发现,在盐度为5 ‰~10 ‰时,精子活力最强[11];严安生等研究发现盐度为5 ‰的 NaCl 浓液中快速运动及寿命最长[8];谷巍等研究发现在 NaCl 浓度为6 ‰时快速运动时间和寿命最长[12];在本试验中,NaCl 浓度为2 ‰时精子 LT 最长,精子 AT 也最长,并且精子激活率也达到了100 %,在 NaCl 溶液浓度为1 ‰时精子快速运动时间最长,其变化规律与上述研究相仿。

在试验中,我们发现,KCl 溶液浓度大于0.04 ‰时中华鲟精子立即死亡;而在其浓度为0.005 ‰时精子活力最强,寿命最长。这说明低浓度的 KCl 做精子稀释液能够激活精子活力,此结果与谷巍[12]等做的高白鲑实验结果相似,即高浓度的 K^+对中华鲟精子有明显的抑制作用,K^+对精子寿命延长的效应只在 K^+含量低于精浆的浓液中才能表现出来。

根据本试验结果以及参考的其他相关资料,作者认为 NaCl,KCl 对鱼类精子活力及激活率的影响都表现在渗透压方面,精浆通过 Na^+,K^+等离子形成高于外界水体的高渗环境。当精液排入水体中后,Na^+,K^+等被稀释使得精子激活。由于外界水体渗透压极低于精子本身内部原生质,使得这整个过程大量消耗精子自身能量。因此在人工繁殖过程中应配置与亲本渗透压相似的 NaCl,KCl 稀释液,尽量减少精子能量的损耗,延长精子活力时间,这是人工繁殖的关键技术,是提高亲本受精率,增加渔业产量的关键所在。这与谷巍等[12]研究的高白鲑结果相似。

3.2 $CaCl_2$,$MgCl_2$ 及 $CuCl_2$ 对中华鲟精子活力的影响

Ca^{2+},Mg^{2+}离子也是精浆中的重要组成成分,但其对精子的作用与 Na^+,K^+离子不同,随着 Ca^{2+},Mg^{2+}浓度的升高,精子活力明显下降。在 $CaCl_2$,$MgCl_2$,$CuCl_2$ 溶液浓度在较低情况下,对精子都产生较强的抑制作用,精子激活率也较低。$CaCl_2$ 浓度为4 ‰以后就只有极少数精子在游动,多数只是微微颤动;而 $MgCl_2$ 溶液在浓度为6 ‰以后,精子就不出现涡动,精子寿命时间很短。表明高浓度的 $CaCl_2$,$MgCl_2$ 溶液对中华鲟精子具有一定的损伤,可能是由于高于精浆中的 Ca^{2+},Mg^{2+}离子浓度引起精子内部发生变化而导致活力下降。这一结果与金属离子对中华倒刺精子的影响研究[13]结果一致。Ca^{2+},Mg^{2+}是衡量水质硬度评定的指标,因此在实际鱼类人工授精过程中,检验水质中 Ca^{2+},Mg^{2+}含量也必不可少的。

$CuCl_2$ 溶液中含有的 Cu^{2+}为常见的重金属离子,较常用于鱼药中,用于杀菌,消毒等,它能够引起蛋白质变性,其对精子破坏性更强,即使 $CuCl_2$ 溶液浓度为0.005 ‰时,精子剧烈运动时间也非常短,为46 s,精子寿命也仅仅维持109 s。在溶液浓度为0.012 5 ‰时就出现了凝聚现象,并且凝结成团,谷巍[12]做高白鲑精子活力测定试验中也曾出现了精子“凝聚”现象的报道,

具体机理还不清楚,这是否是由于重金属离子导致中华鲟精子蛋白质变性缩水而导致精子凝集还不得而知;这两种凝聚现象的诱发机理是否存在相同之处,还有待研究。

3.3 单糖对中华鲟精子活力及寿命的影响

果糖、葡萄糖是精浆中主要的糖类的重要组成部分,Phronen 和 Hyvarinen 就曾在白鲑、硬头鳟等鱼类的精浆测定中发现,其葡萄糖浓度是果糖的 5 倍[14];Gardinerl 发现虹鳟的精子可以利用外源性葡萄糖维持较长时间的活动[15];苏德学等人也发现适宜的葡萄糖可以明显地延长白斑狗鱼[16]和丁鲹[17]鱼精子活动时间,延长其寿命;李胜忠等人也发现加入适量的葡萄糖能够明显延长凹目白鲑精子运动时间[18]。从本试验看,使用单糖物质以后,其精子剧烈运动时间(AT)和寿命(LT)明显延长,说明单糖在维持精子活力中的作用非常显著,这与上述研究观点一致。表明了体外受精鱼类的精子具有利用外源性葡萄糖的能力,来补偿自身运动所需的能量。

参考文献

[1] 四川省长江水产资源调查组. 长江鲟鱼类生物学及人工繁殖研究. 成都:四川科学技术出版社. 1988:1-112.

[2] 危起伟,柯福恩. 论长江口中华鲟幼鱼的保护//绿满东亚:第一届东亚地区国家公园与自然保护区会议暨 CNPPA/ IUCN 第 41 届工作会议文集. 北京:中国环境科学出版社. 1994:786-793.

[3] 傅朝君,刘宪亭,鲁大椿,等. 葛洲坝下中华鲟人工繁殖. 淡水渔业,1985(1):1-5.

[4] 舒琥,黄燕,张海发,等. 盐度及温度对红鳍笛鲷精子活力的影响. 广州大学学报,2005,4(1):29-32.

[5] 喻达辉,陈竞春,苏天凤,等. 合浦珠母贝精子的试验生物学初步研究. 热带海洋,1998,17(3):83-87.

[6] 江世贵,李加儿,区又君,等. 四种鲷科鱼类的精子激活条件与其生态习性的关系. 生态学报,2000,20(3):468-473.

[7] 鲁大椿,傅朝君,刘宪亭,等. 我国主要淡水养殖鱼类精液生物学特性. 淡水渔业,1989,19(2):34-37.

[8] 严安生,王其和,李诗模. 渗透压和钾对鲤、团头鲂精子活力的影响. 淡水渔业,1993,23(3):19-21.

[9] 鲁大椿,刘宪亭,方建萍,等. 我国主要淡水鱼类精浆的元素组成. 淡水渔业,1992,22(2):10-12.

[10] 潘德博,许淑英,叶星,等. 广东鲂精子主要生物学特性的研究. 中国水产科学,1999,6(4):111-112.

[11] 江世贵,苏天凤,喻达辉,等. 中华乌塘鳢精子的生物学特性及超低温保存. 水产学报,2000,24(2):120-122.

[12] 谷巍,李胜忠,蔡林钢,等. Na^+,K^+ 对高白鲑精子活力影响的初步研究. 水利渔业,2005,25(2):8-10.

[13] 张春霖,李英文,邹奇志,等. 金属离子及低温对中华倒刺精子活力的影响. 西南农业大学学报,1999,21(3):274-278.

[14] Phronen J,Hyvarinen H. Composition of the milt of some Teleost fish. J Fish Biol,1983,22:351-361

[15] Gardiner D M. Ut lisation of extracellular glucose by spermatozoa of two viviparous fishes. Comp Biochem Physiol,1978,59A:165-168.

[16] 苏德学,严安生,田永胜,等. 钠、钾、钙和葡萄糖对白斑狗鱼精子活力的影响. 动物学杂志,2004,39(1):16-20.

[17] 苏德学,严安生,田永胜,等. 阳离子、葡萄糖及渗透压对丁精子活力的影响. 水利渔业,2004,24(1):7-8.

[18] 李胜忠,郭焱,蔡林钢,等. Na^+、葡萄糖、果糖对凹目白鲑精子活力的影响. 动物学杂志,2005,40(4):82-85.

Effects of Different Concentration of Irons and Monosaccharides on Sperm Motility of Chinese Sturgeon (*Acipenser sinensis*)

Liu Jianyi[1,2,3] Gan Fang[1,2,3] Wei Qiwei[1,2,3]
Du Hao[1,2,3] Zhu Yongjiu[1,2,3]

(1. Yangtze Rwer Fisherise Research Institude, Chinese Academy of Fishery Science, Jingzhou, 43400;
2. Key Lab of Freshwater Fish Germplasm Resources and Biotechnology by Ministry of Agriculture, Jingzhou, 434000;
3. Freshwater Fisheries Research Center, Chinese Academy of Fisheries Science, Wuxi 214081, China)

Abstract The movement of Chinese sturgeon spermatozoon in various concentration of Na^+, K^+, Ca^{2+}, Mg^{2+}, Cu^{2+} and glucose, fructose solution was studied in this paper. The results showed that the longest lifetime was 347 s when the concentration of Na^+ was 2 ‰, while the longest acute movement time was 98.67 s when the concentration of Na^+ was 1 ‰. The activation rate of sperm was 100 % when the concentration was no more than 2 ‰. Compared with Na^+ solution, when the concentration of K^+ was 0.005 ‰, the motility of sperm was the highest, while the acute movement time and lifetime were the longest, which were 80 s and 174 s. The activation rate increased with the increase of the concentration of K^+. When the concentration was 0.005 ‰, the motility was the highest, which was 60 %. And then it decreased sharply. In the three solution of Ca^{2+}, Mg^{2+} and Cu^{2+}, Along with the increase of the concentration of Ca^{2+}, Mg^{2+} and Cu^{2+}, both the acute movement time and lifetime decreased, and the activation rate of sperm also decreased gradually. The situation in the solution of glucose and fructose was absolutely different from the above 3 solution. Along with the increase of the concentration of glucose and fructose, the acute movement time and lifetime increased gradually. The results indicated that the appropriate concentration of Na^+ and K^+ could increase the acute movement time and lifetime, and activated the sperm, while Ca^{2+}, Mg^{2+} and Cu^{2+} under low concentration had obvious inhibiting effect on the acute movement time, lifetime and activation rate; while glucose and fructose can effectively increased the acute movement time and lifetime of sperm.

Key words Chinese sturgeon Acute movement time Spermic life time Activation rate Sperm motility Monosaccharides

计算机辅助对几种鲟鱼冻精激活液的比较*

柳　凌[1, **]　Otomar Linhart[2]　危起伟[1]　郭　峰[1]　张洁明[1]　刘鉴毅[1]

(1. 中国水产科学研究院长江水产研究所 农业部淡水鱼类种质资源与生物技术重点实验室　湖北荆州　434000

2. Joint Laboratory of Genetics, Physiology and Reproduction of Fish, Institute of Animal Physiology and Genetics, Academy of Sciences of the Czech Republic, University of South Bohemia, Research Institute of Fish Culture and Hydrobiology, 38925 Vodnany, Czech Republic)

摘　要　首次在我国采用鱼类彩色精子图文分析系统(FSQAS-2000)对中华鲟、史氏鲟、西伯利亚鲟和匙吻鲟经超低温冷冻保存后的精子,在不同激活液条件下,平均曲线运动速度(VCL)、平均直线运动速度(VSL)、平均路径运动速度(VAP)、平均移动角度(MAD),以及冷冻前后精子活率的变化等多项参数进行了统计学比较研究。研究分析了不同渗透压、pH 以及含 Mg^{2+} 的激活液对这些参数的影响。研究结果表明:几种鲟鱼激活液的渗透压均较低,其中西伯利亚鲟最低,为 10 mOsm/kg;中华鲟和史氏鲟为 20 mOsm/kg;匙吻鲟最高,为 30 mOsm/kg。不同的渗透压对冷冻精子的 VCL, VSL, VAP,以及精子活率均产生显著($P \leqslant 0.05$)或极显著($P \leqslant 0.01$)影响。激活液的 pH 除了对冷冻精子的上述参数产生影响外,最显著的是对精子的运动轨迹,即平均移动角度(MAD(°)/s)产生影响($P \leqslant 0.01$)。pH 越低,精子的 MAD 越小。几种鲟鱼的最适 pH 分别为西伯利亚鲟 pH 7.5,中华鲟、史氏鲟和匙吻鲟 pH 8.5。当 4 种鲟鱼的冷冻精子激活液中加入 5 mmol/L $MgCl_2$ 时,冷冻精子的 3 种运动速度显著提高($P \leqslant 0.01$)。综合分析后认为,西伯利亚鲟的冷冻精子激活液应为 10 mmol/L Tris-HCl + 5 mmol/L $MgCl_2$, pH 7.5;中华鲟和史氏鲟的冷冻精子激活液为 20 mmol/L Tirs-HCI + 5 mmol/L $MgCl_2$, pH 8.5;匙吻鲟的冷冻精子激活液为 30 mmol/L Tris HCI + 5 mmol/L $MgCl_2$, pH 8.5。用此激活液对冷冻前后 4 种鲟鱼精子活力的相关性进行线性回归分析发现,西伯利亚鲟的 $R^2 = 0.8296$ ($P < 0.01$);中华鲟的 $R^2 = 0.9860$ ($P < 0.01$);史氏鲟的 $R^2 = 0.962$ ($P < 0.01$);匙吻鲟 $R^2 = 0.947$ ($P < 0.01$)。说明冷冻前后 4 种鲟鱼精子的平均活率存在着极显著的线性正相关性。

关键词　鲟鱼　计算机辅助分析系统　超低温保存　精子活力

在我国,鱼类精子活力检测一直未引起水产界的高度重视,其方法主要采用传统的估计法。即在显微镜同一视野下,依据观察者的经验,对激活后精子的运动状态、精子的活率以及快速运动时间等参数进行评价。这种方法人为的估计成分较大,精确性较差,缺乏可重复性,所获得的结果也无法进行统计学分析[1-2]。目前,国际上已逐渐开展对鱼类的计算机辅助精子活力检测(computer-aided spem ranalysis, CASA)相关研究[3-5]。我国虽然也能见到少数几篇有关采用计算机辅助分析的方法对鱼类或其他水生生物的精子活力分析的研究报道[6-9],但分析的基础主要是采用医学软件和哺乳类精子的分析方法。由于鱼类精子和哺乳类精子在生物学特性上存在很大差异[10],其研究的结果有很大的局限性。

作者依据鱼类精子的生物学特性,参照医学对人类精子的检测方法,在世界卫生组织对精子检测标准的基础上[11],与武汉千屏影像技术有限公司合作,开发了一款具有自主知识产权的鱼类精子彩色图文分析系统(FSQAS-2000),并开展了利用该软件对多种鱼类精子进行活力

* 基金项目:国家基础公益项目(2004DIB3J099)。

** 第一作者,男,1959 年出生,湖南长沙人,研究员,主要从事鱼类生殖生理及发育生物学研究,E-mail:Liul@yfi.ac.cn

检测的相关研究。本论文研究了4种鲟鱼精子经超低温冷冻保存后，冻精激活液的成分对精子活力的影响。其目的一方面是为鱼类精子的超低温冷冻保存效果进行数字化评价，为以后的研究提供精确的、可重复的参考依据；另一方面，为我国鱼类精子活力数字化评价方法探索一条行之有效的途径。

1 材料和方法

1.1 精液采集

试验中所用精液取自西伯利亚鲟（*Acipenser baeri*）、中华鲟（*A. sinesis*）、匙吻鲟（*Polydon gladius*）和史氏鲟（*A. schrenekii*）4种鲟鱼（表1）。

表1 4种鲟鱼精液采集相关信息

Tab.1 The information for collecting sperm of four sturgeons

种类 species	采集地 collecting site	采集时间（年-月）collecting time	亲本类型 source of fish	亲鱼数/尾 number of male fish	保存精液量/ml quantity of sperm
中华鲟 *A. sinensis*	湖北宜昌 Yichang	2005-10	野生 wild	3	100
西伯利亚鲟 *A. baerii*	湖北十堰 Shiyan	2006-04	养殖 farmed	4	30
史氏鲟 *A. schrenckii*	湖北十堰 Shiyan	2006-04	养殖 farmed	3	30
匙吻鲟 *P. sputhulu*	湖北荆州 Jinzhou	2006-05	养殖 farmed	5	50

虽然精液来源不尽相同，但均采用类固醇激素人工催情。待亲鱼发情后，用干毛巾擦净亲鱼生殖孔周围，挤压腹部，收集无粪便和杂物的干净精液于塑料精液存放袋中，充氧后密封，4 ℃存放。精液运回实验室后，除用少量精液进行镜检外，其余立即进行超低温冷冻保存。

1.2 冷冻保存

冷冻方法采用修正后的Linhart等[12]和Liu等[13]方法：鲟精与稀释液按1∶3混匀。西伯利亚鲟的稀释液为50 mmol/L蔗糖+20 mmol/L Tris-HCI+0.5 mmol/L KCI；中华鲟和史氏鲟的稀释液为75 mmol/L蔗糖+20 mmol/L Tris-HCI+0.5 mmol/L KCI；匙吻鲟的稀释液为100 mmol/L蔗糖+20 mmol/L Tris-HCl+0.5 mmol/L KCl。所有稀释液的pH均为8。混合液置于4 ℃冰箱平衡30 min后，加入终浓度为8%~10%的甲醇。混匀，立即用0.5 ml麦管（IMV，法国）分装，放入预先设置起始温度为0 ℃的程序降温仪（kryo. 550-16，英国）中。降温程序为0开始，3 ℃/min降至−5 ℃；5 ℃/min降至−15 ℃；10 ℃/min降至−25 ℃；20 ℃/min降至−80 ℃；最后平衡5 min。降温完成后，直接将精液从程序降温仪中取出，立即放入液氮（−196 ℃）中保存。保存1个月以上后，取出解冻、激活、镜检。

1.3 解冻镜检

精液在液氮中保存1个月以上后，取出，在38~40 ℃水浴中8 s即可解冻，解冻后的精液在4 ℃冰箱存放，待镜检。

在定量板上涂抹一层1 mg/ml PBS，加入50 μl待测激活液，然后加入0.25 μl冻精，检测最终稀释比为1∶600，立即混匀，在10×10倍显微镜（Leica DM 2500 德国）下镜检，同时录像（JVC TK-U890EG，日本）。待测激活液按如下内容和顺序进行。

渗透压对冻精活力的影响 分别用5 mmol/L，10 mmol/L，20 mmol/L和30 mmol/L Tris-

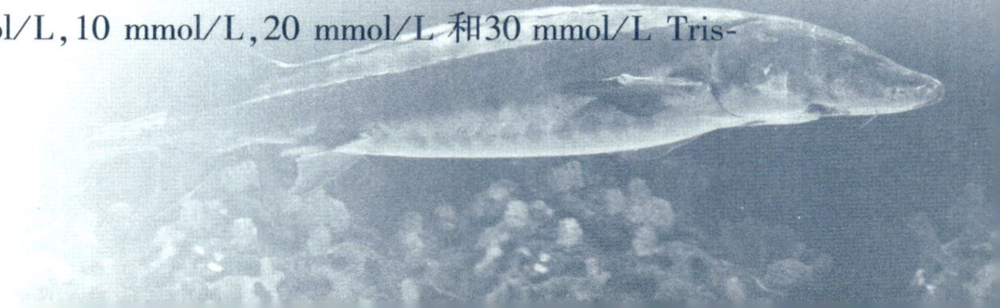

HCl 缓冲液激活冻精，pH 均为 8。对照组用蒸馏水激活。

PH 对冻精活力的影响　用 Tris-HCI 调 pH(Delta320，中国)分别为 6.5，7.5，8.5 和 9.5 激活冻精。其中，西伯利亚鲟为 10 mmol/L Tisr-HCl；中华鲟、史氏鲟为 20 mmol/L Tris-HCI；匙吻鲟为 30 mmol/L Tris-HCl。对照组用蒸馏水激活。

Mg^{2+} 对冻精活力的影响　分别用 Tris-HCI + 2.5 mmol/L，5 mmol/L，7.5 mmol/L 和 10 mmol/L $MgCl_2$ 激活冻精。其中，西伯利亚鲟用 10 mmol/L Tisr-HCl，pH 7.5；中华鲟、史氏鲟用 20 mmol/L Tris-HCl，pH 8.5；匙吻鲟用 30 mmol/L Tirs-HCI，pH 8.5 激活对照组。

在上述试验中，每解冻一支麦管可完成 1 种激活液的试验，约 20 min 完成，平行重复 3 次。所有试验的对照组均用蒸馏水作激活液。

1.4　计算机辅助检测冻精活力

精子激活后的录像采用鱼类彩色精子图文分析系统(FSQAS-2000)(中国水产科学研究院长江水产研究所和武汉千屏影像技术有限公司联合研制)进行分析。精子激活后每隔 30 s 进行一次序列图像采集，直到激活后 3 min。每个序列图像为 3 s，72 帧。系统检测参数设置为：最小步距 25 像素；当前阈值为 150；A 级精子最小直线速度 20 μm/s；B 级精子最小直线速度 5 μm/s；C 级精子最小直线速度 1 μm/s。然后对每个序列图像分别进行检测和报告输出。

1.5　数据分析

实验所得数据采用 SPSS 进行分析。两组平均数间的比较用 Paired-Sample *T* Test，多重比较采用 oen-way ANOVA。$P<0.05$ 为显著差异，$P<0.01$ 为极显著差异。

2　结果

2.1　渗透压对冻精活力的影响

鱼类精子在低温保存时，由于稀释液、抗冻剂，以及降温过程的影响，其精子的渗透压将发生一定的改变。表 2 列出了不同渗透压的激活液对 4 种鲟鱼冻精活力参数的影响。

表 2　不同渗透压的激活液对几种鲟鱼冷冻精子活力参数的影响

Tab. 2　The post-thaw sperm motility of several sturgeons stimulated by activating medium with different osmolalities

种类 Species	激活液 Activating medium	60s 时的冻精活力/% Rate of moving sperm at the 60 second after stimulated				平均曲线运动速度 /(μm/s) VCL	平均直线运动速度 /(μm/s) VSL	平均路径运动速度 /(μm/s) VAP	平均移动的角度 /((°)/s) MAD
		A 级	B 级	C 级	D 级				
西伯利亚鲟 *A. bareii*	5 mmol/L Tris	53.26	9.67	19.87	16.17	89.94 ± 13.74[b]	43.89 ± 13.62[b]	49.55 ± 14.35[b]	42.40 ± 4.40[b]
	10 mmol/L Tris	65.87	19.94	11.02	3.18	112.96 ± 13.18[a]	65.61 ± 20.61[a]	62.60 ± 8.57[a]	58.68 ± 6.39[a]
	20 mmol/L Tris	55.93	13.50	14.79	15.78	92.64 ± 12.66[b]	48.74 ± 15.76[b]	53.78 ± 9.97[b]	43.31 ± 6.52[b]
	30 mmol/L Tris	46.18	25.63	18.34	9.85	79.96 ± 14.72[b]	42.83 ± 14.34[b]	45.15 ± 12.33[c]	46.09 ± 7.28[b]
	蒸馏水	41.32	11.36	26.92	20.40	62.28 ± 12.48[c]	37.34 ± 13.25[c]	42.42 ± 10.53[c]	44.74 ± 8.63[b]

续表

种类 Species	激活液 Activating medium	60 s 时的冻精活力/% Rate ofmoving sperm at the 60 second after stimulated				平均曲线运动速度/(μm/s) VCL	平均直线运动速度/(μm/s) VSL	平均路径运动速度/(μm/s) VAP	平均移动的角度/((°)/s) MAD
		A 级	B 级	C 级	D 级				
中华鲟 *A. sinensis*	5 mmol/L Tris	63.03	19.29	7.61	10.07	81.88 ±9.26'	23.27 ±4.90^{b}	32.06 ±6.78^{b}	39.86 ±4.29^{b}
	10 mmol/L Tris	63.42	18.28	5.97	12.33	84.14 ±11.52^{b}	25.28 ±4.13^{b}	34.23 ±6.08^{b}	40.09 ±5.26
	20 mmol/L Tris	64.18	20.92	9.14	5.76	104.62 ±14.37^{a}	29.37 ±4.23^{a}	40.97 ±6.85^{a}	42.96 ±3.61
	30 mmol/L Tris	57.26	15.57	17.64	9.53	72.18 ±12.75^{c}	22.23 ±4.87^{c}	30.02 ±4.32^{c}	40.66 ±5.20
	蒸馏水	55.25	13.69	14.38	16.68	74.22 ±10.29^{c}	22.85 ±3.16^{c}	31.71 ±6.45^{c}	40.20 ±5.32
史氏鲟 *A. schrenckii*	5 mmol/L Tris	45.92	17.79	21.78	14.51	103.70 ±13.45^{c}	49.19 ±6.46^{c}	59.99 ±5.13^{c}	51.85 ±4.62^{a}
	10 mmol/L Tris	48.95	21.40	21.73	7.90	118.74 ±13.49^{b}	57.09 ±6.23^{b}	71.87 ±5.55^{b}	51.38 ±3.64^{a}
	20 mmol/L Tris	57.68	20.88	14.85	6.59	139.32 ±12.10^{a}	68.99 ±7.65^{a}	82.91 ±5.90^{a}	59.19 ±3.38^{a}
	30 mmol/L Tris	46.54	18.69	20.06	14.71	115.10 ±13.13^{b}	56.06 ±6.29^{b}	68.64 ±5.82^{b}	55.79^{a} ±4.61^{b}
	蒸馏水	44.37	16.51	22.71	16.41	100.16 ±13.75^{c}	45.02 ±6.72^{c}	56.66 ±5.45^{c}	52.12 ±4.98^{a}
匙吻鲟 *P. spathula*	5 mmol/L Tris	50.31	15.72	19.69	14.28	61.86 ±11.45^{c}	22.61 ±5.81^{b}	27.05 ±6.16^{c}	53.72 ±7.14^{a}
	10 mmol/L Tris	52.10	34.13	3.59	10.18	64.02 ±10.17^{c}	21.75 ±5.25^{b}	28.34 ±7.13^{c}	54.68 ±8.13^{a}
	20 mmol/L Tris	61.85	16.10	8.47	13.56	94.00 ±16.86^{b}	42.51 ±10.50^{a}	52.34 ±13.92^{b}	44.86 ±9.09^{b}
	30 mmol/L Tris	66.17	17.29	3.76	12.78	130.44 ±12.48^{a}	44.07 ±7.33^{a}	69.47 ±8.23^{a}	58.99 ±4.32^{a}
	蒸馏水	50.96	13.79	16.51	8.74	62.30 ±10.77^{c}	21.77 ±5.00^{b}	27.99 ±7.29^{c}	56.38 ±5.20^{a}

注:表中同一栏内,右上角标有不同英文字母的数据,相邻字母间的比较(如 a 与 b;b 与 c 的比较)具有显著差异($P<0.05$),间隔 字母间的比较(如 a 与 c 的比较)具有极显著差异($P<0.01$)。

Notes:The continuously different letters in the sameline show significant difference($P<0.05$),and the uncontinuously different letters show highly significant difference($P<0.01$)。

西伯利亚鲟 所有实验组的精子平均曲线运动速度(VCL)、平均直线运动速度(VSL)和平均路径运动速度(VAP)均高于对照组。其中,10 mmol/L Tris 组的 VCL,VSL 和 VAP 最高,与其他实验组和对照组比较存在显著($P<0.05$)或极显著($P<0.01$)差异;而 10 mmol/L Tris 组的精于平均移动角度(MAD)最大($P<0.05$),说明该实验组的精子的运动轨迹更趋于直线。30 mmol/L Tris 组的 VCL,VSL 和 VAP 在所有实验组中最低,但高于对照组,说明适当的提高西伯利亚鲟冻精激活液的渗透压可以提高精子的运动速度,但过高反而会降低精子的运动速度。60 s 时的冻精活力测定同样证明了这一点。10 mmol/L Tris 组的 A 级精子所占比例最高,而 D 级最低。虽然 30 mmol/L Tris 组的 B 级精子在所有实验组和对照组中最高,但 A + B 的值则低于 10 mmol/L Tris 组,而高于对照组。

中华鲟 所有实验组精子的 VCL,VSL 和 VAP 均高于对照组。其中,20 mmol/L Tris 组的 VCL,VSL 和 VAP 最高,与其他实验组和对照组比较存在显著($P<0.05$)或极显著($P<0.01$)差异。MAD 与其他实验组和对照组比较略有升高,但无显著差异,说明在试验设计的渗透压变化范围内,渗透压的改变对中华鲟冻精的运动轨迹无影响。30 mmol/L Tris 组的各项参数与对照组比较无显著差异。60 s 时的冻精活力测定仍然是 20 mmol/L Tris 的 A 级最高,D 级最低。

史氏鲟 所有实验组精子的 VCL,VSL 和 VAP 均高于对照组。其中,20 mmol/L Tris 组的 VCL,VSL 和 VAP 最高,与其他实验组和对照组比较存在显著($P<0.05$)或极显著($P<0.01$)

差异;5 mmol/L Tris 组的 VCL,VSL 和 VAP 与对照组比较无显著差异;除 30 mmol/L Tris 组的 MAD 与其他实验组和对照组比较存在显著差异($P<0.05$)外,其他实验组和对照组间均无显著差异。60 s 时的冻精活力测定仍然是 20 mmol/L Tris 的 A 级最高,D 级最低。

匙吻鲟 30 mmol/L Tris 组的 VCL,VSL 和 VAP 在所有实验组和对照组中最高。而 5 mmol/L Tris 组和 10 mmol/L Tris 组的 VCL,VSL 和 VAP 与对照组比较无差异。30 mmol/L Tris 组的 MAD 也最高,虽然与 5 mmol/L Tris 组、10 mmol/L Tris 组和对照组无显著差异,但与 20 mmol/L 组存在显著差异($P<0.05$)。60 s 时的冻精活力测定仍然是 30 mmol/L Tris 的 A 级最高,D 级最低。

2.2 pH 对冻精活力的影响

不同种的鱼精子的 pH 不尽相同。精子在冷冻过程中,由于稀释液和抗冻剂的影响,其 pH 也可能发生一些改变。表 3 列出了不同 pH 的激活液对 4 种鲟鱼冻精活力参数的影响。

表 3 不同 pH 的激活液对几种鲟鱼冻精活力参数的影响

Tab. 3 The post-thaw sperm motility of several sturgeons stimulated by activating medium with different pH

种类 Species	激活液 Activating medium	60s 时的冻精活力/% Rate ofmoving sperm at the 60 second after stimulated				平均曲线运动速度 /(μm/s) VCL	平均直线运动速度 /(μm/s) VSL	平均路径运动速度 /(μm/s) VAP	平均移动的角度 /((°)/s) MAD
		A 级	B 级	C 级	D 级				
西伯利亚鲟 *A. bareii*	pH 6.5	60.38	17.42	11.34	10,86	99.96 ± 14.36	42.83 ± 14.34^{b}	45.15 ± 12.33^{b}	46.09 ± 7.28^{b}
	pH 7.5	67.17	20.63	8.32	3.88	112.96 ± 13.18^{a}	65.61 ± 20.61^{a}	62.60 ± 8.57^{c}	58.68 ± 6.39^{c}
	pH 8.5	62.93	17.54	9.79	9.74	92.64 ± 12.66^{b}	48.74 ± 15.76^{b}	53.78 ± 9.97^{b}	53.31 ± 6.52^{b}
	pH 9.5	58.22	19.67	9.87	12,24	89.94 ± 13.74^{b}	43.89 ± 13.62^{b}	49.55 ± 14.35^{b}	52.40 ± 4.40^{b}
	蒸馏水	41.32	11.36	26.92	20.40	62.28 ± 12.48^{c}	37.34 ± 13.25^{c}	42.42 ± 10.53^{c}	47.74 ± 8.63^{b}
中华鲟 *A. sinensis*	pH 6.5	63.03	19.29	7.61	10,07	99.48 ± 15.21^{b}	23.27 ± 4.90^{c}	32.06 ± 6.78^{b}	39.86 ± 4.29^{c}
	pH 7.5	03.42	18.28	5.97	12.33	102.14 ± 10.52^{ab}	25.28 ± 4.13^{b}	33.03 ± 6.08^{b}	43.91 ± 5.77^{b}
	pH 8.5	64.18	20.92	9.14	5.76	104.62 ± 14.37^{a}	29.37 ± 4.23^{c}	40.97 ± 6.85^{a}	58.96 ± 3.61^{a}
	pH 9.5	57.20	15.57	17.64	9.53	94.54 ± 5.75^{b}	28.03 ± 2.81^{a}	38.87 ± 4.32^{a}	58.27 ± 3.29^{a}
	蒸馏水	55.25	13.69	14.38	16.68	74.22 ± 10.29^{c}	22.85 ± 3.16^{c}	31.71 ± 6.45^{b}	40.20 ± 5.32^{a}
史氏鲟 *A. schrenckii*	pH 6.5	53.92	17.79	16.78	11.51	125.30 ± 14.45^{ab}	49.19 ± 6.46^{c}	73.50 ± 9.52^{b}	39.21 ± 4.62^{c}
	pH 7.5	55.90	21.40	4.73	7.79	132.78 ± 13.49^{a}	57.09 ± 6.23^{b}	82.37 ± 7.51^{a}	55.18 ± 6.34^{b}
	pH 8.5	57.68	20.88	14.85	6.59	139.32 ± 12.10^{a}	68.99 ± 7.65^{c}	82.91 ± 5.90^{a}	69.19 ± 3.38^{a}
	pH 9.5	46.54	18.69	20.06	14.71	120.86 ± 17.11^{b}	56.06 ± 6.29^{b}	77.64 ± 3.82^{ab}	60.76 ± 5.61^{a}
	蒸馏水	44.37	16.51	22.71	16.41	100.16 ± 13.75^{c}	45.02 ± 6.72^{c}	56.06 ± 5.45^{c}	52.12 ± 4.98^{b}
匙吻鲟 *P. spathula*	pH 6.5	61.37	15.72	10.69	12.22	120.04 ± 17.39^{a}	23.54 ± 6.46^{b}	31.17 ± 6.49^{c}	43.72 ± 9.11^{c}
	pH 7.5	65.00	14.13	8.52	12.35	119.68 ± 17.25^{a}	28.10 ± 5.25^{b}	65.42 ± 7.85^{b}	54.68 ± 10.13^{b}
	pH 8.5	66.17	17.29	3.76	12.78	130.44 ± 12.48^{a}	44.07 ± 7.33^{c}	69.47 ± 8.23^{a}	64.86 ± 9.09^{a}
	pH 9.5	61.85	16.10	8.47	13.56	104.74 ± 14.56^{b}	42.51 ± 10.50^{a}	68.00 ± 11.17^{c}	59.23 ± 4.17^{a}
	蒸馏水	50.96	13.79	16.51	18.74	62.30 ± 10.77^{c}	21.77 ± 5.00^{b}	27.99 ± 7.29^{c}	46.38 ± 5.20^{c}

注:表中同一栏内,右上角标有不同英文字母的数据,相邻字母间的比较(如 a 与 b;b 与 c 的比较)具有显著差异($P<0.05$),间隔字母间的比较(如 a 与 c 的比较)具有极显著差异($P<0.01$)。

Notes:The continuously differentletters in the sameline show significant difference($P<0.05$),and the uncontinuously different letters show highly significant difference($P<0.01$)。

西伯利亚鲟 所有实验组精子的VCL,VSL和VAP均高于对照组。其中,pH 7.5组的VCL,VSL和VAP最高,与其他实验组和对照组比较存在显著($P<0.05$)或极显著($P<0.01$)差异。说明适当的pH激活液可显著提高精子的运动速度。而除了pH 7.5组的MAD显著高于其他实验组和对照组外($P<0.05$),其他试验组与对照组比较,MAD无显著差异。因此认为激活液在该试验范围内的pH对精子的运动轨迹无显著影响。同时,60 s时的冻精活力测定是pH 7.5组的A级最高,D级最低。所以,pH 7.5应该是西伯利亚鲟冻精激活液的最适pH。

中华鲟 所有实验组精子的VCL,VSL和VAP均高于对照组。其中,pH 8.5组的VCL,VSL和VAP最高,除了与pH 9.5组比较,VSL和VAP无差异外,与其他实验组和对照组比较存在显著($P<0.05$)或极显著($P<0.01$)差异。值得注意的是,pH 8.5组的MAD与pH 9.5组的无差异,但与pH 7.5组和对照组存在显著差异($P<0.05$),而与pH 6.5组存在极显著差异($P<0.01$)。图1为pH 6.5组在冻精被激活60 s时的运动轨迹图,图2为pH 8.5组在冻精被激活60 s时的运动轨迹图。从图中可看出,pH 6.5组的精子几乎在原地旋转,而pH 8.5则呈较正常的运动状态。所以,MAD值越小,说明精子的旋转程度越大。低pH的激活液会使精子做旋转运动,从而影响精子的迅速扩散。60 s时的冻精活力测定是pH 8.5组的A级最高,D级最低。所以pH 8.5应该是中华鲟冻精激活液的最适pH。

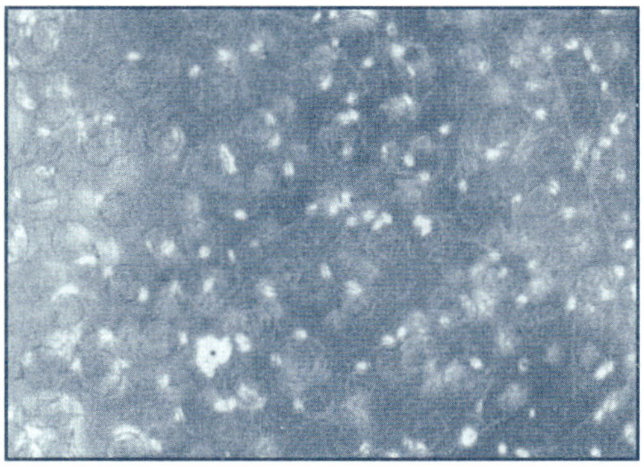

图1 中华鲟pH 6.5组冻精在激活60 s时的运动轨迹

Fig. 1 The moving track of post-thaw sperm of *A. sinensis* in pH 6.5 test group at the time of activated 60 s

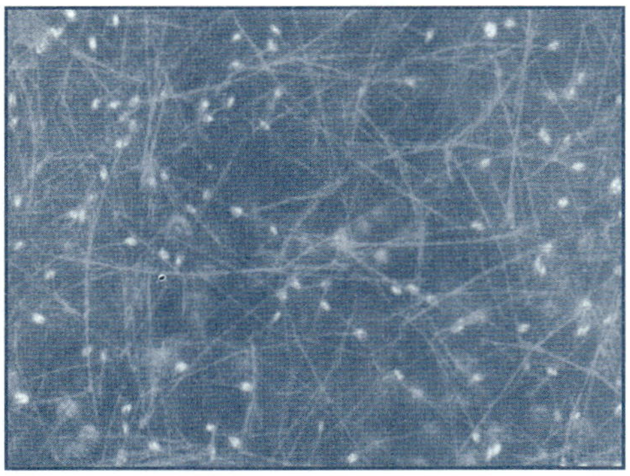

图2 中华鲟pH 8.5组冻精在激活60 s时的运动轨迹

Fig. 2 The moving track of post-thaw sperm of *A. sinensis* in pH 8.5 test group at the time of activated 60 s

史氏鲟 所有实验组精子的VCL,VSL和VAP均高于对照组。其中,pH 8.5组的VCL,VSL和VAP最高,与对照组比较有极显著差异($P<0,01$);pH 6.5组的MAD最低,与pH 8.5,pH 9.5和对照组比较有极显著差异($P<0.01$)。pH 6.5组的精子运动轨迹也呈旋转运动状。60 s时的冻精活力测定是pH 8.5组的A级最高,D级最低。所以pH 8.5应该是匙吻合鲟冻精激活液激活液的最适pH。

匙吻鲟 所有实验组精子的VCL,VSL和VAP均高于对照组。其中,pH 8.5组的CL,VSL和VAP最高,与对照组比较VCL和VAP呈极显著差异($P<0,01$),VSL呈显著差异($P<0.05$);pH 6.5组和对照组的MAD最低,与pH 8.5和pH 9.5组比较有极显著差异($P<0.01$),与pH 7.5比较有显著差异($P<0.05$)。pH 6.5组的精子运动轨迹也呈旋转运动状。60 s时的冻精活力测定是pH 8.5组的A级最高,D级最低。所以pH 8.5应该是匙吻鲟冻精激活液的最适pH。

2.3 Mg^{2+}对冻精活力的影响

试验发现在4种鲟鱼冻精激活液中加入适当浓度的Mg^{2+},可显著增加冻精被激活后的运动速度。表4列出了不同浓度Mg^{2+}的激活液对4种鲟鱼冻精活力参数的影响。从表4可看出4种鲟鱼的冻精激活液中加入5 mmol/L $MgCl_2$可显著提高冻精的VCL,VSL和VAP。所有实验组与对照组比较均呈显著($P<0.05$)或极显著($P<0.01$)差异。但Mg^{2+}对精子的运动方式无影响。除中华鲟和匙吻鲟的5 mmol/L $MgCl_2$组与其他实验组和对照组有显著差异($P<0.05$)外,其他2种鲟鱼的实验组和对照组均无差异。4种鲟鱼60 s时的冻精活力测定均是5 mmol/L $MgCl_2$组的A级最高,D级最低。所以,4种鲟鱼的冻精激活液中加入5 mmol/L $MgCl_2$应是较适宜的浓度。

表4 不同浓度的$MgCl_2$激活液对几种鲟鱼冻精活力参数的影响

Tab.4 The post-thaw sperm motility of several sturgeons stimulated by activating medium with different concentration of $MgCl_2$

种类 Species	激活液 Activating medium	60 s时的冻精活力/% Rate of moving sperm at the 60 second after stimulated				平均曲线运动速度/(μm/s) VCL	平均直线运动速度/(μm/s) VSL	平均路径运动速度/(μm/s) VAP	平均移动的角度/((°)/s) MAD
		A级	B级	C级	D级				
西伯利亚鲟 *A. bareii*	2.5 mmol/L $MgCl_2$	60.38	20.63	11.34	7.65	118.86 ±11.61^{b}	65.01 ±14.55^{b}	64.97 ±12.13^{b}	50.89 ±7.08
	5.0 mmol/L $MgCl_2$	66.26	18.79	11.13	3.82	126.74 ±10.42^{d}	74.03 ±10.13^{a}	76.92 ±9.40^{a}	57.38 ±6.09
	7.5 mmol/L $MgCl_2$	62.93	17.54	9.79	9.74	121.32 ±12.32^{a}	64.14 ±15.12^{b}	62.63 ±9.87^{b}	58.51 ±6.53
	10 mmol/L $MgCl_2$	58.22	19.07	9.87	12.24	109.48 ±13.79^{b}	60.93 ±13.20^{b}	60.80 ±14.05^{b}	59.44 ±4.48
	10 mmol/L Tris pH 7.5	65.87	19.94	11.02	3.18	112.96 ±13.18^{b}	65.61 ±20.61^{b}	62.60 ±8.57^{b}	58.68 ±6.39
中华鲟 *A. sinensis*	2.5 mmol/L $MgCl_2$	63.03	19.29	7.61	10.07	108.48 ±15.17^{b}	37.37 ±6.05^{a}	41.06 ±6.78^{b}	59.86 ±4.29^{b}
	5.0 mmol/L $MgCl_2$	70.15	17.34	8.52	3.99	122.34 ±18.38^{a}	38.40 ±6.78^{a}	49.79 ±6.62^{d}	67.55 ±4.00^{a}
	7.5 mmol/L $MgCl_2$	63.42	18.28	5.97	12.33	102.14 ±10.52^{b}	33.80 ±7.18^{ab}	40.23 ±6.18^{b}	58.27 ±3.29^{b}
	10 mmol/L $MgCl_2$	57.26	15.57	17.64	9.53	94.54 ±15.75^{c}	27.63 ±6.87^{b}	36.48 ±7.91^{b}	53.91 ±5.77^{b}
	20 mmol/L Tris pH 8.5	64.18	20.92	9.14	5.76	104.62 ±14.37^{b}	29.37 ±4.23^{b}	40.97 ±6.85^{b}	58.96 ±3.61^{b}
史氏鲟 *A. schrenckii*	2.5 mmol/L $MgCl_2$	53.92	17.79	16.78	11.51	141.30 ±14.49^{b}	69.86 ±6.46^{b}	88.56 ±6.52^{ab}	59.21 ±4.62
	5.0 mmol/L $MgCl_2$	62.41	18.85	15.52	3.22	151.78 ±15.23^{d}	77.61 ±7.49^{a}	94.17 ±6.90^{a}	68.61 ±4.44
	7.5 mmol/L $MgCl_2$	59.90	19.43	14.70	7.77	142.78 ±13.49^{b}	67.59 ±6.23^{b}	85.37 ±7.51^{b}	60.18 ±6.34
	10 mmol/L $MgCl_2$	46.54	18.69	20.06	14.71	120.86 ±17.11^{c}	56.06 ±6.29^{c}	77.64 ±6.82^{c}	60.76 +5.61
	20 mmol/L Tris pH 8.5	57.89	20.85	14.68	6.58	139.32 ±12.10^{b}	68.99 ±7.65^{b}	82.91 ±5.90^{b}	69.19 ±3.38
匙吻鲟 *P. spathula*	2.5 mmol/L $MgCl_2$	65.00	14.13	8.52	12.35	139.68 ±17.25^{b}	43.10 ±5.25^{b}	69.92 +7.85^{b}	54.68 ±10.13^{b}
	5.0 mmol/L $MgCl_2$	71.02	16.91	7.45	4.62	142.82 ±17.58^{a}	53.73 ±4.44^{a}	78.92 ±6.72^{a}	63.39 ±8.57^{a}
	7.5 mmol/L $MgCl_2$	61.37	15.72	10.69	12.22	120.04 ±17.39^{b}	42.51 ±10.50^{b}	76.17 ±6.49^{a}	53.72 ±9.11^{b}
	10 mmol/L $MgCl_2$	61.85	16.10	8.47	13.56	104.74 ±14.56^{c}	23.54 ±6.46^{c}	68.00 ±11.17^{b}	55.23 ±4.17^{b}
	30 mmol/L Tris pH 8.5	66.17	17.29	3.76	12.78	130.44 ±12.48^{b}	44.07 ±7.33^{a}	69.47 ±8.23^{a}	64.86 ±9.09^{a}

注:表中同一栏内,右上角标有不同英文字母的数据,相邻字母间的比较(如a与b;b与c的比较)具有显著差异($P<0.05$),间隔字母间的比较(如a与c的比较)具有极显著差异($P<0.01$)。

Notes:The continuously differentletters in thesameline show significant difference($P<0.05$),and the uncontinuously different letters show highly significant difference($P<0.01$)。

2.4 精子冷冻保存效果比较

图 3 和图 4 分别为冷冻后和冷冻前 4 种鲟鱼精子在各检测时段的平均活率。从图中可看出冷冻前中华鲟的精子寿命最长，达到了 420 s 以上；其次是西伯利亚鲟为 390 s；史氏鲟的精子寿命为 360 s；匙吻鲟的精子寿命最短，为 300 s 左右。240 s 前，4 种鲟鱼的精子平均活率均在 60 %以上；但 240 s 以后，匙吻鲟精子的平均活率急速下降。其他 3 种鲟鱼精子在 330 s 后，平均活率才相继低于了 60 %。

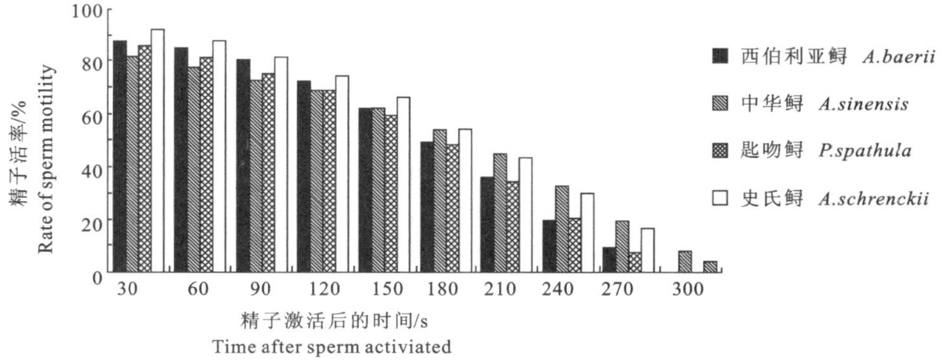

图 3　4 种鲟鱼冻精被激活后各时间段的精子活率

Fig. 3　The rate of post-thaw sperm motility at the each time after activated

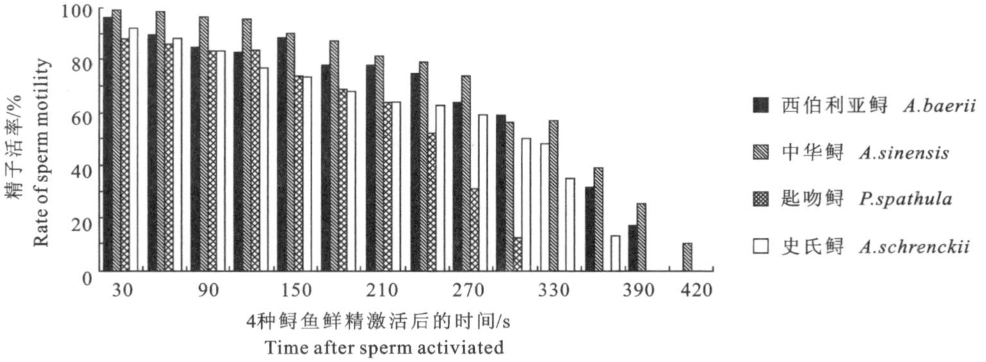

图 4　4 种鲟鱼鲜精被激活后各时间段的精子活率

Fig. 4　The rate of frsh sperm motility at the each dme after activated

冷冻后 4 种鲟鱼精子的平均寿命都不同程度地缩短了。中华鲟和史氏鲟为 300 s，西伯利亚鲟和匙吻鲟为 270 s。4 种鲟鱼精子在 180 s 以后均低于了 60 %。

对 4 种鲟鱼冷冻前后精子的平均活率相关性作线性回归分析，结果显示西伯利亚鲟的 $R^2 = 0.8296$（$P < 0.01$）；中华鲟的 $R^2 = 0.9860$（$P < 0.01$）；匙吻鲟的 $R^2 = 0.9622$（$P < 0.01$）；史氏鲟 $R^2 = 0.9477$（$P < 0.01$）。说明冷冻前后 4 种鲟鱼精子的平均活率存在着极显著的线性正相关性。

3 讨论

3.1 鱼类精子活力检测方法

计算机辅助精子活力分析系统(CASA)是在包括显微摄影、显微图像分析和荧光显微技术等多项技术和设备的基础上发展起来的新技术[4]。该系统利用硬件(显微镜、录像机、频闪仪、计算机等)和软件(Micro Image v 5.0、SQIAS-1000,以及 FSQAS-2000 等)对精子激活后的录像进行数字化处理和分析。此项技术最初是应用在男性生殖学研究中,以后随着技术的发展和完善逐渐被应用到其他哺乳类精子活力的检测中[15-18]。

鱼类精子的活力检测最初是 1985 年由 Cosson 等[19]对鲑鱼的研究开始的。当时是通过对鲑鱼精子激活后的录像,进行慢镜头回放,并在显示器上加入了预先设置好的定标网格,通过精子在各网格中的运动,用秒表记录时间,实现了半数字化鱼类精子活力检测。真正的计算机辅助鱼类精子活力检测技术的研究是最近 10 年才开展起来的[14,20-25]。CASA 技术的引入克服了传统鱼类精子活力检测无法克服的难题,如精子在激活后快速运动的时间很短,精子的密度很高,以及鞭毛摆动的频率很快等[5]。使鱼类精子活力检测在一个更精确、重复性更强、数字化程度更高,并快速、简便的基础上进行。

FSQAS-2000 是作者依据鱼类精子的生物学特性,参照医学对人类精子的检测方法,在世界卫生组织对精子检测标准的基础上,与武汉千屏影像技术有限公司合作开发的,一款具有自主知识产权的鱼类精子彩色图文分析系统。该系统对激活后的精子录像,按时间分段进行采集,每个时间段为 3 秒 72 帧图像。系统会依据每个精子在图像中像素点所在的位置进行自动编号,然后计算机根据每个精子在每一帧图像中的像素点来分析该精子的运动状态。该系统与国外同类系统相比,省掉了频闪仪、时间打卡器等造价很高的硬件设备,而用精子的像素点移动原理和计算机本身的时间记录功能取代。因此,该系统比国外同类系统更简便、更经济,同时由于精子与像素点的关系是一对一的关系,所以也更精确。

3.2 精子活力的相关参数

无论是本论文所采用的 FSQAS-2000,还是其他形式的 CASA,均能对鱼类精子的运动状态提供 20 多项参数的分析结果。但在对美洲鮰(*Noturus flavipinnis*)、鲤(*Cyprinus carpio*)、欧洲鲈(*Dicentrarchus labrax*)、鲑(*Oncorhynchus mykiss*)和湖鲟(*Acipenser fulvescens*)等的许多研究中发现,最重要的参数是精子的 VCL(精子在其实际运动轨迹上的速度。)和 VSL(精子开始运动的位置与最后所处位置间的直线距离/所花的相应时间。)[24,26-29]。我们对 4 种鲟鱼的研究也同样证明,不同的激活液可以使鲟鱼精子的 VCL 和 VSL 发生很大变化。

另一个与精子运动速度有关的参数是精子的 VAP(精子沿其空间实际运动轨迹所做的平滑曲线的距离/所花的相应时间。),一些学者认为它在鱼类精子分析中不重要,原因是鱼类精子总是在一个方向上做平滑的曲线运动。因此,认为鱼类的 VAP 和 VCL 是基本相同的[24,26-29]。但另一些学者则认为 VAP 和 VCL 同样重要。在研究棘鱼(*Acanthodii aculeatus*)和狼鱼(*Anarhl minor*)时发现,由于这些鱼的精子要在胶体渗透压很高的激活液中运动,激活液所产生的阻力迫使精子的运动轨迹呈飘逸状。因此,VAP 和 VCL 产生了很大差异[30,31]。作者在对 4 种鲟鱼精子的研究中也发现 VAP 和 VCL 有很大差异。分析认为,鱼类精子的 VAP 和 VCL 是否有差异主要取决于精子的体积和形状、精子的运动速度、精子的运动方式以及精子在激活液中所受到的阻力。试验中的 4 种鲟鱼精子体积均比一般鱼类大 20 ~ 50 倍,形状呈梭形,速度比一般淡水鱼类精子快(鲤科鱼类平均曲线速度可达到 100 μm/s 以上[32])。最主要的是 4 种鲟鱼精子的运动方式,精子在激活后,不仅沿自身长轴做旋转运动,而且还沿运动方向的轴做旋转运动。因此,精子在运动过程中所受的阻力很大,迫使精子做不规则的飘逸运动。另一种可能是,试验主要是针对冷冻保存后的精子进行冻精激活试验。精子在冷冻过程中,其膜结构可能受到了损伤,从而改变了精子的正常运动状态,—这种可能性是否存在尚有待进一步的研究。

参考文献

[1] 邓岳松,林浩然. 鱼类精子活力研究进展[J]. 生命科学,1999,3(4):271-278.

[2] Billard R, Cosson J. Some problems related to the assessment of sperm motility in freshwater fish[J]. J Exp Zool, 1992,261:122-131.

[3] Rurangwaa E, Kimeb D E, Ollcviera F, et al. The measurement of sperm motility and factors affecting sperm quality in cultured fish [J]. Aquaculture, 2004,234:1-28.

[4] Boyer S P, Davis R O, Katz D F. Automated semen analysis current problems in obstetrics[J]. Gynccol Fertil, 1989 (12):165-200.

[5] Billard R, Cosson J, Crim L W, et al. Sperm physiology and quality [A]//Bromage N, Roberts R. Broodstock management and egg and larval quality[C]. Oxford: Blackwell, 1995:25-52.

[6] 周定刚,温安祥. 黄鳝精子活力检测和精子入卵早期过程观察[J]. 水产学报,2003,27(5):398-402.

[7] 管卫兵,王桂忠,李少菁. 甲壳动物精子质量和活力评价[J]. 海洋通报,2003,22 (2):83 -88.

[8] 苏德学,严安生,田永胜. 钠、钾、钙和葡萄糖对白斑狗鱼精子活力的影响[J]. 动物学杂志,2004,39(1):16-22.

[9] 胡家会,张士璀,张永忠,等. 青岛文昌鱼精子运动特征及计算机辅助分析[J]. 动物学报,2006,52(4):706 -711.

[10] Linhart O, Kudo. Surface ultrastructure of paddlefish eggs before and after fertilization[J]. J Fish Biol, 1997,51:573-582.

[11] Medea A V, Gmb H. Manual for medeaLAB CASA (Computer Aided Sperm Analysis) Version 5. 1 [M]. Am Weichselgarten 23 D-91058 Erlangen Germany, 1997:15-31.

[12] Linhart O, Rodina M, Cosson J. Cryopreservation of sperm in common carp *Cyprinu scarpio*: spermmotility and hatching success of embryos [J], Cryobiology, 2000,41:241-250.

[13] Liu L, Wei Q W, Guo F, et al. Cryopreservation of Chinese sturgeon (*Acipenser sinensis* Gray) sperm [J]. J Appl Ichthyol ,2006,22:55-59.

[14] Gallis J L, Fedrigo E, Jatteau P, et al. Siberian sturgeon, *Acipenser baeri*, spermatozoa: effects of dilution, pH, osmotic pressure, sodium and potassium ions on motility [A]//Williot P. Acipenser[C]. Cemagref Publ, 1991:143-151.

[15] Vantman D, Banks S M, Koukoulis G, et al. Asse ssment of sperm motion characteristics from fertile and infertile men using a fully automated computer assisted semen analyzer[J]. Fertil Steril, 1989,51:151-161.

[16] Farrell P B, Presicce G A, Brockett C C, et al. Quantification of bull sperm characteristics measured by computer-assisted sperm analysis(CASA) and the relationship to fertility [J]. Theriogenology, 1998,49:871-879.

[17] Moore H D M , Akhondi M A. Fertilizing capacity of rat spermatozoa is correlated with decline in straight-line velocity measured by continuous computer-aided sperm analysis: epididymal rat spermatozoa from proximal cauda have a greater fertilizing capacity in vitro than those from the distal cauda or vas deferens [J]. J Androl, 1996,17:50-60.

[18] Hirano Y, Shibahara H, Obara H, et al. Relationships between sperm motility characteristics assessed by computer-aided sperm analysis (CASA) and fertilisation rates in vitro [J]. J Assist Reprod Genet, 2001,18:213-218.

[19] Cosson M P, Billard R, Gatti J L, et al. Rapid and qualitative assessment of trout sperm motility using stroboscopy [J]. Aquaculture, 1985,46:71-75.

[20] Toth G P, Christ S A, McCarthy H W, et al. Computer-assisted motion analysis of semen from the common carp [J]. J Fish Biol, 1995,47:986-1003.

[21] Toth G P, Ciereszko A, Christ S A, et al. Objective analysis of sperm motility in the lake sturgeon, *Acipen*, ser

fulvescens: activation and inhibition conditions [J]. Aquaculture, 1997, 154: 337-348.
[22] Christ S A, Toth G P, McCarthy H W, et al. Monthly variation in sperm motility in common carp assessed using computer-assisted sperm analysis (CASA) [J]. J Fish Biol, 1996, 48: 1210-1222.
[23] Kime D E, Ebrahimi M, Nysten K, et al. Use of computer assisted sperm analysis (CASA) for monitoring the effects of pollution on sperm quality of fish; application to effects of heavy metals [J]. Aquat Toxicol, 1996, 36: 223-237.
[24] Kime D E, van Look K J W, McAllister B G, et al. Computer assisted sperm analysis (CASA) as a tool for monitoring sperm quality in fish [J]. Comp Biochem Physiol, 2001, 130: 425-433.
[25] Ravinder K, Nasaruddin K, Majumdar K C, et al. Computerized analysis of motility, motility patterns and motility parameters of spermatozoa of carp following short-term storage of semen [J]. J Fish Biol, 1997, 50: 1309-1328.
[26] Ciereszko A, Toth G P, Christ S A, et al. Effect of cryopreservation and theophylline on motility characteristics of lake sturgeon (*Acipemer fulvescens*) spermatozoa [J]. Theriogenology, 1996, 45: 665-672.
[27] Rurangwa E, Volckaert F A M, Huyskens G, et al. Quality control of refrigerated and cryopreserved semen using computer-assisted sperm analysis (CASA), viable staining and standardized fertilisation in African catfish (*Clarias gariepinus*) [J]. Theriogenology, 2001, 55: 751-769.
[28] Rurangwa E, Biegniewska A, Slominska E, et al. Effect of tributyltin on adenylate content and enzyme activities of teleost sperm: a biochemical approach to study the mechanisms of toxicant reduced spermatozoa motility [J]. Comp Biochem Physiol, 2002, 131: 335-344.
[29] Jobling S, Coey S, Whitmore J G, et al. Wild inter-sex roach (*Rutihts rutilus*) have reduced fertility [J]. Biol Reprod, 2002, 67: 515-524.
[30] Kime D E, Tveiten H. Unusual motility characteristics of sperm of the spotted wolfish [J]. J Fish Biol, 2002, 61: 1549-1559.
[31] Elofsson H, McAlister B G, Kime D E, et al. Long lasting sticklebacks sperm; is ovarian fluid a key to success in freshwater [J]. J Fish Biol, 2003, 63: 240 -253.
[32] Perchec G, Jeulin C, Ander K, et al. Relationship between sperm ATP content and motility of carp spermatozoa [J] Cell Sci, 1995, 108: 747-753.

Comparative Study of Activating Mediums for the Cryopreserved Sperm of Several Sturgeons Using CASA

Liu Ling[1] Otomar Linhart[2] Wei Qiwei[1]
Guo Feng[1] Zhang Jieming[1] Liu Jianyi[1]

(1. Key Luboratory of Freshuater Fish Germplasm Resources and Biotechnology, Yangtse River Fisheries Research Institute, Chinese Academy of Fisheries, Sciences, Jinzhou 434000, China
2. Joint Laboratory of Genetics, Physiology and Reproduction of Fish, Institute of Animal Physiology and Genetics, Academy of Sciences of the Czech Republic, University of South Bohemia, Research Institute of Fish Culture and Hydrobiology, 38925 Vodnany, Czech Republic)

Abstract The digital analysis technique of sperm quality assessment in fish has provided reliable reference evidences for the effects and the damage mechanism of cryopreservation, genetics and breeding, etc. From the study using fish color sperm image analysis systems (FSQAS-2000) conducted on the cryopreserved sperm of Chinese sturgeon (*Acipenser sinensis*), Amur sturgeon (*A. schrenckii*), Siberian sturgeon (*A. baerii*) and paddlefish (*Polyodon spathula*) under different activating mediums, the most parameters, including the curvilinear velocity (VCL), the straight line velocity (VSL), the angular path velocity (VAP), angular displacement (MAD) and percentage of motile sperm, etc. were compared for the first time in China. The influences of activating mediums with different osmolality, pH and concentration of Mg^{2+} on the parameters were also analyzed in the study. Our results indicated that: First, the osmolality of activating medium for the sturgeons mentioned above was quite low, among them, the lowest for Siberian sturgeon (10 mOsm/kg), the highest for paddlefish (30 mOsm/kg), the middle for Chinese sturgeon and Amur sturgeon (20 mOsm/kg). Second, it had a significant ($P \leqslant 0.05$ or $P \leqslant 0.01$) influence on VCL, VSL, VAP and percentage of motile sperm in cryopreserved samples. Third, pH of activating medium had the most significant difference on MAD ($P \leqslant 0.01$) among the parameters, and it was lower and MAD was lager. Fourth, the most appropriate pH was 7.5 for Siberian sturgeon and 8.5 for Chinese sturgeon, Amur sturgeon and paddlefish. And last, the velocities (VCL, VSL, VAP) had significant increases ($P \leqslant 0.01$), when $MgCl_2$ (5 mmol/L) was added into activating medium. In conclusion, the appropriate activating mediums for the cryopreserved sperm of Siberian sturgeon, Chinese sturgeon and Amur sturgeon, paddlefish were 10 mmol/L Tris-HCl + 5 mmol/L $MgCl_2$, pH 7.5, 20 mmol/L Tris-HC1 + 5 mmol/L $MgCl_2$, pH 8.5, 30 mmol/L Tris-HCl + 5 mmol/L $MgCl_2$, pH 8.5, respectively. Using the mediums to activate sperm, there were highly positive linear correlations ($P \leqslant 0.01$) in percentage of motile sperm between prefreeze and post-thaw sperm. R^2 was 0.83, 0.99, 0.96, 0.95 for Siberian sturgeon, Chinese sturgeon, Amur sturgeon and paddlefish, respectively.

Key words Sturgeon Computer-aided sperm analysis (CASA) Cryopreservation Sperm motility

Na^+，K^+，Mg^{2+}，Ca^{2+}和葡萄糖溶液作为授精-激活介质对中华鲟精子受精率的影响*

甘 芳[1,2,3,**] 杜 浩[2,3] 危起伟[1,2,3,***]
陈细华[2,3] 刘鉴毅[2,3] 郑跃平[1,2,3]

（1.农业部淡水生物多样性保护与利用重点开放实验室 华中农业大学水产学院 湖北武汉 430070
2.中国水产科学研究院长江水产研究所 湖北荆州 434000
3.中国水产科学研究院淡水渔业研究中心 江苏无锡 214081）

摘 要 以不同浓度的 NaCl，KCl，$MgCl_2$，$CaCl_2$ 溶液和葡萄糖溶液作为授精介质，研究了中华鲟（*Acipenser sinensis*）的受精效果。结果显示，适量的阳离子和葡萄糖作为激活授精介质时中华鲟卵受精率都有所提高。在实验设置浓度范围内 25 mmol/L NaCl 溶液、0.1 mmol/L KCl 溶液、1 mmol/L $MgCl_2$溶液、1 mmol/L$CaCl_2$溶液和 50 mmol/L 葡萄糖溶液浓度下，受精率分别可达到最高值，依次为 87.72％，86.82％，82.24％，89.76％，80.92％。随着实验浓度继续增加，受精率反而呈下降趋势。结果显示，作为人工配制的中华鲟精子授精－激活介质，最适 NaCl 溶液浓度在 25 mmol/L 附近，最适葡萄糖溶液浓度在 25 mmol/L 附近，最适 KCl 溶液浓度≤0.1 mmol/L，最适 $MgCl_2$溶液浓度≤1 mol/L，最适 $CaCl_2$溶液浓度≤1 mmol/L。

关键词 中华鲟 精子活力 受精率

中华鲟（*Acipenser sinensis*）为鲟形目鲟科鲟属鱼，是典型的江海洄游性鱼类，主要分布于我国东海、黄渤海大陆架水域及汇入其中的大型河流中[1]。在长江中，性成熟后的中华鲟要上溯三千多公里到达金沙江下游及长江上游产卵。葛洲坝水利工程修建后阻隔了中华鲟产卵洄游的通道，中华鲟在葛洲坝下游形成了新的产卵场[2]。由于大坝阻隔及过度捕捞和环境退化等因素，中华鲟自然繁殖群体资源大幅下降[2]。1988 年国家颁布《野生动物保护法》，将中华鲟列为国家一级保护野生动物，经国家渔政行政主管部门审批每年有少量的科研捕捞指标。葛洲坝下游中华鲟人工繁殖自 1983 年首次获得成功[3]，以后每年即采取人工放流的方法对中华鲟种群进行增殖，放流工作主要由葛洲坝集团中华鲟研究所和长江水产研究所实施，至 2006 年长江水产研究所共向长江中投放各种规格中华鲟苗约 191 万尾。伴随着中华鲟目前的生存压力逐渐增加，中华鲟自然繁殖种群的性腺发育质量有明显下降的趋势[4]，雌雄性比失调及种群结构的变化和产卵场环境的改变对中华鲟自然繁殖将有重要的影响[5]，因此人工增殖放流显得很重要。受精率大小是影响中华鲟人工繁殖效果的重要因素。研究表明，阳离子、渗透压、温度、pH、单糖等都可能对精子活力和受精率产生影响[6-8]，合适的人工激活液作为授精介质能够极大地提高精子的受精率。作者曾对阳离子和单糖对中华鲟精子活力的影响进行了初

＊基金项目：国务院三峡办重点项目（SX97-25/HB）；国家自然科学基金重大项目专题（30490231）；湖北省科技攻关项目（2006AA203A02）。

＊＊第一作者，女，硕士研究生，研究方向为濒危鱼类生理生态学，E-mail：ganfang@yfi.ac.cn

＊＊＊通讯作者（Corresponding author），E-mail：weiqw@yfi.ac.cn

步实验[9]，本实验在原有基础上，对阳离子和单糖对受精效果的影响进行单因子系统实验和分析。旨在找出中华鲟人工授精激活液中各组分的最佳配制浓度，提高中华鲟人工繁殖效果，为同类研究提供借鉴以及参考。

1 材料与方法

1.1 实验材料

中华鲟亲本是由国家渔政主管部门审批后，由渔政部门组织本地专业渔民于2006年10～11月份从宜昌葛洲坝下游捕获。雌鲟由中华鲟专用运输车运到荆州中华鲟增殖放流基地，并进行人工催产获得卵子。雄鲟在宜昌葛洲坝集团中华鲟研究所进行人工催产并获得精液，低温保存运输至荆州中华鲟增殖放流基地，放入2 ℃冰箱充氧保存，每4 h换气一次。实验所用卵子、精液分别采自同一亲本，选择已成熟的卵子和活力强的精液进行授精实验。

1.2 实验方法

选择性腺发育良好、人工催产顺利的同一尾雌鱼的卵，每个浓度实验设3个平行组，每个平行组取100颗卵放入培养皿内。用注射器滴入2 μl的精子，混匀后添加适量的不同化学激活液授精，5 min后再加入同种溶液清洗2次。清洗后放入装有循环水装置的白桶内孵化，每隔4 h观察一次，挑出坏(死)卵，胚胎发育至原肠中期时(图1)计算受精率。同时各实验共同设置一个对照组，采用蒸馏水作为激活液授精，然后在循环水装置内孵化。人工控制及观察方法与实验组相同。配制精卵激活-授精介质所用药品均为分析纯，用蒸馏水稀释。NaCl配制10 mmol/L，25 mmol/L，50 mmol/L，75 mmol/L和100 mmol/L 5个浓度梯度；KCl配制0.1 mmol/L，0.3 mmol/L，0.5 mmol/L，0.7 mmol/L，0.9 mmol/L和1.0 mmol/L 6个浓度梯度；$MgCl_2$配制1 mmol/L，3 mmol/L，5 mmol/L，10 mmol/L，15 mmol/L和20 mmol/L 6个浓度梯度；$CaCl_2$配制1 mmol/L，3 mmol/L，5 mmol/L，10 mmol/L 15 mmol/L和20 mmol/L 6个浓度梯度；葡萄糖配制25 mmol/L，50 mmol/L，100 mmol/L，125 mmol/L和150 mmol/L 5个浓度梯度。以空白蒸馏水作为对照组。

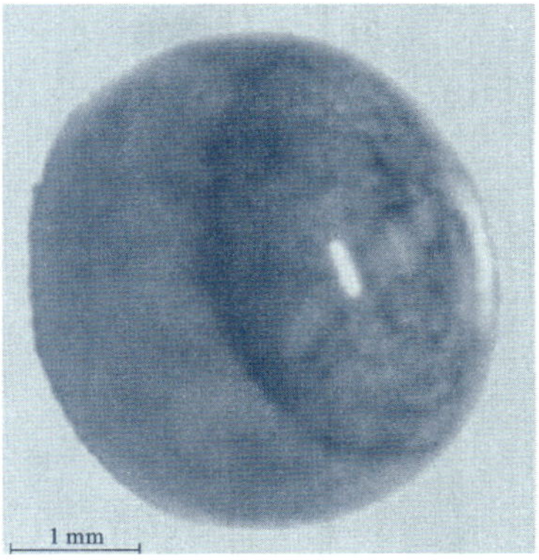

图1 中华鲟的中期原肠胚

Fig. 1 Chinese sturgeon embryo at mid gastrula stage

1.3 数据分析

采用SPSS 13.0单因素方差分析Duncan多重检验比较各浓度组及对照组之间的差异。

2 结果与分析

2.1 不同阳离子溶液对中华鲟精子受精率的影响

实验结果显示，适量的阳离子溶液作为激活授精介质时的受精率比蒸馏水直接作为授精介质都有一定程度增加，有利于中华鲟卵受精率的提高(图2)。实验范围内NaCl溶液在低于25 mmol/L时，随浓度的增加受精率有所增加，NaCl溶液浓度在25 mmol/L时，中华鲟精子受精率达到最高，为87.72 %；而高于25 mmol/L时随浓度增加受精率呈不明显的下降趋势(*P* >

0.05)，至 NaCl 溶液为 100 mmol/L 时，受精率显著降低($P<0.05$)。结果表明，NaCl 溶液作为中华鲟精子授精介质，其最佳浓度在 25 mmol/L 附近，这一浓度虽与对照组无显著差异，但明显比对照组中华鲟精子受精率高，说明在精子授精液中适当加入 NaCl 溶液有利于提高受精率，然而高于或低于一定浓度都会降低受精率，高浓度甚至明显抑制精子的激活。

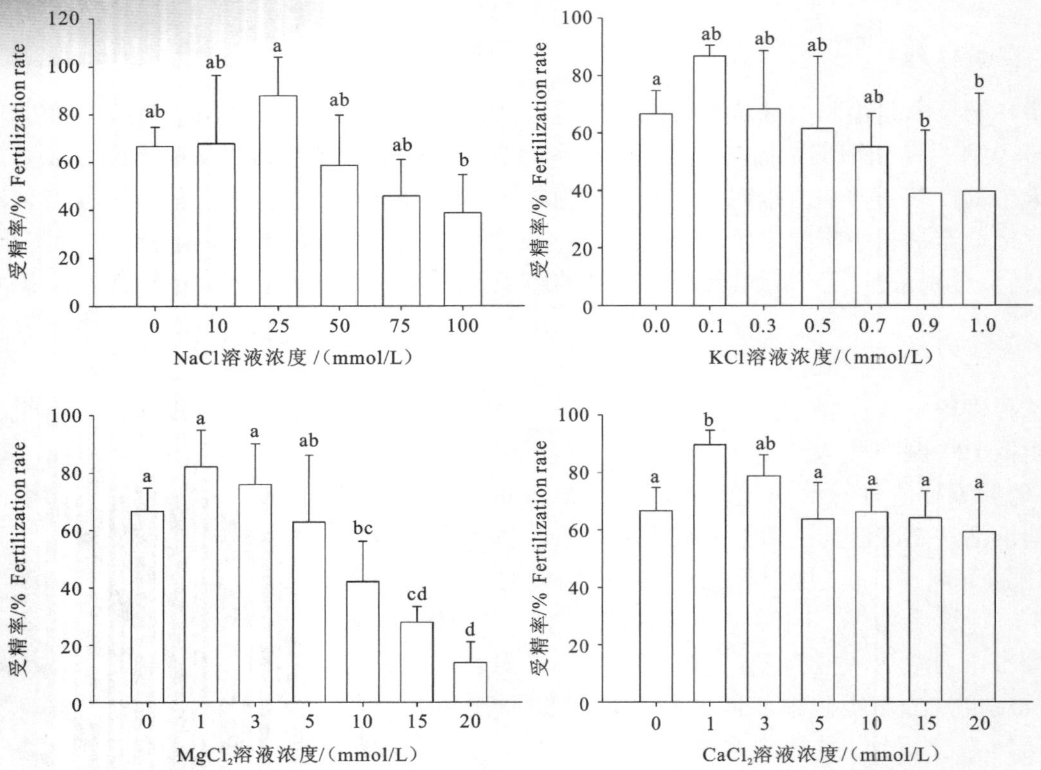

图2 不同浓度的 Na^+，K^+，Mg^{2+}，Ca^{2+} 溶液对中华鲟精子受精率的影响(Mean ± SD)
不含有相同字母的各浓度组的受精率有显著差异($P<0.05$，Duncan 多重比较)。

Fig. 2 The effect of Na^+, K^+, Mg^{2+}, or Ca^{2+} solution as insemination-activation medium on the fertilization rate of Chinese Sturgeon milt. Columnswithout same small letters were significantly different from each other by Duncan's in Spss ($P<0.05$).

与 NaCl 溶液相同，KCl 溶液浓度低于 0.3 mmol/L 时受精率随浓度的提高呈不明显增高($P>0.05$)，在 KCl 溶液浓度为 0.1 mmol/L 时，中华鲟精子受精率最大，为 86.82 %。随着 KCl 溶液浓度的升高，中华鲟精子受精率逐渐下降，至 KCl 溶液浓度达 0.9 mmol/L 时，受精率显著低于对照组，说明在人工激活授精液中配制低浓度的 KCl 溶液可激发中华鲟精子活力，提高受精率。高浓度的 KCl 溶液影响中华鲟精子活力，从而导致受精率下降。

$MgCl_2$溶液与 $CaCl_2$溶液浓度对中华鲟精子受精率作用较相似。当 $MgCl_2$，$CaCl_2$溶液浓度为 1 mmol/L 时，中华鲟精子受精率最高，为 82.24 %和 89.76 %；但随着 $MgCl_2$，$CaCl_2$溶液浓度升高，中华鲟精子受精率呈下降趋势。$MgCl_2$溶液浓度为 1 mmol/L 时中华鲟精子受精率与对照组之间虽无显著差异，但明显高于对照组，在达到 10 mmol/L 高浓度以后的 $MgCl_2$溶液对精子的激活作用受到明显抑制($P<0.05$)。而 $CaCl_2$溶液为 1 mmol/L 时中华鲟精子受精率与对照组之间存在显著差异($P<0.05$)，高于 3 mmol/L $CaCl_2$溶液作为授精激活介质时中华鲟卵的受精率呈不显著下降的趋势。实验结果说明，在中华鲟授精液中加入低浓度的 $MgCl_2$，$CaCl_2$溶

液有利于激活中华鲟精子活力,提高精子受精率,高浓度 $MgCl_2$,$CaCl_2$ 溶液不利于精子受精,对精子受精率有一定抑制作用,比对照组(蒸馏水)的受精率还要低。

2.2 葡萄糖溶液对中华鲟精子受精率的影响

以葡萄糖作为授精介质时中华鲟卵的受精率随着葡萄糖溶液浓度的增加呈先上升后下降趋势(图3)。在葡萄糖溶液为50 mmol/L 时,中华鲟精子受精率最大,为80.92 %。100 mmol/L 以上的葡萄糖溶液使中华鲟受精率随浓度增加呈不明显下降趋势($P>0.05$),至125 mmol/L 时相对于对照组有显著下降($P<0.05$)。结果表明,中华鲟精子授精激活液中加入低浓度的葡萄糖溶液,有利于提高精子受精率,高浓度的葡萄糖溶液对中华鲟精子受精的激发作用均受到不同程度的抑制。

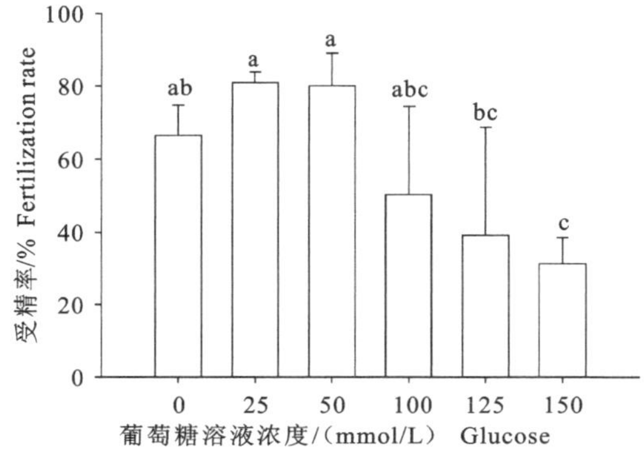

图3 不同浓度的葡萄糖溶液作授精介质对中华鲟精子受精率的影响(Mean ± SD)
不含有相同字母的各浓度组的受精率有显著差异($P<0.05$,Duncan 多重比较)。

Fig. 3 The effect of glucose solution as the insemination-activation medium on the fertilization rate of Chinese Sturgeon milt. Columns without same small letters were significantly different from each other by Duncan's in Spss ($P<0.05$).

3 讨论

3.1 NaCl,KCl 溶液对中华鲟精子受精率的影响

精子活力与受精率有着密切的联系,为了得到高的受精率,精子必须具有较强的活力[6]。鱼类精子在精巢和精浆中是不活动的,进入低渗透压水环境中则被激活,其重要的抑制因素就是渗透压。构成鱼类渗透压的主要离子是鱼类血浆、精浆中的 Na^+,K^+ 等组分。Alavi 和 Cosson 的研究表明,鲟鱼精子对 Na^+ 和其他离子的生物敏感性主要由精浆的成分和 Na^+ 含量决定[10]。本实验表明,适宜浓度的 Na^+ 可以增强中华鲟精子活力,从而提高精子受精率,且精子活力激发的最适宜 Na^+ 浓度为25 mmol/L,明显高于精浆中的 Na^+ 浓度(10 ~ 20 mmol/L)。

K^+ 离子对鲟鱼精子的影响却与 Na^+ 离子的作用有很大差别。K^+ 主要起抑制鲟鱼精子活力的作用[11]。本实验表明,随着 K^+ 浓度的逐渐升高,中华鲟精子的活力降低,从而受精率也依次递减,这与其他人的研究得出的结论一致[10,12]。根据郑跃平的试验检测,中华鲟精浆中的 K^+ 浓度为(7.18 ± 0.47) mmol/L 和(4.46 ± 0.02) mmol/L[13],远高于实验中精子活力受到抑制时的浓度,所以,K^+ 是抑制中华鲟精子在精浆中活动的主要因子。在预实验中也发现,KCl 溶液浓度低于0.1 mmol/L 时,其受精率变化幅度很小,Na^+,K^+ 离子在精浆中形成高于外界水体的渗透压,当精液排入水体,其浓度稀释即对精子有激活作用,提高精子受精率。在体

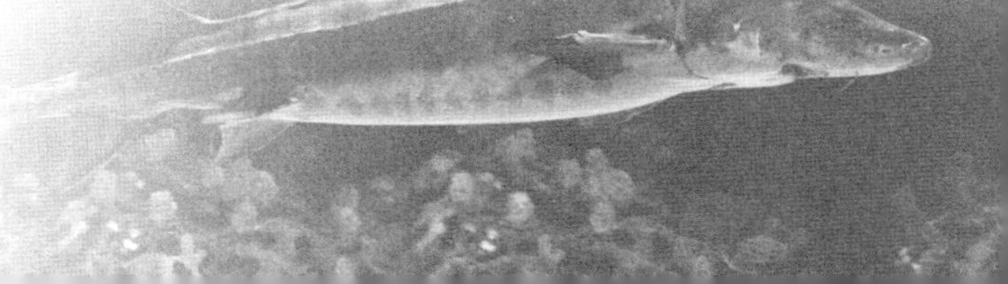

外环境下,精子活力是随 K^+ 浓度升高而下降,体外环境中 K^+ 浓度较高时就会导致其精子受精率的降低。

3.2 $MgCl_2$,$CaCl_2$溶液对中华鲟精子受精率的影响

Ca^{2+},Mg^{2+} 作为鱼类精浆主要离子,其含量较 Na^+,K^+ 低得多,鲟鱼精浆中的 Ca^{2+},Mg^{2+} 含量通常都低于 1 mmol/L,但是在精子活力的调节中具有重要作用。Ca^{2+} 是调节鱼类精子活力的重要因子,外界环境的 Ca^{2+} 通过 Ca^{2+} 通道进入精子细胞,使精子细胞内的 Ca^{2+} 增加,这是精子激活所必需的[10]。Ca^{2+} 还可解除 K^+ 对鲟鱼精子的抑制作用,0.1 mmol/L Ca^{2+} 即可部分解除 1 mmol/L K^+ 对中华鲟精子的抑制,Ca^{2+} 0.25 mmol/L 时即可完全解除 K^+ 的抑制作用[11]。刘鉴毅的研究表明,Ca^{2+} 浓度高于 20 mmol/L 可明显抑制中华鲟精子的活力,在 40 mmol/L 时精子活力完全被抑制[9]。本实验也表明,Ca^{2+} 浓度范围从 1 ~ 20 mmol/L,中华鲟精子受精率依次降低。因此,虽然 Ca^{2+} 是精子激活所必需的,但浓度并不是越高越好,在中华鲟中以≤1 mmol/L 为适宜。另外,中华鲟在 Ca^{2+} 浓度为 5 mmol/L 时受精率出现巨降,较 15 mmol/L 的受精率还低,可能是由于实验过程中的误差造成。受精率大小是评价人工繁殖效果的重要指标,但实际操作可能带入误差。因此,在实际鱼类人工繁殖过程中需要参照精子活力合理配制激活授精液。

有关 Mg^{2+} 对鲟鱼精子活力影响的报道很少。Alavi 等研究显示,Mg^{2+} 对中华鲟精子受精率的影响效果与 Ca^{2+} 的效果相似[14]。本实验印证了这一观点,与 Ca^{2+} 作用相似,中华鲟精子受精率在 Mg^{2+} 浓度≤1 mmol/L 为宜,随着 Mg^{2+} 浓度升高,中华鲟精子受精率下降,精子活力应是逐渐受到了抑制。

3.3 葡萄糖溶液对中华鲟精子受精率的影响

葡萄糖是精浆中主要糖类的重要组分,它的作用主要是参与精子代谢,维持精子活力[8]。白斑狗鱼(*Esox lucius*)的精子在葡萄糖溶液中的活动规律与在 NaCl,KCl 溶液中基本一致,但在葡萄糖中的活力较二者高[7]。而本实验表明,中华鲟精子在葡萄糖溶液中活力与 KCl 溶液不同,与 NaCl 溶液类似,精子受精率先升高后呈下降趋势,且在浓度 50 mmol/L 时受精率最高。可见,葡萄糖作为细胞能源物质对精子激活有一个最适浓度,在这个浓度下可达到中华鲟精子的最高受精率。

致谢

本实验过程中得到长江大学刘劲松同学的协助;论文撰写中得到胡佳、石莹等人的帮助,在此一并表示感谢。

参 考 文 献

[1] 四川省长江水产资源调查组. 长江鲟鱼类生物学及人工繁殖研究. 成都:四川科学技术出版社,1988:284.

[2] Wei Q, Ke F, Zhang J, et al. Biology, fisheries, and conservation of sturgeons and paddlefish in China. Environmental Biology of Fishes, 1997, 48: 241-255.

[3] 傅朝军,刘宪亭,鲁大椿,等. 葛洲坝下中华鲟人工繁殖. 淡水渔业,1985(1):1-5.

[4] 刘鉴毅,危起伟,陈细华,等. 葛洲坝下中华鲟繁殖生物学特性及其人工繁殖效果. 应用生态学报,2007,18(6):1397-1402.

[5] 危起伟,陈细化,杨德国,等. 葛洲坝截流 24 年来中华鲟产卵群体结构的变化. 中国水产科学,2005,12(4):

452-457.
[6] 阮国良,冯家斌,杨代勤. 渗透压,pH 和温度对泥鳅精子活力及受精率的影响. 湖北农学院学报,2004,24(1):22
[7] 苏德学,严安生,田永胜,等. 钠、钾钙和葡萄糖对白斑狗鱼精子活力的影响. 动物学杂志,2004,39(1):16-20.
[8] 严安生,宋贵文,闫拥军. 鲤、团头鲂精子生理特性的研究-Ⅲ. 单糖和渗透压对精子活力的影响. 淡水渔业,1995,25(2):3-5.
[9] 刘鉴毅,甘芳,危起伟,等. 几种不同浓度的离子及单糖对中华鲟精子活力的影响. 水生生物学报,2007,31(6):95-100.
[10] Alavi S M H, Cosson J, Karami M, et al. Chemical composition and osmolality of seminal fluid of *Acipenser persicus* their physiological relationship with sperm motility. Aquaculture Research, 2004, 35(13):1238-1243.
[11] Alavi S M H, Cosson J. Sperm motility in fishes (II) eff ects of ions and osmolality: a review. Cell Biology International, 2006, 30(1):1-14.
[12] 谷魏. Na^+、K^+、葡萄糖和果糖对高白鲑精子活力影响的研究[D]. 乌鲁木齐:新疆农业大学,2005:30-32.
[13] 郑跃平. 中华鲟精子生理生态特性研究[D]. 武汉:华中农业大学,2007:26-28.
[14] Alavi S M H, Cosson J, Karami M, et al. Spermatozoa motility in the Persian sturgeon, Acipenser persicus: effects of pH, dilution rate, ions and osmolality. Reproduction, 2004, 128(6):819-828.

Effect of Na^+, K^+, Mg^{2+}, Ca^{2+} and Glucose Solutions as Insemination-Activation Media on the Fertilization of *Acipenser sinensis* Milt

Gan Fang[1,2,3]   Du Hao[2,3]   Wei Qiwei[1,2,3]
Chen Xihua[2,3]   Liu Jianyi[2,3]   Zheng Yueping[1,23]

(1. Key Laboratory of Freshwater Biodiversity Conservation and Utilization, Ministry of Agriculture of China, College of Fisheries, Huazhong Agricultural University, Wuhan 430070;
2. Yangtze River Fisheries Research Institute, Chinese Academy of Fisheries Science, Jingzhou 434000;
3. Freshwater Fisheries Research Center, Chinese Academy of Fishery Sciences, Wuxi 214081, China)

Abstract The fertilization of Chinese Sturgeon (*Acipenser sinensis*) milt was assessed by using different concentrations of NaCl, KCl, $MgCl_2$, $CaCl_2$ or glucose solution as the insemination-activation medium. The results showed that the fertilization rates of Chinese Sturgeon were all improved when suitable concentrations of NaCl, KCl, $MgCl_2$, $CaCl_2$ and glucose solutions were used, and that the fertilization rates reached the highest when the following concentrations of solutions were used: 87.72% in 25 mmol/L NaCl, 86.82% in 0.1 mmol/L KCl, 82.24% in 1 mmol/L $MgCl_2$, 89.76% in 1 mmol/L $CaCl_2$, and 80.92% in 50 mmol/L glucose. Further increase in the salt or glucose concentration caused gradual decrease in fertilization rate. The results indicate that, as insemination-activation medium of Chinese Sturgeon milt, the most suitable concentration for NaCl, KCl, $MgCl_2$, $CaCl_2$ and glucose solution were 25 mmol/L, 0.1 mmol/L, 1 mmol/L, 1 mmol/L, and 50 mmol/L, respectively.

Key words Chinese Sturgeon (*Acipenser sinensis*)   Sperm motility   Fertilization rate

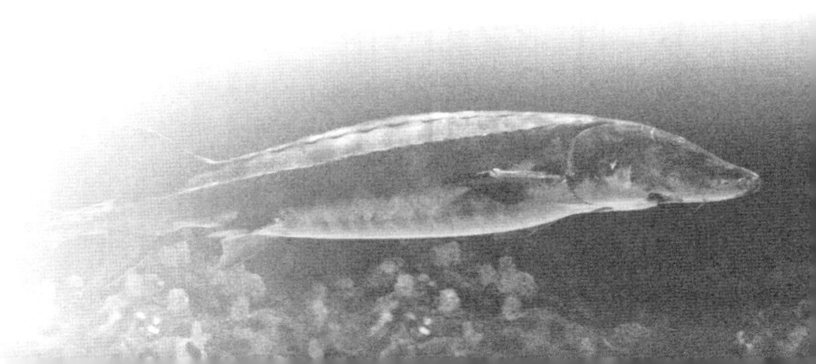

不同水深和光照强度对中华鲟受精卵孵化率的影响*

柴　毅[1,**]　谢从新[2]　危起伟[3]　李罗新[3]

（1. 长江大学动物科学学院　湖北荆州　434025　2. 华中农业大学水产学院　湖北武汉　430070
3. 中国水产科学院长江水产研究所 农业部淡水鱼类种质资源与生物技术重点开放实验室　湖北荆州　434000）

摘　要　试验研究了不同水深和光照强度对中华鲟受精卵孵化率的影响。不同水深试验结果表明，在0～40 cm水层中，随水深增加，中华鲟受精卵孵化率逐渐升高，其中在水深40 cm时孵化率达85.2％；此后孵化率又有所下降，在水深60 cm时孵化率为82.6％。可见中华鲟受精卵最佳孵化水深为40 cm，最适孵化水深范围为40 ～ 60 cm。不同光照强度试验结果表明，中华鲟受精卵在强光750 lx以上和弱光100 lx以下孵化率均较低，受精卵孵化最适光照强度范围为250 ～500 lx，其中在250 lx时可获得最佳孵化率，达83.9％。

关键词　中华鲟　受精卵　水深　光照　孵化率

中华鲟（*Acipenser sinensis*）1996年被国际自然保护联盟（IUCN）列入红皮书，划分到濒危动物等级中[1]。增殖放流是现阶段对中华鲟采取的有效保护措施之一。中华鲟苗种培育模式为集约化养殖，但在受精卵孵化阶段往往由于达不到最佳孵化条件，造成孵化率较低和初孵仔鱼活力较差。目前关于中华鲟早期生活史阶段的研究主要集中在发育、养殖等方面[2-6]，而对其孵化率的相关研究较少[7,8]。本试验研究了水深和光照强度对中华鲟孵化率的影响，确定了获得最佳孵化率的水深和光照，以期为中华鲟基础研究积累资料，并为更好地集约化养殖提供理论参考依据。

1　材料与方法

1.1　材料来源与处理

2006年11月上旬在葛洲坝下采捕野生中华鲟亲鱼，在长江水产研究所中华鲟繁育基地进行人工催产授精。挑选同一批的4 000粒受精卵，平均放入40个孵化筛（40cm × 40 cm × 20 cm）中。在圆形孵化池（直径1 m，深度0.8 m）内流水孵化。

1.2　试验方法

1.2.1　水深

用钢尺测量孵化筛网片离水面高度以设定水深，精确至0.1 cm。分别设置2 cm，10 cm，20 cm，40cm，60 cm 5个试验组。将孵化筛准确固定。每组水深放置4个孵化筛，即每个试验组有受精卵400粒，共计2 000粒。

1.2.2　光照强度

用ST-80C数字式水下照度计测定水中光照强度。通过调节光照度，使各试验组放卵区光

---

* 基金项目：科技部国家社会公益研究专项（2004DIB3J099）。

** 第一作者，女，1978年出生，山东临沂人，博士，讲师，研究方向为鱼类生理生态学，E-mail：chaiyi123456@126.com

照强度分别为 1 000 lx,750 lx,500 lx,250 lx,100 lx。每天光照时间为 12 h(06:00 ~ 18:00)。每组光照强度放置 4 个孵化筛,即每个试验组有受精卵 400 粒,共计 2 000 粒。

1.3 孵化管理

孵化筛使用前在 K_2MnO_4溶液(15 mg/L)中浸泡消毒 48 h,再用清水浸泡、冲洗干净后固定在孵化池内。淋浴喷头表层供水,水交换量 6 ~ 8 L/min,溶氧 >7 mg/L,水温 20 ~ 22 ℃。孵化用水为过滤消毒后的湖水。每隔 2 h 轻摇孵化筛避免受精卵堆积,及时去除死卵并记录死卵数量。

1.4 受精卵孵化率的计算

孵化出膜时及时吸出出膜仔鱼,统计每个孵化筛内孵出的中华鲟苗总数并计算孵化率。孵化率 = 出膜苗数/孵化卵数 ×100 %。

1.5 数据处理

使用 Excel 2000 对数据进行统计分析并绘图。

2 结果

2.1 水深对中华鲟受精卵孵化率的影响

在不同水深条件下,中华鲟受精卵的孵化率如图 1 所示。在水深 0 ~ 40 cm 范围内,随着水深的增加,孵化率逐渐升高。其中水深 2 cm 时孵化率最低,为 53.3 %。水深 40 cm 时孵化率最高,达 85.2 %。而水深 60 cm 时孵化率又有所下降,为 82.6 %。可见中华鲟受精卵孵化最适水深范围为 40 ~ 60 cm,其中在水深 40 cm 时可获得最佳孵化率。

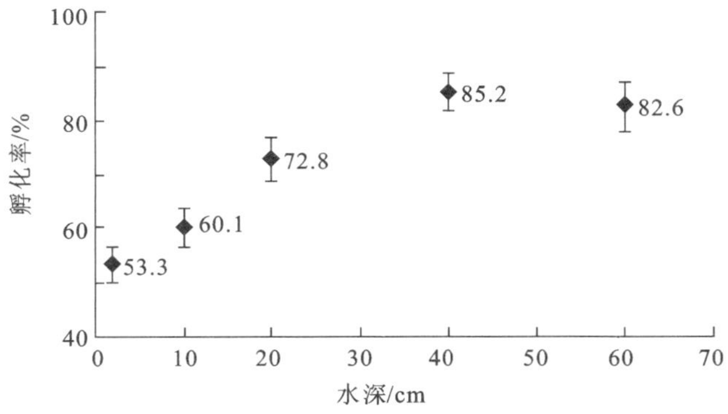

图 1 水深对中华鲟孵化率的影响(n =4)

2.2 光照对中华鲟受精卵孵化率的影响

在不同光照强度条件下,中华鲟受精卵的孵化率如图 2 所示。光照强度在 100 lx 时孵化率较低,为 67.1 %;250 lx 时孵化率最高,达 83.9 %。此后随着光照强度增加,孵化率逐渐降低,500 lx 时孵化率为 80.6 %,1 000 lx 时孵化率降至 60.2 %。可见中华鲟受精卵孵化最适光照强度范围为 250 ~ 500 lx,其中在 250 lx 时可获得最佳孵化率。

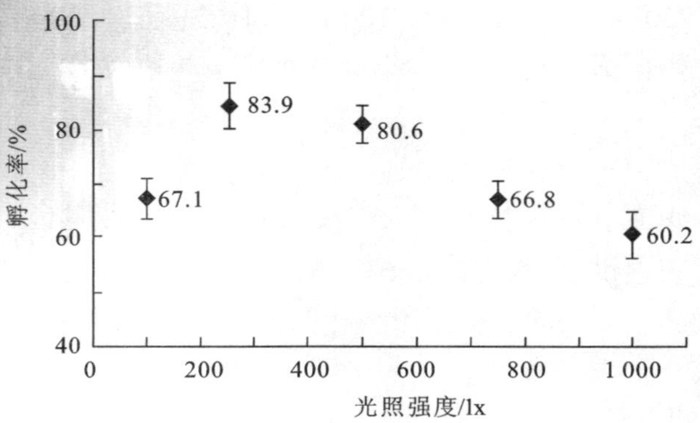

图2 光照强度对中华鲟孵化率的影响($n=4$)

3 讨论

合适的水深可以保证充足的溶氧、稳定的水温和流速,从而获得较高孵化率[9]。中华鲟产漂浮性卵,黏附于江底岩石上孵化[10],表明孵化需要一定的水深,这可能是中华鲟长期在较深水域产卵的适应结果。从本试验结果可知,中华鲟受精卵孵化最适水深范围为40 ~ 60 cm,其中在水深40 cm时孵化率可达85.2 %,为最佳孵化率。但在自然环境中,中华鲟受精卵孵化所处的水深通常远远大于40 cm,这也可能是自然条件下孵化率较低的原因之一。

光照也是影响孵化率的重要因素之一[9]。本试验结果表明,中华鲟受精卵孵化最适光照为250 ~500 lx,其中在250 lx时孵化率可达83.9 %。此后随着光照强度的增加孵化率逐渐降低,1 000 lx时孵化率已降至60.2 %。这与自然条件孵化所需较弱光照强度相似。因此在人工孵化过程中应避免阳光直射,采取较好的遮阳措施,且在室内孵化时也要避免强光长时间照射。

此外,影响中华鲟受精卵孵化率的因素还有很多,如卵质、水压、水温、氨氮、pH、溶氧、有害动物等[9]。本试验仅研究了水深和光照强度对其孵化率的影响,其他因素对孵化率的影响还有待于进一步研究。

参考文献

[1] 汪松. 中国濒危动物红皮书:鱼类[M]. 北京:科学出版社,1998:6-7.
[2] 柴毅. 中华鲟感觉器官的早期发育及其行为机能研究[D]. 武汉:华中农业大学,2006.
[3] 厉萍,危起伟,鲁大椿,等. 中华鲟精卵特征及精子入卵早期过程观察[J]. 水利渔业,2007,27(2):9-11.
[4] 曹志华,邓伦飞,李萍. 不同开口饵料对中华鲟仔鱼生长及成活率的影响[J]. 湖北农业科学,2007,46(2):278-280.
[5] 徐雪峰. 中华鲟消化系统的发育及消化酶活性变化的研究[D]. 武汉:华中农业大学,2006.
[6] 李罗新,危起伟,朱永久,等. 中华鲟苗种集约化培育技术[J]. 淡水渔业,2006,36(2):52-54.
[7] 程高潮. 环道孵化中华鲟效果观察[J]. 水利渔业,1987(5):23-24.
[8] 刘鉴毅,危起伟,杜浩,等. 中华鲟人工繁殖关键技术的改进效果研究[J]. 经济动物学报,2006(2):96-100.
[9] 殷名称. 鱼类生态学[M]. 北京:中国农业出版社,1999:136-139.
[10] 四川省长江水产资源调查组. 长江鲟鱼类生物学及人工繁殖研究[M]. 成都:四川科学技术出版社,1988:90-91.

Effects of Water Depth and Illumination on the Hatching Rate of Zygote of Chinese Sturgeon (*Acipenser sinensis*)

Chai Yi[1] Xie Congxin[2] Wei Qiwei[3] Li Luoxin[3]

(1. College of Animal Science, Yangtze University, Jingzhou 434025, Hubei, China;
2. Fishery College, HuaZhong Agricultural University, Wuhan 430070, Hubei, China;
3. Key Laboratory of Freshwater Fish Germplasm Resources and Biotechnology of Agriculture Ministry, Yangtze River Fisheries Research Institute, Chinese Academy of Fishery Sciences, Jingzhou 434000, Hubei, China)

Abstract Effects of water depth and light intensities on the hatching rate of zygote of Chinese sturgeon were researched. Results indicated that in the stage of 0-40 cm, with the increase of water depth, the hatching rate of zygote of Chinese sturgeon increased. And the hatching rate was 85.2 % when water depth was 40 cm. The hatching rate decreased to 82.6 % when water depth was 60 cm. The optimal water depth range for zygote hatching was during 40 cm to 60 cm. And the optimal hatching rate, 85.2 %, was obtained when water depth was 40 cm. The hatching rate of zygote of Chinese sturgeon was low when light intensities extra 100 lx and 750 lx. The optimal light intensities range for zygote hatching was 250-500 lx. And the optimal hatching rate, 83.9 %, was obtained when light intensities was 250 lx. Therefore in the process of artificial incubation should avoid direct sunlight, good sunshade measures, and to avoid strong light for a long time at the indoor exposure. In addition, there are many factors influencing the Chinese sturgeon egg hatching rate, such as egg quality, water pressure, water temperature, pH, dissolved oxygen, ammonia nitrogen, harmful animals, etc. This experiment studied the only water depth and light intensity on its hatching rate, the influence of other factors on the hatching rate has yet to be further research.

Key words *Acipenser sinensis* Zygote Water depth Light intensities Hatching rate

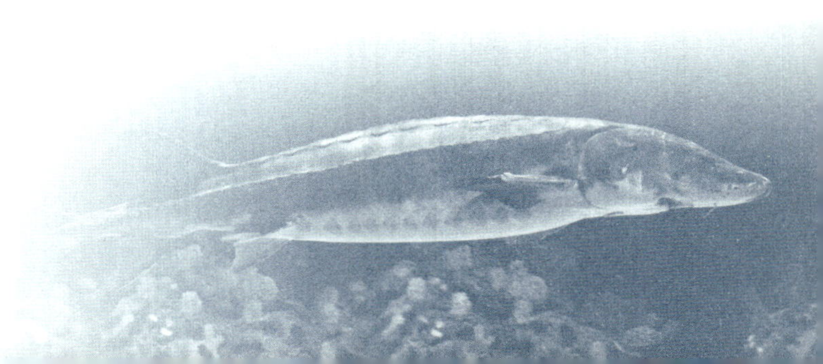

养殖中华鲟四种天然饵料的细菌分布及灭菌效果*

张艳珍[1,2,**]　张晓雁[1]　王彦鹏[1]　危起伟[2,***]

（1. 北京信沃达海洋科技有限公司（北京海洋馆）　北京　100081

2. 农业部淡水生物多样性保护重点实验室 中国水产科学研究院长江水产研究所　湖北武汉　430223）

摘　要　为预防北京海洋馆养殖中华鲟（*Acipenser sinensis*）因饵料引起的细菌性疾病暴发，对鲜活大泷六线鱼（*Hexagrammos otakii*）、冷冻斑尾刺虾虎鱼（*Acanthogobius ommaturus*）、南美白对虾（*Penaeus Vannamei*）和多春鱼（*Malltus villosus*）的细菌分布及不同方式灭菌效果进行了研究。结果显示，饵料优势菌类型为致病力较强的2类条件致病菌，即气单胞菌和不动杆菌，但菌种有差别，活饵优势菌主要为嗜水气单胞菌和两种非发酵革兰阴性菌（其中一种为鲍曼不动杆菌），冷冻饵料为鲁氏不动杆菌、嗜水气单胞菌和维罗纳气单胞菌。分别采用消毒、暂养、冷冻、冷冻后紫外灭菌方式处理鲜活、冰冻饵料，对于活饵大泷六线鱼，消毒处理效果不明显，条件适宜池体暂养3 d后，菌群结构趋于稳定，采用紫外灭菌方法可降低细菌总数但不能杀灭优势致病菌。冷冻处理冰鲜斑尾刺虾虎鱼，延长处理时间，从1 d到4 d，40 d，110 d，可使细菌种类明显变化，但优势菌稳定，且紫外灭菌后优势致病菌不能去除。而冷冻饵料南美白对虾和多春鱼紫外灭菌后细菌种类和数量均有减少。

关键词　中华鲟（*Acipenser sinensis*）　天然饵料　细菌　灭菌方法

长期以来，对饵料和饲料的消毒被我国广大渔民作为水产养殖“四消”原则中的内容之一而在水产养殖生产中普遍应用[1]。人工养殖中华鲟（*Acipenser sinensis*）既可以使用人工配合饲料，也可以用鲜活或冷冻的鱼类等水产品，但水产品饵料自身可能携带多种致病菌，加工处理过程中也容易感染其他菌。对于其携带微生物及致病性未见有研究报道。此外，对饵料投喂前的灭菌力度未做过深度分析。北京海洋馆是中华鲟重要的易地保护和保育科教基地之一，我们于2012年1月27日至9月28日对中华鲟天然饵料的安全性进行了分析，研究了活饵的检疫暂养效果、冷冻饵料的细菌分布动态以及各种饵料制作后灭菌效果，旨在为有效预防和控制中华鲟细菌性疾病的暴发、控管饵料安全提供依据。

1　材料和方法

1.1　饵料种类及灭菌方式

实验中华鲟饵料主要为鲜活和冷冻的天然饵料，均购自北京水产品市场，鲜活饵料为大泷六线鱼（*Hexagrammos otakii*），冷冻饵料包括斑尾刺虾虎鱼（*Acanthogobius ommaturus*）、南美白对虾（*Penaeus Vannamei*）和多春鱼（*Malltus villosus*）。

---

* 基金项目：公益性行业（农业）科研专项（201203086）。

** 第一作者，女，1985年出生，硕士，研究方向为濒危物种保护生理生态学，E-mail：zhyanzh1985@163.com

*** 通讯作者（Corresponding author），男，1960年出生，研究员，博导，主要研究方向为珍稀鱼类保护，E-mail：weiqw@yfi.ac.cn

灭菌方式：新引进大泷六线鱼经常规消毒后暂养 3 d，水温设定 16 ℃，盐度 30 ‰，养殖密度 4.2 kg/m^3，使用当日致死去鳃及内脏；斑尾刺虾虎鱼常温解冻，去鳃及内脏；南美白对虾去头部和尾部；多春鱼不处理。投喂前均经紫外灯(功率 20 W，紫外线波长 275 nm)灭菌 15 min 或 30 min。

1.2 采样方法

随机选取大泷六线鱼、斑尾刺虾虎鱼、南美白对虾和多春鱼各 3 ~ 4 尾，要求黏液透明、鳞片完整，无外伤。其中，大泷六线鱼体重 300 ~ 400 g，全长 30 ~ 40 cm；斑尾刺虾虎鱼体重 150 ~ 200 g，全长 15 ~ 20 cm；多春鱼体重 20 ~ 30 g，全长 10 ~ 15 cm；南美白对虾体重 10 ~ 15 g，全长8 ~ 10 cm。

大泷六线鱼在引进时、消毒后、暂养 2 d 和 3 d 后分别取鳃表黏液及养殖水样；使用当日致死后取灭菌前、灭菌 15 min 和 30 min 后的肌肉和内脏组织。斑尾刺虾虎鱼分别在冷冻 1 d，4 d，40 d 和 110 d 时取灭菌前后肌肉样本。南美白对虾和多春鱼在灭菌前后取肌肉样本。

鳃黏液采样时提起鱼体头部、吸干流下来的水后，采用拭子轻轻刮取 2 cm × 2 cm 面积的鳃丝表面黏液。肌肉、内脏样本各采集 10 g，装入无菌瓶。水样采用无菌注射器抽取 20 ml，装入无菌瓶中，常温保存送检。

样本采集在无菌条件下操作。

1.3 微生物培养与鉴定

样品经细菌培养、鉴定及分类计数，增菌时间为 7 d。黏液样品直接放入培养基中培养；组织样品经研磨与灭菌生理盐水制成悬浮液，稀释后进行培养；水样经稀释后进行培养。

培养基选用麦康凯培养基、巧克力培养基、沙包罗氏培养基、土豆培养基和察氏培养基培养，培养基采用常规方法配制或购买。采用法国梅里埃 VITEK2 COMPACT 全自动微生物分析仪进行细菌鉴定。

1.4 数据处理

实验数据采用 Excel 软件分析处理。

2 结果与分析

2.1 鲜活饵料细菌分布和灭菌效果

由表 1 可见，引进时大泷六线鱼携带多种细菌，样本间菌种类型复杂且不一致，菌落计数值高，菌种与运输鱼袋水体部分菌种一致。消毒对菌种类型影响小，个别菌种可被除去，数量比例发生明显变化，无明显优势菌。同时间水样菌与鱼体菌种一致，表明引进饵料携带大量细菌与其养殖水体相互影响，安全性差，不可直接使用。暂养 2 d，细菌总数大幅降低，细菌类型发生明显变化，出现优势菌种，包括嗜水气单胞菌、鲍曼不动杆菌和另一种非发酵革兰氏阴性菌，且携带菌种与养殖水体不一致。

表 1　活饵大泷六线鱼各样本细菌检查结果(单位 cfu/ml)

Tab. 1　The bacteria in the live *Hexagrammos otakii*(unit:cfu/ml)

样本		大泷六线鱼			水样
		1#鱼	2#鱼	3#鱼	
鳃表黏液	I	>100 000,A10 %,B20 %,C70 %	>100 000,A20 %,C70 %,D10 %	>100 000,D30 %,B40 %,E10 %,F20 %	>100 000,G40 %,D10 %,E30 %,F20 %
	II	>100 000,A30 %,B30 %,C40 %	>100 000,B40 %,D10 %,H30 %,F20 %	>100 000,D85 %,B15 %	>100 000,A40 %,I20 %,B30 %,C10 %
	III	400,J50 %,K50 %,	8 000,L40 %,J60 %,	1 200,A17 %,M41 %,N25 %,D17 %,	100,O
	IV	>100 000,L60 %,J20 %,K20 %	>100 000,L60 %,J10 %,K30 %	>100 000,L50 %,J50 %	—
肌肉	i	>100 000,K	>100 000,P95 %,K5 %	>100 000,J10 %,L90 %	100,J
	ii	>100 000,K	—	>100 000,C	
	iii	—	700,J	G	
内脏	i	300,J	—	>100 000,J60 %,L40 %	
	iii	200,J	400,K	100,J	

注:I 为引进时,II 为消毒后,III 为暂养 2 d,IV 为暂养 3 d;i 为灭菌前,ii 为灭菌 15 min,iii 为灭菌 30 min;A 为克氏库克菌,B 为莫拉菌属,C 为少动鞘氨醇杆菌,D 为藤黄微球菌,E 为奇异变形杆菌,F 为肠球菌,G 为维多纳气单胞菌,H 为棒状杆菌,I 为类志贺邻单胞菌,J 为鲍曼不动杆菌,K 为嗜水气单胞菌,L 为非发酵革兰阴性菌,M 为腐败希瓦菌,N 为霍乱弧菌,O 为杀鲑气单胞菌,P 为类鼻疽伯克霍尔德菌,—为普通培养无细菌生长。

暂养 3 d,优势菌仍为非发酵革兰氏阴性菌和嗜水气单胞菌,比较稳定,当日养殖水体培养无细菌生长。当日饵料鱼肌肉及内脏组织结果与鳃部优势菌种一致,灭菌 15 min 后,菌种类型减少,除嗜水气单胞菌外其他细菌可被杀灭;灭菌 30 min 后菌种数量与类型均有减少,嗜水气单胞菌数量也明显减少。结果显示,灭菌时间的延长,使灭菌效果提升。

2.2　冷冻饵料细菌分布和灭菌效果

对冷冻斑尾刺虾虎鱼分别在冷冻 1 d,4 d,40 d 和 110 d 时取样结果见表 2。不同冷冻时间样本,灭菌前菌落计数值均高,检出细菌种类多;灭菌后菌种类型减少,细菌总数降低。灭菌前后优势菌均为鲁氏不动杆菌和嗜水气单胞菌。冷冻 1 d,灭菌前菌种不一致,灭菌后只有鲁氏不动杆菌。冷冻 4 d,灭菌前后均为类志贺邻单胞菌和嗜水气单胞菌,但比例变化。冷冻 40 d,灭菌前粪产碱杆菌和少动鞘氨醇杆菌占多数,灭菌后细菌总数减少,粪产碱杆菌和稍动鞘氨醇杆菌不能全部杀灭。冷冻 110 d,灭菌前种群结构复杂,灭菌后鲁氏不动杆菌和嗜水气单胞菌不能被杀灭。综合比较,冷冻斑尾刺虾虎鱼灭菌前按频次依次为鲁氏不动杆菌(5 次),嗜水气单胞菌(3 次)和粪产碱杆菌(2 次)及其他菌种;灭菌后按频次依次为鲁氏不动杆菌(7 次),嗜水气单胞菌(4 次),类志贺邻单胞菌(3 次)及其他菌种。随着冷冻时间延长,灭菌前细菌种类明显变化,但优势菌比较稳定,菌落计数结果持续较高,灭菌后优势菌不变,但数量显著降低。该结果也显示,灭菌过程容易引发一些菌种的暴发,可能满足了这些细菌生长的环境条件。

表2 斑尾刺虾虎鱼冷冻期间细菌检查结果

Tab. 2 The bacteria in *Acanthogobius ommaturus* during the freezing

冷冻时间	灭菌前		灭菌后	
	总数/(cfu/ml)	结果	总数/(cfu/ml)	结果
1 d	>100 000	A,B	>100 000	C
	>100 000	D	>100 000	C
4 d	>100 000	E10 %,C20 %,F70 %	>100 000	E10 %,F90 %
	>100 000	E20 %,F40 %,A40 %	>100 000	E30 %,F10 %,C50 %,G10 %
40 d	2800	H10 %,I90 %	700	I100 %
	>100 000	I90 %,J10 %	3 000	K20 %,I60 %,J20 %
110 d	>100 000	L40 %,C60 %	2 000	E30 %,C70 %
	>100 000	M20 %,C10 %,F70 %	800	C40 %,F60 %
	>100 000	L20 %,N20 %,C60 %	1 000	L10 %,Q10 %,C80 %
	>100 000	A30 %,O20 %,P20 %,C30 %	600	C70 %,F30 %

注:A为维罗纳气单胞菌,B为睾酮丝毛单胞菌,C为鲁氏不动杆菌,D为枯草芽孢杆菌,E为类志贺邻单胞菌,F为嗜水气单胞菌,G为粘金黄杆菌,H为缺陷/泡囊短波单胞菌,I为粪产碱杆菌,J为少动鞘氨醇杆菌,K为微球菌,L为解藻朊酸弧菌,M为弗劳地枸橼酸杆菌,N为副溶血弧菌,O为嗜麦芽窄食单胞菌,P为琼氏不动杆菌,Q为鲍曼不动杆菌。

由表3可见,冷冻南美白对虾和多春鱼在灭菌前菌落总数值高,灭菌后细菌总数显著降低。灭菌前虾和多春鱼共检出5种细菌,每个样均检出维罗纳气单胞菌,鲁氏不动杆菌出现3次,产吲哚金黄杆菌和表皮葡萄球菌各检出1次;灭菌后仍检出维罗纳气单胞菌3次,表皮葡萄球菌检出2次,鲁氏不动杆菌检出1次,无其他菌种。

表3 冷冻南美白对虾和多春鱼细菌检查结果

Tab. 3 The bacteria in the frozen *Penaeus vannamei* and *Malltus villosus*

类型	灭菌前		灭菌后	
	总数/(cfu/ml)	结果	总数/(cfu/ml)	结果
虾	>100 000	A	600	A50 %,B50 %
	>100 000	A25 %,C25 %,D25 %,E25 %	1 000	B80 %,A20 %
多春鱼	>100 000	A40 %,E27 %,B33 %	800	B20 %,E80 %
	>100 000	E,A	7 000	A40 %,B60 %

注:A为维罗纳气单胞菌,B为表皮葡萄球菌,C为尿放线杆菌,D为产吲哚金黄杆菌,E为鲁氏不动杆菌。

3 讨论

3.1 饵料细菌的致病性

中华鲟为江海洄游性鱼类,是国家一级保护动物,且已有人工养殖群体[2]。本次实验中华鲟的养殖条件及密度依据报道设定[3],饵料细菌作为外源因子,对中华鲟的安全养殖至关重要。人工养殖中华鲟极易感染细菌性疾病,已有研究报道中华鲟感染细菌性疾病案例,包括细菌性败血症、肠炎病、肿嘴病、烂鳃病以及胃充气病等;感染细菌种类包括嗜水气单胞菌、点状气单胞菌以及柱状屈桡杆菌等[4,5]。其他鲟鱼感染嗜水气单胞菌报道也较多,包括俄罗斯鲟

(*Acipenser gueldstaedti*)、史氏鲟(*Acipenser schrencki*)、杂交鲟(Bester)等[6-9]。

本次检出的嗜水气单胞菌、维罗纳气单胞菌、两种非发酵革兰氏阴性(其中一种为鲍曼不动杆菌)、鲁氏不动杆菌和类志贺邻单胞菌为水产养殖中常见的致病菌。嗜水气单胞菌为水生生态系统中的一种重要病原菌,可引起多种水产养殖动物发病且常造成严重的经济损失,致病的主要原因是水质不良,引种及对所使用的生物活饵料检疫、消毒不严[10,11]。维罗纳气单胞菌已有报道是罗非鱼病害主要病原[12],也是欧洲鳗鲡(*Anguilla anguilla*)红头病的病原之一[13]。鲍曼不动杆菌以及非发酵革兰氏阴性菌均为条件致病菌,致病性较强。鲁氏不动杆菌也为具有致病性的不动杆菌之一,在胡子鲶(*Clariasfuscus lacepede*)"吊头病"等病例中曾分离出[14],也是欧洲鳗鲡红头病的病原之一[13]。

天然饵料由于来源地及运输方式等不同,携带不同菌种,在正常情况下细菌、鱼体与环境保持动态平衡并不致病。其致病性取决于水产养殖动物易感程度,直接受外界自然环境条件的影响,同时与人工养殖的整体条件有关。当鱼体免疫力或环境发生改变,导致某种致病菌增殖,造成鱼类细菌感染。饵料携带条件致病菌是否会引发细菌性疾病需引起重视,通过强化饵料灭菌以提高饲养的安全性。

3.2 饵料的灭菌效果

根据本实验结果,紫外灯照射灭菌方式可减少或杀灭部分细菌,但不能完全灭菌,延长时间杀菌率大大提高。该方法对多春鱼和南美白对虾等小型饵料灭菌效果好于个体较大的饵料,这与紫外线杀菌照射深度和照射时间有较大关系,而长时间的紫外灭菌对肉质的蛋白质的影响程度,还需要进一步的研究。水产品灭菌方法还包括低温粉碎技术和高温加热灭菌等方法。根据中华鲟饵料需求的具体情况,可进一步摸索适合其饵料消毒灭菌的方式。

对于活饵大泷六线鱼易感菌群以及如何灭菌的研究较少,有报道暴发链球菌病以及采用中药和西药对大泷六线进行治疗比较[15]。对新鱼的消毒灭菌方法有一些报道,如用四环素或金霉素类抗生素对新进龙鱼进行消毒灭菌[16],氟苯尼考已广泛用于防治水产养殖中细菌性疾病[17],也有学者采用高浓度臭氧水、二氧化氯浸泡等方法对大马哈鱼(*Oncorhynchus keta*)等进行消毒灭菌效果显著[18]。结合以往结果,通过对大泷六线鱼的菌种进行药敏试验,从而筛选出安全适用的药物,提高消毒灭菌效果。

参考文献

[1] 陈超然,孟长明,陈昌福,等.用虾蟹康给鱼类鲜活饵料和饲料消毒的方法[J].渔业致富指南,2002(7):50.

[2] 陈细华.鲟形目鱼类生物学与资源现状[M].北京:海洋出版社,2007.

[3] 张晓雁,李罗新,危起伟,等.养殖密度对中华鲟行为、免疫力和养殖环境水质的影响[J].长江流域资源与环境,2011(11):1348-1354.

[4] 朱永久.鲟鱼病害防治计数[J].现代农业科技,2009(5):234-237.

[5] 潘连德,刘建,陈景辉,等.中华鲟细菌性烂鳃病和胃充气并发症的临床诊断和控制[J].水产科技情报 2008,35(5):258-260.

[6] 杨志国,孙国秀.一例鲟鱼细菌性败血症的诊断和治疗[J].中国水产科学,2001(3):47-48.

[7] 曹海鹏,杨先乐,王玉洁,等.鲟源致病性豚鼠气单胞菌的分离及其生长特性[J].动物学杂志,2007,42(6):1-6.

[8] 韩英,李天龙,郑秀华.史氏鲟常见疾病的诊断与防治[J].中国水产,2004(3):52.

[9] 储卫华,于勇.鲟鱼嗜水气单胞菌的分离与鉴定[J].淡水渔业,2003,33(2):16-17.

[10] 王潮林,邢华. 水产养殖动物嗜水气单胞菌病的免疫技术介绍[J]. 中国水产,2006(7):88.
[11] 吕爱军,胡秀彩,李槿年,等. 水产动物致病性嗜水气单胞菌的分离鉴定[J]. 安徽农业科学,2008(2):172-174.
[12] 黎炯,叶星,卢迈新,等. 罗非鱼维氏气单胞菌的分离鉴定和药敏试验[J]. 水生态学杂志,2011,32(3):132-136.
[13] 宋学宏,李蒙英,蔡春芳,等. 欧洲鳗鲡红头病病原及组织病理研究[J]. 苏州大学学报:自然科学版,2001,17(3):85-91.
[14] 李桂峰,李海燕,毕英佐. 胡子鲇"吊头病"病原的研究[J]. 中国水产科学,2001,8(2):72-75.
[15] 冯春明,景宏丽,李莹,等. 大泷六线鱼链球菌病的病原及其防治[J]. 大连水产学院学报,2010(3):84-88.
[16] 陈思行,邱卫华. 龙鱼的饲养与疾病防治[J]. 水产科技情报,2006,33(5):229-231.
[17] 高俊岭,高启平. "氟苯尼考"在水产养殖中的应用[J]. 中国水产,2004(3):52.
[18] 李归浦,夏会龙. 二氧化氯在食品灭菌消毒上的应用研究进展[J]. 食品科技,2006:21-25.

The Bacteria Distribution and Sterilization Effect of Four Kinds of Natural Baits for Cultured *Acipenser sinensis*

Zhang Yanzhen[1,2]　Zhang Xiaoyan[1]
Wang Yanpeng[1]　Wei Qiwei[2]

(1. Beijing Xinwoda Marine Technology Co. Ltd(Beijing Aquarium), Beijing 100081, China;
2. Key Laboratory of Freshwater Biodiversity Conservation, Ministry of Agriculture of China, Yangtze River Fisheries Research Institute Chinese Academy of Fishery Sciences, Wuhan 430223, China)

Abstract To prevent the bacterial disease caused by the bait in cultured Chinese sturgeon(*Acipenser sinensis*) in Beijing aquarium, the bacterial distributions and sterilization effects of four kinds of live and frozen baits, *Hexagrammos otakii*, *Acanthogobius ommaturus*, *Penaeus Vannamei* and *Malltus villosus*, were studied. The results showed that two types of dominant bacteria were discovered with serious pathogenicity, belong to the Aeromonas and *Acinetobacter*. The species of bacteria in live baits and frozen baits were different. The dominant bacteria in live baits were *Aeromonas hydrophila* and two kinds of Non-Fermentative Gram-negative Bacteria, one was *Acinetobacter baumanii*. The dominant bacteria in frozen baits were *Acinetobacter lwoffii*, *Aeromonas hydrophila* and *Aeromonas Verona*. The treatment methods of live and frozen baits included disinfection, temporarily raised, frozen and ultraviolet sterilization. For *Hexagrammos otakii*, the sterilization effect was not obvious. After 3days raising in a suitable pool, the flora structure tended to be stable, while the ultraviolet sterilization could reduce the total number of bacteria, but could not kill the dominant bacterial pathogens. For *Acanthogobius ommaturus*, the species of bacteria changed obviously after 1 d, 4 d, 40 d and 110 d freezing. The dominant bacteria were stable in different freezing days. Also, the ultraviolet sterilization could be stable, but could not kill the dominant bacterial pathogens. For *Penaeus Vannamei* and *Malltus villosus*, the ultraviolet sterilization could reduce the numbers of bacteria and kill part of the dominant bacterial pathogens.

Key words Chinese sturgeon　Natural baits　Bacterium　Sterilization method

5月龄、7月龄中华鲟子二代光照偏好性研究*

王 恒[1,2] 危起伟[1,2,3,**] 李 伟[1,2] 刘 猛[2]
李 雷[1,2] 杜 浩[1,2,3] 李罗新[2,3]

(1. 华中农业大学水产学院 湖北武汉 430070
2. 中国水产科学研究院长江水产研究所 农业部淡水生物多样性保护重点实验室 湖北武汉 430223
3. 中国水产科学研究院淡水渔业研究中心 江苏无锡 214081)

摘 要 为获取全人工繁殖中华鲟后代的光照偏好及其可能适应人工环境的遗传行为改变,通过黑白底质颜色选择、趋光性和最适光照选择实验,并采用个体与群体实验相结合的方式对5月龄、7月龄中华鲟子二代幼鱼光照强度偏好进行了探讨。黑白底质实验结果表明,5月龄和7月龄幼鱼偏好白色底质。趋光性个体实验结果表明,5月龄对光亮区(200~250 lx)的选择较为显著($P<0.05$),7月龄对过渡区(20~50 lx)选择较为显著($P<0.05$);群体实验结果表明,5月龄和7月龄幼鱼均对过渡区(20~50 lx)的选择较为显著($P<0.05$),此现象可能与"信号-适应"假说和最适照度范围有关。最适光照实验结果表明,5月龄幼鱼对45~135 lx照度区选择显著高于其他区域($P<0.05$);7月龄幼鱼较多的选择23~202 lx照度区,表明随着年龄增大,中华鲟子二代幼鱼对光照的可接受范围变大。研究表明,中华鲟子二代5月龄和7月龄幼鱼偏好白色底质,最适照度在20~200 lx。实验结果可以为中华鲟子二代人工养殖和放流提供参考。

关键词 中华鲟子二代 幼鱼 黑白底质颜色选择 趋光性 最适光照 鱼类行为

中华鲟(*Acipenser sinensis*)为国家一级保护动物,近年来由于葛洲坝、三峡大坝等水利工程建设和环境污染等原因,其产卵场和野生种群数量受到严重威胁[1-3],2010年被IUCN列为极危(CR)等级。在1981年,研究人员提出用人工繁殖放流补充中华鲟资源量来缓解葛洲坝水利枢纽工程对中华鲟的影响[4]。1983年11月,中华鲟首次突破葛洲坝下"半人工繁殖"[5],1984年开始对中华鲟鱼苗进行人工放流[6];但人工繁殖放流的数量无法完全弥补自然繁殖的不足[7],中华鲟野生群体数量仍在衰减,长江中华鲟群体呈现出明显的高龄化现象[8]。中华鲟全人工繁殖的突破[9],为人工增殖放流恢复中华鲟野生种群带来了转机。

中华鲟行为方面的研究多集中在野生和子一代个体,庄平等[10-11]首次在洄游、趋光性、栖息水深选择等方面对子一代中华鲟仔稚鱼进行了系统的研究;顾孝连等[12-14]对长江口野生幼鱼底质类型选择、趋光性、藏匿性等做了报道,填补了中华鲟幼鱼行为学研究的空白;柴毅等[15-16]研究了中华鲟感觉器官的早期发育,主要侧重视觉的发育和早期趋光行为;何绪刚[17]得出了环境盐度的持续刺激可改变中华鲟幼鱼盐度喜好的结论。其中,大部分研究涉及光照,体现了光照选择在鱼类行为中的重要性。关于子二代幼鱼的研究仅张建明等[18]报道了慢性拥挤胁迫对子二代中华鲟幼鱼生长、摄食行为的影响,中华鲟子二代幼鱼的研究比较匮乏。

被饲养的动物经历若干世代后,一些遗传性的行为可能丢失或改变[19-26],世代间隔较长的

* 基金项目:国家自然科学基金(31172413);公益性行业(农业)科研专项(201203086)。

** 通信作者(Corresponding author),E-mail:weiqw@yfi.ac.cn

中华鲟可能在子二代就发生细微改变。目前已经发现，人工养殖中华鲟子一代在性成熟雌性个体大小方面产生了显著改变，仅为长江野生雌性最小成熟个体体质量的40%~50%，这与养殖空间不足有关[9]。因此，子二代个体存在较大可能产生适应人工养殖的环境偏好改变。本研究从光照偏好角度进行探索，依据野生中华鲟入海时间(5月龄处于纯淡水环境，7月龄进入长江口处于淡咸水环境)[27-29]可能带来的月龄差异，选择5月龄、7月龄中华鲟子二代进行研究，通过黑白底质颜色选择、趋光性、最适光照3个实验相互结合来探讨其对光照强度的偏好，为子二代幼鱼的人工养殖环境优化和后代演化提供基础资料。

1　材料与方法

1.1　实验用鱼

实验鱼为长江水产研究所荆州太湖中华鲟保育和增殖放流中心2012年11月全人工繁殖的中华鲟子二代。5月龄、7月龄实验阶段各随机抽取20尾实验。实验鱼在实验前10天放入实验装置进行适应，并测量体长和体质量，减少实验时的应激反应，减少误差。5月龄体长为(28.7±1.9) cm，叉长为(30.6±2.1) cm，体质量为(152.0±41.8) g；7月龄体长为(43.0±3.1) cm，叉长为(46.0±3.2) cm，体质量为(536.7±127.0) g。

实验鱼每天早、中、晚三次投喂鲟鱼饲料。实验场所为室内，自然光周期，水下照度为0~(100±30) lx，水源为经过曝气、过滤的地下水，实验环境和实验前人工养殖环境一致。实验期间每2天测一次水温，5月龄阶段为17.8~19.1℃，7月龄阶段为19.2~20.2℃。实时观察实验鱼适应期摄食情况。经观察，实验鱼从第2天起开始摄食，昼夜摄食行为强度相似；第5天起摄食行为达到最大，摄食行为明显，基本恢复到实验前状态，可以认为已经完全适应实验环境。实验鱼在实验开始前2 h内不喂食，避免饱腹对实验产生干扰。

1.2　实验装置

实验水槽　实验水槽为水泥池，规格为3.6 m×2.4 m×0.5 m(长×宽×高)，水深33 cm，以2块隔板分成面积相等的3个区域(2.4 m×1.2 m)，养殖区、实验区和暂养区(图1)。隔板为双层铁网片，外围包裹100目尼龙纱网并嵌入卡槽，仅水流能通过，投饵区域远离实验区并靠近出水口，及时清理残饵，防止气味扩散影响实验鱼选择。养殖区为实验鱼适应期和实验期养殖区域；实验区为主要实验区域，正上方有红外摄像机进行录像记录；暂养区为防止重复抽样的实验鱼暂养区域。进水口在暂养区右上方边缘，垂直向下水流，将其控制在中等水流时，整个水体基本处于无流速状态。水池采用24 h不间断不循环持续供水，保证溶氧的充足和水质优良。此设计使实验鱼养殖和实验期间达到同水同流、周边环境相似，相比将养殖和实验水槽分开，应激性更小，环境适应时间更短。实验地点为单独实验室，减少环境干扰。

黑白底质选择实验装置　在实验区用黑色、白色瓷砖铺设成左右面积相等的2个区域[10,12](图2)。每组实验结束后更换黑白底质位置，消除位置适应带来的误差。

趋光性实验装置　在实验区一端以2只45 W白炽灯提供光源，以遮光板对光源遮挡，使实验区形成面积相等的3个不同照度区域[11,13]，用水下照度计ZDS-10测照度(上海嘉定学联仪表厂，量程0~10 000 lx，精确度0.01 lx)。暗区为1~8 lx，过渡区为20~50 lx，亮区为180~250 lx(图3)。每组实验后调换亮区暗区位置，消除误差。

最适光照实验装置　与趋光性实验装置类似，去掉遮光板，以水平照明方式形成10~250 lx

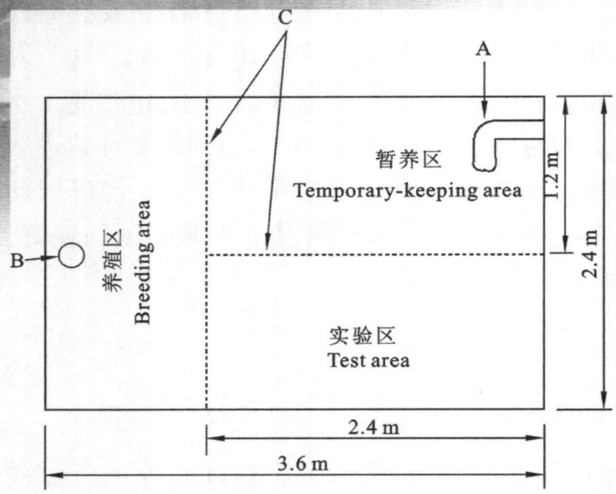

图 1　实验水槽俯视图

A:进水管(垂直水流);B:出水口;C:隔片

Fig. 1　Top view of test tank.

A:Inlet(vertical flow);B:Outlet;C:Septa.

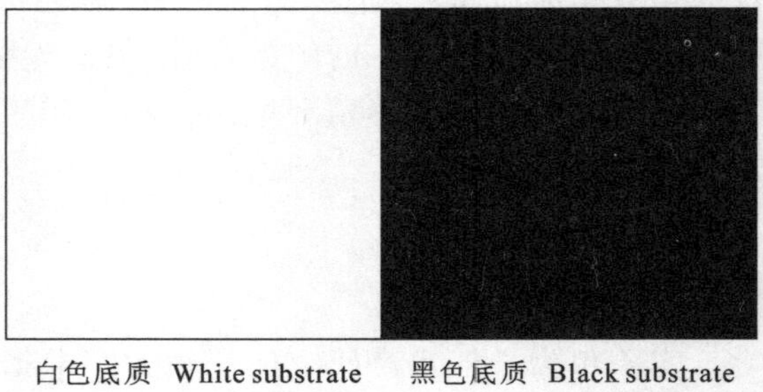

图 2　黑白底质选择实验示意图

Fig. 2　Sketch map for substrate color preference test

的均匀光照度梯度。通过在水槽正上方平行悬挂细棉线,使其在录像画面中形成成 6 个面积相等的照度区,以棉线正下方对应界线的平均照度为区域间分界,从亮到暗为 202 ~ 250 lx, 135 ~ 202 lx, 74 ~ 135 lx, 45 ~ 74 lx, 23 ~ 45 lx, 10 ~ 23 lx,记为 A ~ F 区(图 4),每组实验后调换光源位置。分区仅为统计实验鱼在整个梯度中分布的范围,分区的多少不影响实验鱼的选择,只影响统计结果的可操作性和精确性。

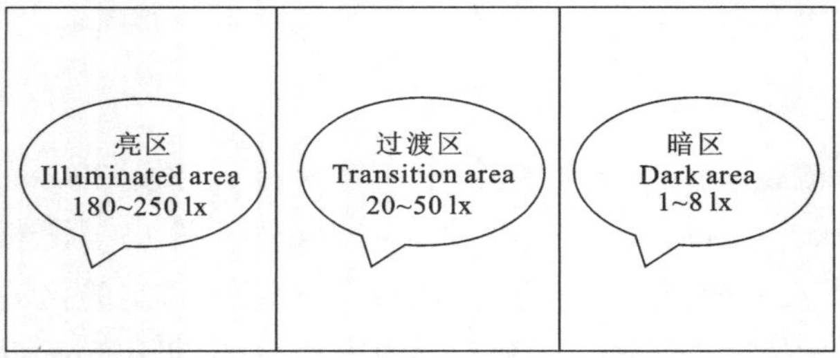

图 3　趋光性实验示意图

Fig. 3　Sketch map for illumination intensity preference test

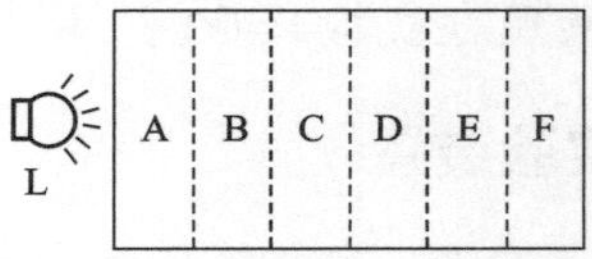

图 4　最适光照实验示意图

L:光源;A:202 ~ 250 lx;B:135 ~ 202 lx;C:74 ~ 135 lx;D:45 ~ 74 lx;E:23 ~ 45 lx;F:10 ~ 23 lx

Fig. 4　Sketch map for the optimum illumination test. L:light source;A:202 ~ 250 lx;B:135 ~ 202 lx;C:74 ~ 135 lx;D:45 ~ 74 lx;E:23 ~ 45 lx;F:10 ~ 23 lx.

1.3　实验方法

黑白底质选择实验　实验分别对 5 月龄、7 月龄鱼分组进行个体实验和群体实验,每个月龄组的实验鱼随机分为 2 组。5 月龄时,随机选择 10 尾进行个体实验,并作为第一组实验鱼,每次从实验水槽中央(黑白底质交界处)用捞网放入 1 尾鱼,适应 5 min 后,利用红外摄像机录像统计随后 5 min 内实验鱼在黑白区域的停留时间,实验结束后把鱼放入暂养区。在不对调黑白底质的情况下,剩余 10 尾进行群体实验,并作为第二组实验鱼。将 10 尾实验鱼同时放入实验水槽,适应 1 h 后,每隔 0.5 h 记录一次黑白区域内的幼鱼数量。第二组鱼的群体实验结束后,对调黑白位置,开始进行第一组实验鱼的群体实验,之后不对调黑白位置进行第二组实验鱼的个体实验。个体实验均在白天进行,实验前后进行照度测定。群体实验增加了夜晚无光状态下(照度 <0.1 lx)的分布作为对比,其中一组从白天到夜晚连续观察,另一组从夜晚到白天连续观察,由于中华鲟子二代幼鱼昼夜摄食强度相同,与野生中华鲟幼鱼一致[12],可排除昼夜节律对实验的干扰。实验期间,实验人员放完鱼后远离实验场所,避免人为干扰。7 月龄组实验方法与 5 月龄组实验方法相同。

这样的设计方法,使单组鱼的个体实验和群体实验的黑白位置条件相反,2 组间个体和群体实验间的黑白位置也相反,且单组个体与群体实验相隔 1 天,符合行为学实验要求,避免了实验鱼产生位置适应影响结果[30],也解决了实验样本获得困难的问题。

趋光性实验 实验与黑白底质实验类似,鱼从过渡区中央放入水槽[13]。由于 7 月龄中华鲟子二代幼鱼较大,7 月龄群体实验每组从 10 尾中抽取 6 尾进行实验。

最适光照实验 和趋光性实验类似,只去掉了遮光板,照度均匀分布。实验只进行个体实验,每组随机选择 6 尾分别从 6 个照度区放入实验水槽,适应 5 min 后,利用红外摄像机记录 5 min内实验鱼在每个光照度区的停留时间。一组实验结束后对调光源位置进行下一组实验。

1.4 数据统计

所有实验以实验鱼胸鳍及胸鳍以上部分全部进入某区域为标准,表示幼鱼处于该区域,开始进行计时或计数。

黑白底质选择实验 个体实验时,通过录像回放统计实验幼鱼在白色(或黑色)底质中活动的时间百分比,将百分比进行反正弦变换(实际进行正态性检验和方差齐性检验,符合正态的不做反正弦变换),求其 95 %置信区间(0.95 CI),如果此区间不包含 50 %,则对此选择具有显著性偏好,反之偏好不显著;群体实验时,同理通过数量百分比的 0.95 CI 判断偏好性[12]。对不同月龄间差异显著性使用非参数方法 Mann-Whitney Test 检验和 Duncan 法多重比较结合,显著水平 $P=0.05$。所有实验数据使用 SPSS 19.0 进行统计。

趋光性实验 个体实验通过录像统计幼鱼在 3 个区域内的活动时间百分比,群体实验统计数量百分比。用非参数方法统计 Kruskal-Wallis Test 检验各区域内活动时间和数量百分比差异显著性[13],利用单因素方差分析(one-way ANOVA)中 Duncan 法多重比较检验两两区域间的差异显著性。

最适光照实验 通过录像统计单尾鱼在某个照度区分布时间占总实验时间(5 min)的百分比,作为在该区的停留频率,利用 Duncan 法多重比较检验不同照度区分布时间的差异显著性,得出 6 个区中实验鱼最偏好的区域,此区域的照度范围即为实验鱼的最适照度区间。分区越多,最适照度范围越精确,但考虑到以实验鱼胸鳍及胸鳍以上部分全部进入区域为统计标准,分区的大小必须大于实验鱼胸鳍至头部的长度(大约 1/3 体长)。综合考虑,本文设立 6 个区。

2 结果与分析

2.1 黑白底质的选择结果

个体实验中,实验照度为 76 ~ 113 lx(5 月龄)和 69 ~ 103 lx(7 月龄),5 月龄中华鲟子二代幼鱼在白色和黑色底质的时间百分比分别为 79.2 % ± 24.5 %和 20.8 % ± 24.5 %,7 月龄为 63.5 % ± 22.1 %和 36.5 % ± 22.1 %;5 月龄幼鱼在白色底质中时间百分比平均值的 0.95 CI 为

72.9%~95.1%,7 月龄为 53.7%~76.4%,均不含 50%(图 5),表明 5 月龄和 7 月龄中华鲟子二代幼鱼个体均对白色底质有显著性偏好。

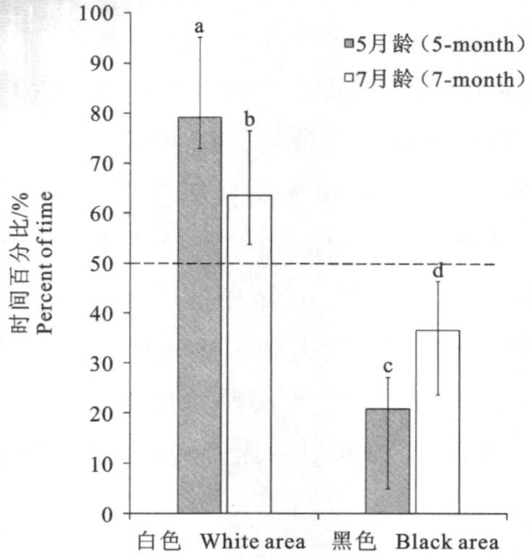

图 5 中华鲟子二代幼鱼对底质颜色的选择(个体实验)

柱状图表示时间(数量)百分比平均数,误差线表示 95% CI;

柱状图上字母相同表示差异不显著,不同表示差异显著(下同)

Fig. 5 Preference for substrate color of juvenile F_2 *A. sinensis*(individual test)

Bars represent mean percent of time(number), error bar means 95% CI. Bars with the same letter are not significant difference, whereas are significant difference, the same below.

群体实验进行白天有光和夜晚无光对比,白天照度为 80~120 lx(5 月龄)和 71~112 lx(7 月龄),5 月龄中华鲟子二代幼鱼在白色和黑色底质的数量百分比分别为 67.5%±20.0%和 32.5±20.0%,7 月龄为 60.0%±18.6%和 40.0%±18.6%;5 月龄幼鱼在白色底质中数量百分比平均值的 0.95 CI 为 58.2%~76.8%,7 月龄为 51.3%~68.7%,均不包含 50%(图 6),表明 5 月龄和 7 月龄子二代幼鱼群体对白色底质同样具有显著偏好性,与个体一致;夜晚对比群体实验中,无光(照度<0.1 lx)状态下,5 月龄中华鲟子二代在白色和黑色底质中的数量百分比分别为 56.0%±10.7%和 44.0%±10.7%,7 月龄为 54.0%±17.6%和 46.0%±17.6%;5 月龄幼鱼在白色底质中数量百分比平均数的 0.95 CI 为 48.3%~63.7%,7 月龄为 45.8%~62.2%,均包含 50%(图 7),表明 5 月龄和 7 月龄子二代幼鱼群体在极低照度下对黑白色底质没有显著偏好性。

对 5 月龄和 7 月龄幼鱼个体在白色底质中的时间百分比进行 Mann-Whitney Test 非参数检验,有显著性差异($P=0.016$),说明 7 月龄对白色底质的偏好低于 5 月龄。对 5 月龄和 7 月龄子二代群体在白天处于白色底质的数量百分比进行 Mann-Whitney Test 非参数检验,无显著性差异($P=0.167$),但从数据上看,7 月龄对白色底质的选择百分比仍小于 5 月龄。综上所述,5 月龄和 7 月龄中华鲟子二代皆偏好白色底质,但 7 月龄的偏好性已经开始减弱。在夜晚进行的群体无光对比实验中,选择 Duncan 法进行多重比较,5 月龄与 7 月龄间对黑白底质的选择均不存在显著偏好($P>0.05$)。

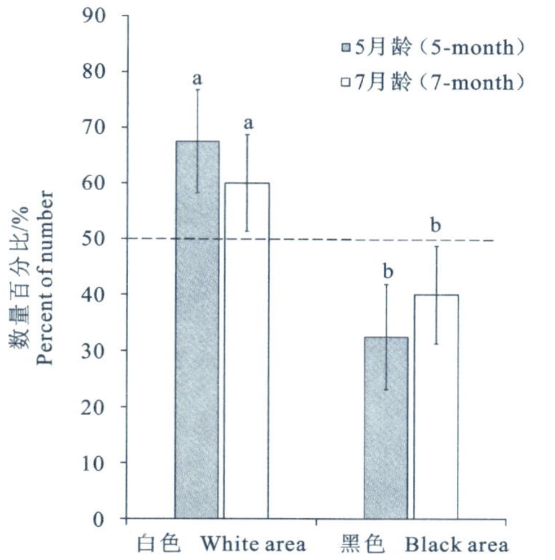

图6　中华鲟子二代幼鱼对底质颜色的选择
（光亮条件下群体实验）

Fig. 6　Preference for substrate color of juvenile F_2
A. sinensis(population test under illuminated situation)

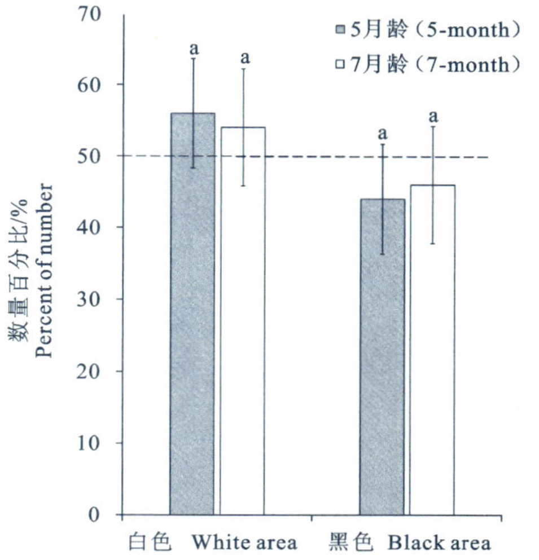

图7　中华鲟子二代幼鱼对底质颜色的选择
（黑暗条件下群体实验）

Fig. 7　Preference for substrate color of juvenile F_2
A. sinensis(population test under dark situation)

2.2　趋光性的选择结果

个体实验中,5 月龄中华鲟子二代幼鱼在 3 个区域(亮区、过渡区、暗区)内的时间百分比依次为 51.7 %±31.7 %,36.0 %±23.3 %和 12.3 %±13.4 %,呈显著性差异(P<0.05),幼鱼偏好亮区;7 月龄依次为 27.6 %±18.9 %,44.2 %±16.3 %和 28.1 %±23.6 %,幼鱼个体在过渡区的时间百分比多于亮区和暗区,且差异显著(P<0.05),亮区和暗区间无显著性差异(P>0.05)(图 8)。群体实验中,5 月龄中华鲟子二代幼鱼在 3 个区域(亮区、过渡区、暗区)内的时间百分比依次为 25.0 %±11.5 %,45.0 %±12.4 %和 30.0 %±12.1 %,7 月龄依次为 24.2 %±19.8 %,44.1 %±20.4 %和 31.7 %±17.0 %,2 个月龄幼鱼在过渡区的时间均多于亮区和暗区,且差异显著(P<0.05),亮区和暗区间无显著性差异(P>0.05)(图 9)。

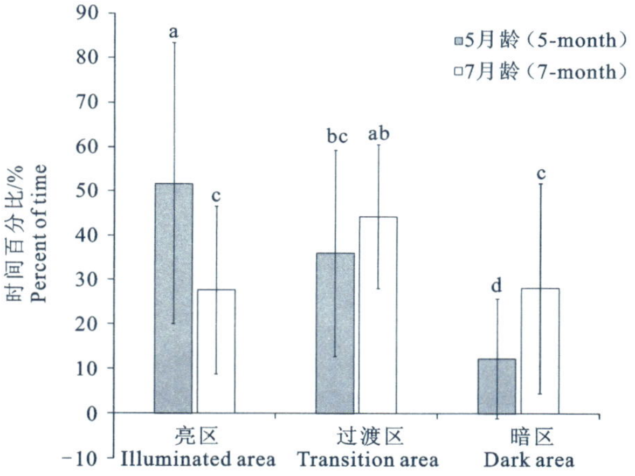

图8　中华鲟子二代幼鱼趋光性(个体实验)

Fig. 8　Preference for illumination intensity of
juvenile F_2 *A. sinensis*(individual test)

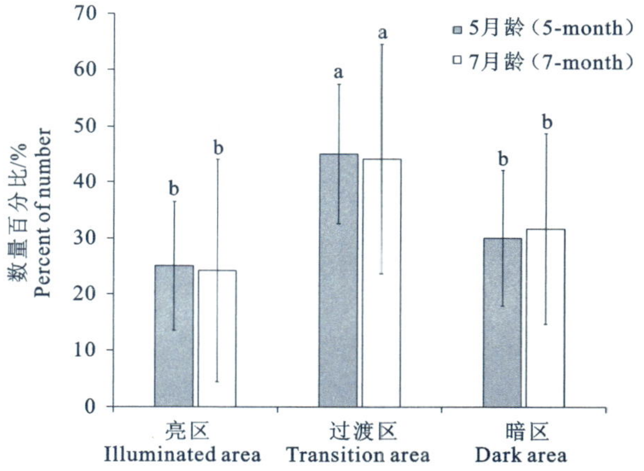

图9　中华鲟子二代幼鱼趋光性(群体实验)

Fig. 9　Preference for illumination intensity of
juvenile F_2 *A. sinensis*(population test)

2.3 最适光照的选择结果

5 月龄中华鲟子二代幼鱼在 6 个光梯度区的时间百分比分别为 5.8 % ± 6.4 %(A 区),17.1 % ± 10.2 %(B 区),30.4 % ± 11.7 %(C 区),25.2 % ± 8.6 %(D 区),13.6 % ± 11.2 %(E 区),7.9 % ± 9.6 %(F 区)(图 10)。结果表明中华鲟子二代幼鱼在 C 区的时间最多,差异显著($P < 0.05$),D 区和 C 区的时间百分比相近,无显著性差异($P > 0.05$),且多于其他区。7 月龄中华鲟子二代在 6 个区的时间百分比为 11.1 % ± 9.7 %(A 区);19.1 % ± 11.9 %(B 区);23.2 % ± 9.3 %(C 区);21.7 % ± 10.3 %(D 区);18.5 % ± 12.2 %(E 区);6.4 % ± 6.6 %(F 区)(图 10)。与 5 月龄相同,7 月龄子二代在 C 区的时间最多,差异显著($P < 0.05$),但其在B ~ E 区的时间百分比无显著性差异($P > 0.05$)。相比 5 月龄,7 月龄中华鲟子二代适合照度范围更广。

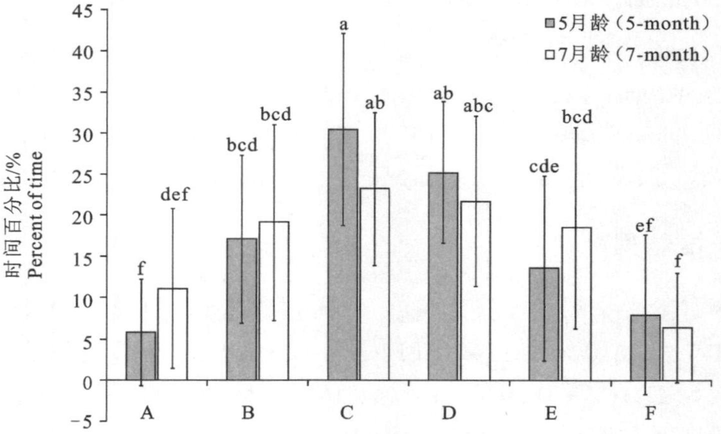

图 10　中华鲟子二代幼鱼最适光照

A 区:202 ~ 250 lx;B 区:135 ~ 202 lx;C 区:74 ~ 135 lx;D 区:45 ~ 74 lx;E 区:23 ~ 45 lx;F 区:10 ~ 23 lx

Fig. 10 Preference for the optimum illumination of juvenile F_2 *A. sinensis*. A area: 202 ~ 250 lx; B area: 135 ~ 202 lx; C area: 74 ~ 135 lx; D area: 45 ~ 74 lx; E area: 23 ~ 45 lx; F area: 10 ~ 23 lx。

3 讨论

3.1 中华鲟子二代幼鱼个体、群体间和不同月龄间的光照偏好差异

除了 5 月龄趋光性实验,其他实验个体、群体间的结果保持一致。本实验结果表明 5 月龄中华鲟子二代个体偏好亮区,而群体偏好过渡区,群体可能干扰了幼鱼对偏好光照区域的选择。联系最适照度实验的结果,5 月龄中华鲟子二代偏好 45 ~ 135 lx 照度区域,处于过渡区和亮区交界处。根据“信号-适应”假说,光照在开始阶段对鱼具有食物和集群的生物学信号意义,一段时间后因为视觉对光的适应更趋向强光区,最后在强光刺激性进行绕光源旋转的圆周运动[30-31]。群体实验的幼鱼暴露在光源下的时间长于个体实验,更容易产生实验鱼在光亮区的圆周运动。亮区进行旋转运动的幼鱼阻挡了其他幼鱼进入亮区,迫使其他幼鱼更多停留在过渡区中的适宜照度区域。野生中华鲟幼鱼群体相比个体,趋光程度也有所下降[13],但未出现如本实验群体转变为趋向过渡区的情况,可能是其趋光程度较强,最适照度区不覆盖过渡区。个体和群体选择不同的现象在墨西哥湾鲟(*Acipenser oxyrinchus desotoi*)中出现过,单尾墨

西哥湾鲟选择卵石,而群体可能由于种内竞争回避卵石或者无选择[32]。

不同月龄实验中,7 月龄对白色底质的选择比 5 月龄有所下降,光照选择范围更宽泛。5 月龄野生中华鲟处于长江中下游,为纯淡水环境,7 月龄野生中华鲟已经来到长江口,处于咸淡水环境,9 月龄左右开始入海进入咸水环境[27-29]。5 月龄和 7 月龄野生中华鲟的栖息地环境改变巨大,趋性行为具有遗传性[33-34],理论上会使 5 月龄和 7 月龄的中华鲟子二代幼鱼产生环境偏好的差异。光照偏好的可接受范围变广,说明从 5 月龄到 7 月龄,光照对中华鲟子二代的限制越来越小,幼鱼可能正从依赖视觉捕食向依赖其他器官捕食转变。

3.2　中华鲟子二代幼鱼与野生中华鲟幼鱼的光照偏好差异

本实验相关野生数据仅有长江口中华鲟幼鱼(7 月龄左右)的行为学研究[12-14]。在黑白底质选择的结果上,本实验与野生数据大致相同,只有在低照度下(照度 <0.1 lx),中华鲟子二代幼鱼对黑白底质无显著选择性,而长江口中华鲟幼鱼在高照度(124.0 ~ 146.0 lx)和低照度(1.2 ~ 1.4 lx)下都显著地选择白色底质[12],原因可能是低照度的选择不同。本实验设计的低照度极低,不足顾孝连[12]设计的 10 %,在这种照度条件下可能无法分辨黑白。低照度下总体黑白选择虽然无显著差异,但白色底质的选择仍稍多于黑色,这是因为其中一组是从有光到无光进行的连续实验,在有光实验阶段对白色区域产生了一定的位置适应[30],间接证明了中华鲟子二代幼鱼对白色底质的选择。

趋光性实验中,长江口 7 月龄中华鲟幼鱼显著趋光[13],本研究结果表明 7 月龄子二代幼鱼更趋向于过渡区,趋光性较弱,但 5 月龄是趋光的。产生这种差异的原因可能是 7 月龄中华鲟不依靠视觉捕食,对光照的偏好更容易受环境影响,子二代中华鲟对人工养殖光照环境产生了适应性改变。根据梁旭方[35]和庄平等[36]的研究,长江口中华鲟幼鱼、1 龄中华鲟及成年中华鲟主要依靠电感觉器官、嗅觉和触觉进行觅食过程,视觉作用不是很大。顾孝连等[13]虽然得出中华鲟幼鱼显著趋光,但同时发现光照对摄食强度无影响,中华鲟幼鱼的捕食效率与光照强度无关,幼鱼不依靠视觉捕食,光线更多作为幼鱼寻找食物丰富的栖息地的信号,信号意义大于利用视觉捕食的意义。本实验鱼所处的养殖环境照度为 10 ~ 100 lx,此照度成为食物丰富信号,决定了本实验鱼对过渡区的偏好。从月龄间差异可以推断,5 月龄到 7 月龄可能是中华鲟视觉作用发生转变的重要时期。

中华鲟子二代来自 10 年以上人工养殖状态下的子一代,在人为条件影响下,子一代中华鲟很可能产生适应人工养殖环境的变异,并遗传后代,使子二代某些野外生存所必需的遗传行为发生改变,如洄游、避敌、捕食等。Darwin[37]在物种起源第一章就提到了家养动物的大量变异情况。这种现象在水生动物也广泛存在,有正面的变异,如抗病性增强,环境适应力增强[22-23],也有生长性能下降、遗传变异水平降低的负面变异[19,21]。本实验的结果与野生中华鲟幼鱼存在一定的偏差,已经产生了微小的变异,可能是暂时的适应,未对遗传行为产生影响;也可能已经产生遗传变异。这还需要进一步研究证明。

3.3　中华鲟子二代幼鱼光照偏好改变的意义

中华鲟子二代幼鱼光照偏好相比野生幼鱼,在 7 月龄已经有所改变,结合中华鲟性成熟个体的小型化趋势[9],人工养殖已经对中华鲟的生物学特征产生了影响。这种改变在子二代放流后可能会对野生种群的遗传多样性产生冲击,因此建议谨慎放流中华鲟子二代。月龄间的

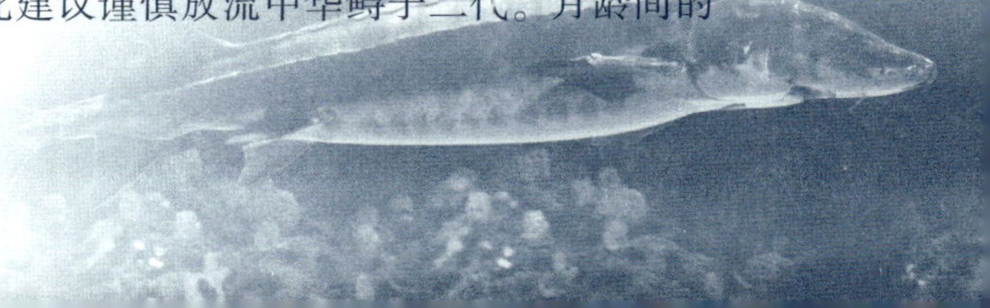

差异可以得出结论，随着年龄的增长，中华鲟子二代幼鱼对光照的可接受范围变大。7月龄最适照度范围比较宽泛，建议从7月龄以后开始进行放流或转移至室外养殖，以获得更好效果。

参考文献

[1] Wei Qiwei, Ke F E, Zhang J M, et al. Biology, fisheries and conservation of sturgeons and paddlefish in China [J]. Developments in Environmental Biology of Fishes, 1997, 48(1/4): 241-255.

[2] 危起伟，陈细华，杨德国，等. 葛洲坝截流24年来中华鲟产卵群体结构的变化[J]. . 中国水产科学，2005，12(4)：452-457.

[3] Wang C Y, Kynard B, Wei Qiwei, et al. Spatial distribution and habitat suitability indices for non-spawning and spawning adult Chinese sturgeons below Gezhouba Dam, Yangtze River: effects of river alterations [J]. Journal of Applied Ichthyology, 2013, 29(1): 31-40.

[4] 中国科学院水生生物研究所. 关于长江葛洲坝水利枢纽救鱼对象和措施的意见[J]. . 中国水利，1981(3)：25-29.

[5] 傅朝君，刘宪亭，鲁大椿，等. 中华鲟人工蓄养和催情试验[J]. 淡水渔业，1983(4)：38-40.

[6] 傅朝君，刘宪亭，鲁大椿，等. 葛洲坝下中华鲟人工繁殖[J]. 淡水渔业，1985(1)：1-5.

[7] 常剑波. 长江中华鲟繁殖群体结构特征和数量变动趋势研究[D]. 武汉：中国科学院水生生物研究所，1999.

[8] 易继舫，唐大明，刘灯红，等. 长江中华鲟繁殖群体资源现状的初步研究[J]. 水生生物学报，1999，23(6)：554-559.

[9] 危起伟，李罗新，杜浩，等. 中华鲟全人工繁殖技术研究[J]. 中国水产科学，2013，20(1)：1-11.

[10] 庄平. 鲟科鱼类个体发育行为学及其在进化与实践上的意义[D]. 武汉：中国科学院水生生物研究所，1999.

[11] Zhuang P, Kynard B, Zhang L Z, et al. Ontogenetic behavior and migration of Chinese sturgeon, *Acipenser sinensis* [J]. Environmental Biology of Fishes, 2002, 65(1): 83-97.

[12] 顾孝连. 长江口中华鲟(*Acipenser sinensis* Gray)幼鱼实验行为生态学研究[D]. 上海：上海海洋大学水产与生命学院，2007.

[13] 顾孝连，庄平，章龙珍，等. 长江口中华鲟幼鱼趋光行为及其对摄食的影响[J]. 水产学报，2009，33(5)：778-783.

[14] 顾孝连，庄平，章龙珍，等. 长江口中华鲟幼鱼对底质的选择[J]. 生态学杂志，2008，27(2)：213-217.

[15] 柴毅. 中华鲟感觉器官的早期发育及其行为机能研究[D]. 武汉：华中农业大学水产学院，2006.

[16] 柴毅，谢从新，危起伟，等. 中华鲟视网膜早期发育及趋光行为观察[J]. 水生生物学报，2007，31(6)：920-922.

[17] 何绪刚. 中华鲟海水适应过程中生理变化及盐度选择行为研究[D]. 武汉：华中农业大学水产学院，2008.

[18] 张建明，郭柏福，高勇. 中华鲟幼鱼对慢性拥挤胁迫的生长、摄食及行为反应[J]. 中国水产科学，2013，20(3)：592-598.

[19] 杨学明，李思发. 长江鲢、草鱼原种-人繁群体生长差异与生化遗传变化[J]. 中国水产科学，1996，3(4)：1-10.

[20] 张志伟，韩曜平，仲霞铭，等. 草鱼野生群体和人工繁殖群体遗传结构的比较研究[J]. 中国水产科学，2007，14(5)：720-725.

[21] 王伟继，孔杰，庄志猛，等. 真鲷野生群体和人工繁殖群体的同工酶遗传差异[J]. 生物多样性，2000，8(4)：391-396.

[22] 纪荣兴，邹文政，鄢庆枇，等. 日本对虾抗白斑病子三代的抗白斑综合征病毒感染能力及免疫特性[J]. 水产学报，2008，32(1)：98-104.

[23] 刘鉴毅，谭永安，谭启森，等. 中国大鲵子二代适应能力及生长优势的研究[J]. 四川动物，2006，25(2)：387-390.

[24] 管云雁，刘文广，何毛贤. 马氏珠母贝选育群体4个世代的遗传变异[J]. 中国水产科学，2013，20(4)：764-770.

[25] Victor B C. Settlement strategies and biogeography of reef fishes[A]// Sale P F. The Ecology of Fishes on Coral Reefs[C]. New York: Academic Press, 1991: 231-260.

[26] Barlow G W. The relevance of behavior and natural history to evolutionarily significant units[A]// Nielsen J L. Evolution and the Aquatic EcosysTEM [C]. Bethesda: American Fisheries Society, 1995: 169-175.

[27] 四川长江水产资源调查组. 长江鲟鱼类生物学及人工繁殖研究[M]. 成都: 四川科学技术出版社, 1988: 1-173.

[28] 王成友. 长江中华鲟生殖洄游和栖息地选择[D]. 武汉: 华中农业大学水产学院, 2012.

[29] 李罗新, 张辉, 危起伟, 等. 长江常熟溆浦段中华鲟幼鱼出现时间与数量变动[J]. 中国水产科学, 2011, 18(3): 611-618.

[30] 何大仁, 蔡厚才. 鱼类行为学[M]. 厦门: 厦门大学出版社, 1998: 1-387.

[31] 殷名称. 鱼类生态学[M]. 北京: 中国农业出版社, 2007: 158-161.

[32] Chan M D, Dibble E D, Kilgore K J. A laboratory examination of water velocity and substrate preference by age-0 Gulf sturgeons [J]. Transactions of the American Fisheries Society, 1997, 126(2): 330-333.

[33] 范志勤. 动物行为[M]. 北京: 科学出版社, 1988: 1-304.

[34] 尚玉昌. 动物行为学[M]. 北京: 北京大学出版社, 2004: 1-384.

[35] 梁旭方. 中华鲟吻部腹面罗伦氏囊结构与功能的研究[J]. 海洋与湖沼, 1996, 27(1): 1-5.

[36] 庄平, 章龙珍, 罗刚, 等. 长江口中华鲟幼鱼感觉器官在摄食行为中的作用[J]. 水生生物学报, 2008, 32(4): 475-481.

[37] Darwin C. On the Origin of Species [M]. London: John Murray, 1866: 7-45.

Light Intensity Preferences of 5-month and 7-month F_2 Chinese Sturgeon (*Acipenser sinensis*)

Wang Heng[1,2] Wei Qiwei[1,2,3] Li Wei[1,2] Liu Meng[2] Li Lei[1,2] Du Hao[1,2,3] Li Luoxin[2,3]

(1. College of Fisheries, Huazhong Agricultural University, Wuhan 430070, China;
2. Key Laboratory of Freshwater Biodiversity Conservation, Ministry of Agriculture of China; Yangtze River Fisheries Research Institute, Chinese Academy of Fishery Sciences, Wuhan 430223, China;
3. Freshwater Fisheries Research Center, Chinese Academy of Fishery Sciences, Wuxi 214081, China)

Abstract After the success of complete controlled propagation in Chinese sturgeon (*Acipenser sinensis*), it is necessary to find out the possible behavior variation between wild group and second generation. We combined individual test and population test in 3 different experiments: test of substrate color preference, illumination intensity preference and the optimum illumination. The results showed that both 5-month and 7-month juvenile F_2 *A. sinensis* significantly chose white substrate (95 % Confidence interval of preference on white substrate does not contain 50 %). 5-Month juvenile F_2 *A. sinensis* individuals significantly chose illuminated area (200 ~ 250 lx) ($P < 0.05$), while 7-month ones significantly chose transition area (20 ~ 50 lx) ($P < 0.05$). Both 5-month and 7-month juvenile F_2 *A. sinensis* populations preferred transition area (20 ~ 50 lx) ($P < 0.05$) in population test. The different choice of illumination intensity preference between individuals and population in 5-month F_2 *A. sinensis* may be related to 'signal -adaptation hypothesis' and the optimum illumination range. 5-month juvenile F_2 *A. sinensis* significantly chose illumination area of 45 ~ 135 lx ($P < 0.05$); 7-month juvenile F_2 *A. sinensis* significantly chose illumination area of 23 ~ 202 lx ($P < 0.05$). It meant that with the increasing age, F_2 Chinese sturgeon strengthened its adaptability to illumination. From all above, both 5-month and 7-month juvenile F_2 *A. sinensis* preferred white substrate, the optimum illumination range was 20 ~ 200 lx. Compared with the exiting wild juveniles data, their phototaxis has already declined.

Key words F_2 *Acipenser sinensis* Juvenile Preference on black and white substrate Illumination intensity The optimum illumination fish behaviors

苯酚，Cu^{2+}，亚硝酸盐和总氨氮对中华鲟稚鱼的急性毒性*

杜 浩[1,2,3,**] 危起伟[1,2,3,***] 刘鉴毅[1,2]
杨德国[1,2] 陈细华[1,2] 朱永久[1,2] 唐国盘[3]

（1. 中国水产科学研究院长江水产研究所 农业部淡水鱼类种质资源与生物技术重点开放实验室 湖北荆州 434000
2. 中国水产科学研究院淡水渔业研究中心 江苏无锡 2140811 3. 华中农业大学水产学院 湖北武汉 430070）

摘 要 研究了苯酚，Cu^{2+}，亚硝酸盐和总氨氮对 8～12 日龄中华鲟（*Acipenser sinensis*）稚鱼的急性毒性，中华鲟稚鱼的全长为（2.36±0.10）cm，平均体重为 0.07 g。结果表明：苯酚对中华鲟稚鱼 24 h，48 h，72 h 和96 h的半致死浓度（LC_{50}）及 95 %可信区间分别为 97（87～109）mg/L，83（73～95）mg/L，75（69～83）mg/L 和 71（64～80）mg/L；Cu^{2+} 对中华鲟稚鱼的 24 h，48 h，72 h 和 96 h 的 LC_{50} 及 95 %可信区间分别为 0.31（0.29 ～0.34）mg/L、0.20（0.15～0.27）mg/L，0.09（0.08～0.12）mg/L 和 0.07（0.06～0.10）mg/L；亚硝酸氮对中华鲟稚鱼的 24 h，48 h，72 h 和 96 h 的 LC_{50} 及 95 %可信区间分别为 7.05（6.22～8.03）mg/L，4.07（1.51～7.88）mg/L，3.10（2.07～4.34）mg/L 和 2.57（2.31～2.85）mg/L；非离子氨对中华鲟稚鱼的 24 h，48 h，72 h 和 96 h 的 LC_{50} 及 95 %可信区间分别为 0.67（0.64～1.07）mg/L，0.52（0.44～0.67）mg/L，0.43（0.40～0.54）mg/L 和 0.39（0.30～0.45）mg/L。126 mg/L 苯酚浓度组中，中华鲟稚鱼的心率（119±10 次/min，$n=10$）明显快于对照组（106±5 次/min，$n=10$）（$P<0.05$）；0.4 mg/L Cu^{2+} 浓度组中，稚鱼的心率与对照组没有显著差别；20.30 mg/L 亚硝酸氮和 29.20 mg/L 总氨氮浓度组中，鱼的心率分别为（112±4）次/min 和（112±7）次/min，也显著高于对照组（$P<0.05$）。

关键词 中华鲟 急性毒性 半致死浓度 苯酚 Cu^{2+} 亚硝酸盐 总氨氮

中华鲟（*Acipenser sinensis*）是中国国家一级保护动物，由于水质污染，修建水利工程及人为捕捞，使得中华鲟的资源量锐减。近 20 年来，我国一直采用人工放流的方法对中华鲟种群进行增殖和保护，但 1999 年和 2002 年在长江河口地区误捕时却发现，中华鲟幼鱼中人工增殖放流苗的比例小于 3 %。可见，中华鲟资源的补充主要还是来源于自然繁殖[1]。这是否与中华鲟的人工繁殖幼苗在天然水体中的生存能力有关，值得进一步研究。纺织、印染、电镀、皮革以及石油等工业废水的排放，农药、化肥的使用等[2]，导致长江部分江段形成了污染带，石油烃类和挥发酚、重金属铜已成为某些地区沿岸排入的主要污染物[3]，这些水域环境已经变得不利于中华鲟的生存。在人工育苗中，池水中的总氨氮和亚硝酸氮浓度若不控制在安全范围之内，也将影响幼苗的正常发育。目前，关于有毒物质对中华鲟稚鱼毒性作用方面的报道较少，仅姜礼燔曾经调查了长江环境污染对中华鲟的影响[4]。

鱼类从胚胎至稚鱼的早期发育时期是其生命周期中对各种污染物最为敏感的阶段[5]。本试验中，笔者用生物测试法，研究了中华鲟稚鱼在卵黄吸收完毕阶段对苯酚，Cu^{2+}，亚硝酸氮和总氨氮的耐受能力，并据此评价各种物质对中华鲟稚鱼的急性毒性作用。

---

* 基金项目：国家自然科学基金资助项目（30490231）；科技部基础性工作重点专项项目资助（2002DEA1004）；中国长江三峡工程开发总公司和国务院三峡办重点项目资助（SX97-25 /HB）。

** 第一作者，男，1981 年出生，硕士，E-mail：duhao@ yfi. ac. cn

*** 通讯作者（Corresponding author），男，研究员，博士，E-mail：weiqw@ yfi. ac. cn

1 材料与方法

1.1 材料

试验用鱼为当年人工繁殖的中华鲟稚鱼（8～12 日龄），全长为（2.36±0.10）cm，平均体重为 0.07 g。试验前 24 h 及试验期间停止喂食。

试验在湖北省荆州中华鲟增殖放流实验基地室内进行，所用容器是容积为 7 L 的塑料盆。试验用水为基地充分曝气的稚鱼培育水，其水化指标如下：pH 6.5，溶氧 8.3 mg/L，总硬度 5.55 mmol/L，总碱度 3.97 mmol/L，亚硝酸氮 0.006mg/L，总氨氮 0.087 mg/L。pH 用 0.01 mol/L HCl 和 NaOH 调节。所用试剂苯酚，$NaNO_2$，NH_4Cl，$Cu_2SO_4\cdot5H_2O$ 均为分析纯。

1.2 方法

1.2.1 毒物浓度的测定

用 pH S-2C 酸度计测定 pH；用 4-氨基安替比林分光光度法测定苯酚的浓度；用重氮偶氮法测定亚硝酸氮的浓度；用纳氏比色法测定总氨氮的浓度；用原子吸收分光光度法测定 Cu^{2+} 浓度。

1.2.2 毒物对中华鲟稚鱼的半致死浓度

根据预试验的结果，确定 4 种毒物的试验浓度见表 1。急性毒性试验采用静水法进行，每塑料盆盛水 5 L，放 20 尾稚鱼。每一试验浓度设三次重复，并设对照组。水浴恒温（20±0.3）℃。总氨氮试验组，pH 为 8.00±0.05，根据总氨氮浓度与非离子氨百分比换算表，求得非离子氨的浓度[10]。按 6 h 间隔测定水温、溶氧、pH，并调解 pH，每 24 h 更换一半毒性溶液。

表 1 4 种毒物的试验设计浓度

Tab.1 The nom inal designed concentrations of the toxicities ρ/(mg/L)

苯酚 Phenol	Cu^{2+}	NO_2^--N	总氨氮 Total-ammonia-N
126	0.40	20.30	29.20
100	0.32	10.15	24.04
75	0.24	5.07	18.95
56	0.16	2.54	13.74
42	0.08	1.27	8.59
32	0.04	—	3.44
0	0	0	0

试验期间，连续 96 h 观察记录，及时捞出死鱼，统计各组受试鱼以及对照组鱼的累积死亡数，计算死亡率。

1.2.3 中华鲟稚鱼心率的测定

分别对在各种毒物的最高浓度组中放置 1 h 的 10 尾鱼进行心率测定。测定心率时，在试验室 40 W 日光灯下进行，用显微镜灯照明，电子秒表计数[8]，每尾鱼测量 60 s 的心跳次数，重复三次。

1.3 数据处理

采用 SPSS 13.0 软件，用概率单位回归法求出 24 h，48 h，72 h，96 h 试验鱼的半致死浓度（LC_{50}）[7]，并给出 LC_{50} 的 95 %可信区间。用 t 检验法检验毒性试验组与对照组鱼心率的差异。

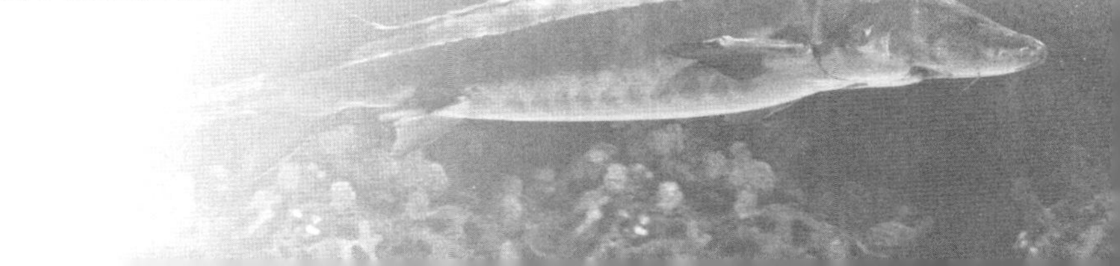

2 结果

2.1 毒物对中华鲟稚鱼的毒性作用

2.1.1 中华鲟稚鱼的中毒症状

试验期间对照组稚鱼无死亡。

在苯酚毒性作用下，大部分鱼游泳速度明显增加。具体表现为，鱼头贴着水面，鱼尾快速摆动，经过一段时间后，有的鱼侧卧水底不游动，敲击盆壁才会窜游；因苯酚致死的鱼身体僵直，一部分鱼尾部扭曲。在 Cu^{2+} 和亚硝酸氮毒物中，80 %的鱼能正常游动，小部分鱼侧卧水底，轻敲盆壁又能够游动；但亚硝酸氮毒物致死的鱼有 80 %身体躯干出现扭曲，有些死鱼尾部上翘。刚放入总氨氮毒物中的鱼就表现出明显的狂游状态，贴着盆壁或者浮于水面窜游；有的沉入水底侧卧，挣扎向上游动。

2.1.2 毒物对中华鲟稚鱼的致死作用

随着毒物作用的持续，中华鲟稚鱼开始出现死亡。图 1 反映了各种毒物不同浓度组累计死

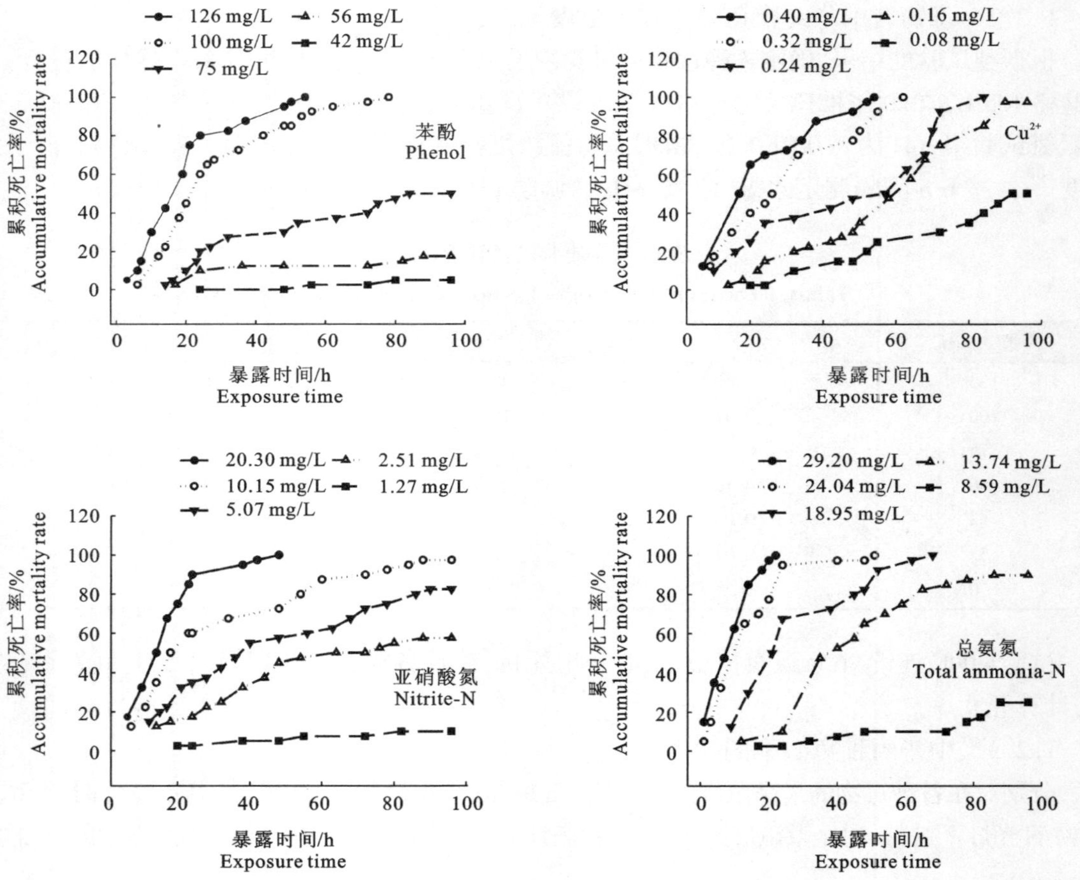

图 1 苯酚，Cu^{2+}，亚硝酸氮和总氨氮各浓度组中华鲟稚鱼累积死亡率随暴露时间的变化

Fig. 1 The changes in accumulative mortality rate of Chinese sturgeon juveniles with exposure time in different concentra tions of phenol, Cu^{2+}, nitrite-N and total ammonia-N

亡率(%)与毒性试验时间(h)间的相互关系。苯酚高浓度组(126 mg/L)鱼在 24 h 内死亡率达 80 %,54 h 后全部死亡;100 mg/L 浓度组鱼 80 h 内也全部死亡;较低浓度组(42 ~ 75 mg/L)鱼 24 h 内死亡率低于 25 %,至 96 h 死亡率低于 50 %。Cu^{2+} 较高浓度组(0.4 mg/L)的鱼在48 h 内全部死亡;0.24、0.32 mg/L 浓度组的鱼在 96 h 内全部死亡;0.16、0.08、0.04 mg/L 低浓度组的鱼在 96 h 内的死亡率分别为 97 %、50 %和 10 %。亚硝酸氮高浓度组(20.30 mg/L)的鱼在 24 h 内死亡率达 90 %,48 h 内全部死亡;其他浓度组的鱼 96 h 内没有全部死亡。总氨氮高浓度组(29.02 mg/L)的鱼在 22 h 内全部死亡;24.04 mg/L 和 18.95 mg/L 浓度组的鱼也在 72 h 内全部死亡;其他较低浓度组的鱼在 96 h 内没有全部死亡。

通过对试验数据进行统计分析,得出各毒性物质对中华鲟稚鱼的 24 h,48 h,72 h,96 h 的半致死浓度及 95 %可信区间见表 2(非离子氨的浓度是根据总氨氮浓度换算得出的)。

表 2 苯酚,Cu^{2+},NO_2^--N,总氨氮及 NH_3-N 对中华鲟稚鱼的半致死浓度及 95 %可信区间

Tab. 2 The median lethal concentrations(LC_{50}) and their 95 % confidence intervals of phenol, copper, nitrite-N, total ammonia-N and unionized ammonia-N to Ch inese sturgeon juveniles

ρ/(mg/L)

时间/h Time	苯酚 Phenol	Cu^{2+}	NO_2^--N	总氨氮 Total ammonia-N	NH_3-N
24	97(87 ~ 109)	0.31(0.29 ~ 0.34)	7.05(6.22 ~ 8.03)	17.61(16.96 ~ 28.27)	0.67(0.64 ~ 1.07)
48	83(73 ~ 95)	0.20(0.15 ~ 0.27)	4.07(1.51 ~ 7.88)	13.65(11.54 ~ 15.65)	0.52(0.44 ~ 0.67)
72	75(69 ~ 83)	0.09(0.08 ~ 0.12)	3.10(2.07 ~ 4.34)	11.44(10.64 ~ 12.65)	0.43(0.40 ~ 0.54)
96	71(64 ~ 80)	0.07(0.06 ~ 0.10)	2.57(2.31 ~ 2.85)	10.40(9.86 ~ 11.43)	0.39(0.30 ~ 0.45)

2.2 毒性物质对中华鲟稚鱼心率的影响

试验期间,对毒性试验最高浓度组及对照组的稚鱼心率进行了测定,结果见表 3。从表 3 可见,中华鲟稚鱼暴露在高浓度的苯酚、亚硝酸氮和总氨氮中 1 h 后,其心率明显加快。t 检验结果表明:苯酚(126 mg/L)、亚硝酸氮(20.3 mg/L)和总氨氮(29.20 mg/L)毒性试验组的鱼,平均心率分别为(119 ± 10)次/min,(112 ± 4)次/min 和(112 ± 7)次/min,均显著高于对照组(106 ± 5 次/min)($P < 0.05$);Cu^{2+}(0.4 mg/L)试验组中的鱼,平均心率比对照组略高,但没有明显差异($P > 0.05$)。

表 3 苯酚、Cu^{2+}、亚硝酸盐和总氨氮 4 种毒物对中华鲟稚鱼心率的影响

Tab. 3 The heart beating rates of Chinese sturgeon juveniles in phenol, copper, nitrite and ammonia nominal concentrations

毒物 Toxicants	ρ/(mg/L)	测试鱼尾数 Numbers of specimens	心率/(次·min^{-1}) Heart rate /(beat·min^{-1})	t	P
苯酚 Phenol	126	10	119 ± 10	4.126	<0.05
Cu^{2+}	0.40	10	108 ± 15	0.439	>0.05
亚硝酸氮 Nitrite-N	20.30	10	112 ± 4	3.669	<0.05
总氨氮 Total ammonia-N	29.20	10	112 ± 7	3.066	<0.05
对照组 Control	0	10	106 ± 5	—	—

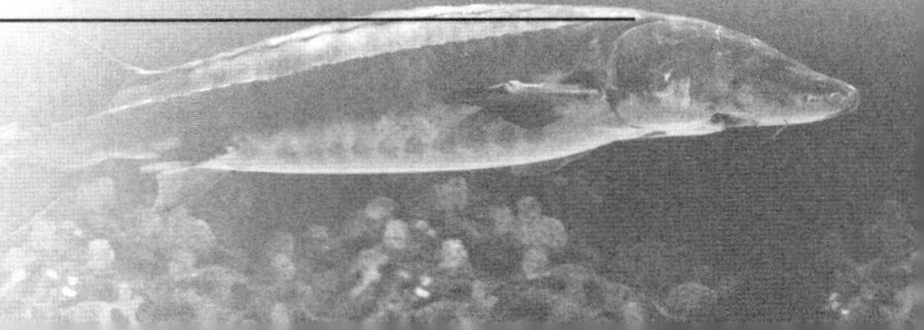

3 讨论

心脏跳动的频率可以反映鱼类呼吸强度的大小，而呼吸系统的变化能迅速反映出机体自身机能的变化，它们对外环境变化敏感，是反映机体机能状态的重要指标[9]。因此，心率的变化可以反映毒物对中华鲟稚鱼机体的影响状况。在苯酚、亚硝酸氮和总氨氮作用下，中华鲟稚鱼的心率明显加快，显著高于对照组($P<0.05$)，表明毒物使得鱼的呼吸频率明显加快。在氧气充足的情况下，呼吸频率的加快反映了机体供氧的不足。有研究表明，苯酚可以破坏红细胞[10]；NO_2^--N 进入鱼的血液后，能将血红蛋白分子中的 Fe^{2+} 氧化成为 Fe^{3+}，抑制血液的载氧能力，导致其呼吸机能下降，中毒严重时可引起鱼类窒息、死亡[11]；非离子氨能够导致鱼鳃对气体的通透性增加，刺激鱼体兴奋，使鱼发生痉挛[12]。本试验中，中华鲟稚鱼在 3 种毒物作用下，心率显著增加，表明这 3 种毒物对鱼体呼吸机能产生明显的影响。因此，在中华鲟育苗水体中，保持较高的溶解氧对于缓解毒物的毒副作用有一定的意义。

有研究表明，苯酚对罗非鱼 96 h 的 LC_{50} 为 28.49 mg/L[13]，比中华鲟 96 h 的 LC_{50} 低(71 mg/L)，表明中华鲟对苯酚的耐受能力要强于罗非鱼。总氨氮和非离子氨对胭脂鱼 96 h 的 LC_{50} 分别为 30.1(24.1 ~ 37.7) mg/L，0.58(0.45 ~ 0.71) mg/L[11]，明显高于中华鲟相应的 96 h 的 LC_{50}，说明中华鲟对总氨氮和非离子态氨更为敏感。育苗池中残剩的饵料常常会导致水泥池底部总氨氮积聚达到一定的有害浓度[13]。在中华鲟幼苗培育过程中，常可见到身体畸形、尾部扭曲的成活个体，类似于本试验中亚硝酸氮致死稚鱼死亡时的状态。活体稚鱼出现畸形是否与亚硝酸氮的毒性作用有关，尚需进一步研究。在不同水温、盐度和 pH 条件下，总氨氮的存在形态将发生变化，NH_3-N 对水生生物的毒性要强于 NH_4^+-N[11,14]。本试验结果表明，NH_3-N 对中华鲟 24 h，48 h，72 h 和 96 h 的 LC_{50} 都明显低于亚硝酸氮，说明非离子氨对中华鲟稚鱼的毒性比亚硝酸氮更强。因此，中华鲟育苗时，须及时清除水体中的残饵，保持良好的水体交换。

Cu^{2+} 对中华鲟稚鱼 72 h 的 LC_{50} 为 0.09 mg/L，这与 Cu^{2+} 对鲫幼鱼 96 h 的 LC_{50} 相同[15]。可见，中华鲟对 Cu^{2+} 的反应要比鲫等一般的淡水鱼类敏感。养殖水体中，常用 0.7 mg/L Cu_2SO_4 (0.4 mg/L Cu^{2+}) 防治车轮虫、口丝虫等原生动物疾病，并可杀死复口吸虫、甲壳类等寄生虫，同时还可杀死青泥苔、水网藻、蓝绿藻等[16]。从本试验结果来看，中华鲟养殖水体中不宜使用 Cu_2SO_4 进行消毒。

陈家长等[3]对长江下游安庆、南京、镇江和南通 4 个江段水质污染状况的监测结果表明，该 4 个江段的挥发酚、重金属铜等污染物都严重超标，沿江的其他江段也很可能有其他污染带存在。这意味着中华鲟自然繁殖幼苗以及人工繁殖放流幼苗降河洄游到长江口的过程中，要经受各种毒物毒害的威胁。因此，对长江沿岸的污染源应该严加管控，为中华鲟等水生生物资源提供良好的生存环境。

参考文献

[1] 危起伟. 中华鲟繁殖行为生态学与资源评估[D]. 武汉：中国科学院水生生物研究所，2003：94-96.
[2] 曲克明，陈民山，马绍赛，等. 3 种工业废水对牙鲆胚胎的毒性效应[J]. 中国水产科学，2003，10(2)：155-159.
[3] 陈家长，孙正中，瞿建宏，等. 长江下游重点江段水质污染及对鱼类的毒性影响[J]. 水生生物学报，2002，26(6)：635-640.
[4] 姜礼燔. 环境污染对中华鲟影响的研究[J]. 现代渔业信息，1996，11(7)：1-7.
[5] Mckia J M. Evalu ation of test with the early life stages of fish for prediction long-term toxicity[J]. Fish Res Board Can，1997，34：1148-1154.
[6] 张宏陶. 水质分析大全[M]. 重庆：科学技术出版社重庆分社，1989：462.

[7] 景体淞,徐镜波. 酚、苯、重金属离子对溞类的毒性作用[J]. 松辽学刊:自然科学版,2000(3):18-22.
[8] 孙修涛,李键,王清印. 中国对虾雌虾的心率及有关影响因素的研究[J]. 海洋与湖沼,1999,30(2):161-166.
[9] 上海水产学院. 鱼类学与海水鱼类养殖[M]. 北京:农业出版社,1979:259-261.
[10] 陈惠群. 苯酚对革胡子鲶血液指标的影响[J]. 环境污染与防治,2002,24(2):104-105,126.
[11] Sampaio L A, Wasielesky W, Camposmiranda-Filho K. Effect of salinity on acute toxicity of ammonia and nitrite to juvenile *Mugilplatanus*[J]. Bu lletin of Env ironmental Contam ination and Toxicology,2002,68:668-674.
[12] Randall D J, Wright P A. Ammonia distribution and excretion in fish[J]. F ish Physiol B iochem,1987,3:107-120.
[13] Saha N C, Bhunia F, Kaviraj A. Toxicity of phenol to fish and aquatic ecosystems[J]. Bullet in of Environmental Contam ination and Toxicology,1999,63:195-202.
[14] 沈成钢,杨凤,刘长发,等. 氨态氮在低温下对鱼的急性毒性研究[J]. 大连水产学院学报,1995,10(3):58-64.
[15] 杨丽华,方展强,郑文彪. 重金属对鲫鱼的急性毒性及安全浓度评价[J]. 华南师范大学学报:自然科学版,2003(2):101-106.
[16] 杨再福,陈立侨,陈华友,等. 重金属铜、镉对蝌蚪毒性的研究[J]. 中国生态农业学报,2003,11(1):102-103.

Acute Toxicity of Phenol, Copper, Nitrite and Total Ammonia to Juvenile Chinese Sturgeon (*Acipenser sinensis*)

Du hao[1,2,3] Wei Qiwei[1,2,3] Liu Jianyi[1,2] Yang Deguo[1,2]
Cheng Xihua[1,2] Zhu Yongjiu[1,2] Tang Guopan[3]

(1. Key Lab of Freshwater Fish Germplasm Resources and Biotechnology, Ministry of Agriculture, Yangtze River Fisheries Research In stitute, Chinese Academy of Fishery S cien ces, Jingzhou 434000, China;
2. Freshwater Fisheries Research Center, Chinese Academy of Fishery Sciences, Wuxi 214081, China;
3. Fisheries College, huazhong Agricultural University, Wuhan 430070, China)

Abstract A cute toxicities of phenol, Cu^{2+}, nitrite and total ammonia to juvenile Chinese sturgeon (*Acipenser sinensis*) fingerling swere determined in static bioassays. The juveniles with (2.36 ± 0.10) cm in totallength and average 0.07 g in body weight were 8-12 day old just before the yolk would be completely exhausted. It was found that the 24 h, 48 h, 72 h and 96 h median lethal concentration (LC_{50}) values (95 % confidence interval) of the phenol to the juveniles were 97 (87-109) mg/L, 83 (73-95) mg/L, 75 (69-83) mg/L and 71 (64-80) mg/L, respectively. The 24 h, 48 h, 72 h and 96 h LC_{50} values (95 % confidence in terval) of Cu^{2+} to the juveniles were 0.31 (0.29-0.34) mg/L, 0.20 (0.15-0.27) mg/L, 0.09 (0.08-0.12) mg/L and 0.07 (0.06-0.10) mg/L, respectively. The 24 h, 48 h, 72 h and 96 h LC_{50} values (95 % confidence in terval) of nitrite-N to the juveniles were 7.05 (6.22-8.03) mg/L, 4.07 (1.51-7.88) mg/L, 3.10 (2.07-4.34) mg/L and 2.57 (2.31-2.85) mg/L, respectively. The 24 h, 48 h, 72 h and 96 h LC_{50} values (95 % confidence interval) of unionized ammonia-N to the juveniles were 0.67 (0.64-1.07) mg/L, 0.52 (0.44-0.67) mg/L, 0.43 (0.40-0.54) mg/L and 0.39 (0.30-0.45) mg/L, respectively. The juveniles had higher heart beating rates (119 ± 10 beat /min, $n = 10$) in 126 mg/L phenol than those in the control group (106 ± 5 beat/min, $n = 10$) ($P < 0.05$). There was no significant difference in heart rates in 0.4 mg/L Cu^{2+} and in the control group ($P > 0.05$). The juveniles had significantly higher heart beating rates in 20.30 mg/L nitrite-N (112 ± 4 beat/min, $n = 10$) and 29.20 mg/L total ammonia-N (112 ± 7 beat/min, $n = 10$) than in the control group ($P < 0.05$).

Key words *Acipenser sinensis* Acute toxicity Median lethal concentration Phenol Cu^{2+} Nitrite Total ammonia

多倍体中华鲟微卫星亲子鉴定体系的建立*

辛苗苗[1,2,**]　张书环[2]　汪登强[2]　汪　珂[2]
张　磊[2]　霍来江[2]　危起伟[2,3,***]

（1. 西南大学生命科学学院　重庆　400715
2. 中国水产科学研究院长江水产研究所 农业部淡水生物多样性保护重点实验室　湖北武汉　430223
3. 中国水产科学研究院淡水渔业研究中心　江苏无锡　214081）

摘　要　应用荧光引物和自动测序技术，检测了 10 个自主研发的微卫星分子标记在 68 个中华鲟(*Acipenser sinensis*)个体中的遗传多态性。结果表明，10 个微卫星标记共检测到 124 个等位基因，等位基因数为 7～22，期望杂合度为 0.730 26～0.873 95，Shannon-Wiener 多样性指数为 1.561 61～2.219 60。基于亲子鉴定排除率不低于 0.99 的标准，最终确定 Asi-75067，Asi-67648，Asi-67123，Asi-73843，Asi-72040，Asi-70421 和 Asi-56700 这 7 个微卫星标记为中华鲟亲子鉴定的核心体系，其单亲、父权和双亲的累积排除率分别为 99.12 %，99.95 %和 99.99 %。该微卫星标记组合为中华鲟准确高效经济的亲子鉴定体系建立提供了科学依据。

关键词　微卫星　中华鲟(*Acipenser sinensis*)　亲子鉴定　多倍体　累积排除率

中华鲟(*Acipenser sinensis*)为典型的大型溯河产卵洄游性鱼类，是我国国家一级保护野生动物。近年来，由于环境污染、过度捕捞、栖息地改变与破坏、葛洲坝修建等人类活动的干扰，中华鲟资源量大幅度锐减，雌雄比例失调，自然种群严重衰退，有灭绝的危险。目前，仅在葛洲坝下发现一处适宜其繁殖和栖息的自然产卵场，有效面积非常有限[1]。因此建立中华鲟人工养殖群体，实现全人工繁殖是保存中华鲟物种和自然资源增殖的重要途径[2]。然而，人工养殖中华鲟亲鱼数量有限，且人工繁殖时通常未进行配种设计，多年累代养殖后，将会造成不同程度的近亲繁殖，导致养殖群体遗传结构破坏，种质退化。全人工繁殖的中华鲟仔鱼受精率低，抗病能力弱，严重影响中华鲟物种保护的效果。为减少近亲繁殖导致的遗传问题以及培养出具有较好生存力的繁殖后代，迫切需要建立科学可靠的遗传系谱来有效指导中华鲟规模化的人工繁殖。

微卫星 DNA(microsatellite DNA)又称简单重复序列(simple sequence repeats，SSRs)，是一种共显性标记，具有多态性丰富、检测简便快捷、PCR 扩增结果重复性好等特点。近年来，随着濒危物种遗传保护研究日益增强，微卫星分子标记在濒危动物保护中也得到越来越多的应用[3]。基于微卫星标记的二倍体动物亲子鉴定体系已经建立[4]，并成功应用于系谱建立和遗传管理。然而国内外关于鲟鱼等多倍体动物亲子鉴定报道还较少。2004 年，Rodzen 等[5]应用微卫星标记对高首鲟(*A. transmontanus*)进行了亲子鉴定分析。多倍体中华鲟[6]因其复杂的遗传特性和未知的染色体数目使其亲子鉴定研究遇到瓶颈。1999 年，朱滨等[7]利用湖鲟(*A. fulvescens*)的 2 对微卫星标记对中华鲟进行了亲子鉴定研究[7]，但只对少量样本进行了简单的分析，并未建立中华鲟亲子鉴定的微卫星标记体系。本试验采用微卫星标记方法，在已有研究的基础上，进一步优化亲子鉴定微卫星标记组合，形成鉴定准确、检测费用低廉的微卫星

---

* 基金项目：公益性行业(农业)科研专项：珍稀水生动物繁育与物种保护技术研究(201203086)。

** 第一作者，女，1989 年出生，硕士研究生，专业方向为渔业资源与环境，E-mail：xinmiao1206@126.com

*** 通讯作者(Corresponding author)，E-mail：weiqw@yfi.ac.cn

亲子鉴定体系，为中华鲟成功高效的人工繁殖提供理论依据和技术指导。

1　材料与方法

1.1　材料

本试验材料取自中国水产科学研究院长江水产研究所农业部中华鲟保育与增殖放流中心，其中包括2012年10月通过人工授精获得的中华鲟家系，子代56尾、真实的亲本2尾及随机挑取与家系无关的10尾后备亲鱼。样品均采集鳍条，保存在无水乙醇中备用。

1.2　基因组DNA的提取

本试验采用高盐法[8]提取基因组DNA，具体步骤如下：将保存的鳍条样本剪取0.5 g左右，放入2 ml离心管中，用TE缓冲液浸泡12 h以充分置换出酒精，然后取出放入352 μl的无菌抽提缓冲液中，加入40 μl 20 % SDS和10 μl 20 mg/ml蛋白酶K混匀；在56 ℃消化4 h以上，每30 min摇晃离心管1次，直到组织完全消化后加入300 μl 6 mol/L NaCl溶液，立刻剧烈震动30 s；12 000 r/min离心10 min，吸上清液，加入等体积的异丙醇；－20 ℃放置1 h，12 000 r/min离心15 min（4 ℃），弃上清，沉淀用70 %的乙醇溶液洗涤1次，37 ℃烘干或自然晾干后，加入50 μl双蒸水，待完全溶解后稀释成100 ng/μl保存备用。

1.3　引物设计与PCR扩增

基于实验室已筛选出的24对多态微卫星标记，选取其中多态性较高的10对（表1）用于亲子鉴定，由上海生工生物工程技术服务有限公司合成5端带有FAM和HEX的荧光引物。

表1　10个微卫星标记引物序列及退火温度

Tab.1　Primer sequences and annealing temperature of 10 microsatellite markers in *A. sinensis*

位点	引物	片段大小/bp	登录号	退火温度/℃
Asi-75067	F:AGAGTTCTCGAAGCGGAAACAG R:CTGGTTCAAACTGGGAGCGAT	293	KM379079	60
Asi-67648	F:TCCGGTACTGGAAACCCTTG R:ATTCGCCTGGAAGAGCACAC	302	KM379090	60
Asi-67123	F:AGCTAACAGCAGTGCATGGTATTT R:CTTGAGATAAAAGGCGCTGTAGAG	276	KM379082	62
Asi-73843	F:GGTGTTTGAAAGACAGCGAGAA R:TTCCTCCAGGACAGAGTTTGC	292	KM379084	58
Asi-72040	F:AGCAGAGTCCACATCCCCCT R:GAGTGTCGCTCGAAAGCCCT	306	KM379089	62
Asi-70421	F:TGCCACAAATAAGATGCAGGAG R:TTTTGCTTTGGAAACTGTACTGC	309	KM379092	56
Asi-56700	F:CAACCTCTTCACTACCGCAAAC R:TGCAAAAAGGAATTGGAATCG	295	KM379094	50
Asi-65194	F:TACAAAATCGGCAGAAAGGCT R:GCAGGCATGATGAGAATAGGC	296	KM379077	58
Asi-77057	F:GGGTCCCGCACAGTTTAAAG R:GACGGCAAGGCAAGATAGGT	300	KM379087	56
Asi-66034	F:CCTACGCCAAGCTCAACCAG R:ACCGCAGAGTCACGGAGTTG	309	KM379072	56

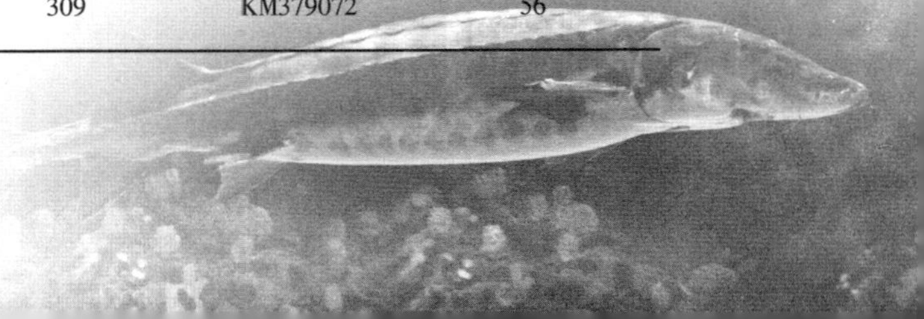

PCR 反应体系总体积为 25 μl：10 × PCR（TaKaRa）缓冲液 3 μl，25 mmol/L Mg^{2+} 离子 0.5 μl，10 mmol/L dNTPs 1 μl，0.4 μl rTaqDNA 聚合酶（TaKaRa）0.5 U，正反向引物各0.5 μl，100 ng 基因组 DNA 以及双蒸水。

PCR 反应条件为：95 ℃预变性 5 min；95 ℃变性 30s，退火（表 1）30 s，72 ℃延伸 1 min，30 个循环；72 ℃延伸 10 min；4 ℃保存。

PCR 产物用 1 %琼脂糖凝胶电泳，EB 染色，自动凝胶成像系统拍照检测，然后送至上海生工生物工程技术服务有限公司测序，在 ABI3730 自动测序仪上进行荧光信号收集，利用毛细管电泳方法进行基因型检测。为降低微卫星基因型测序成本，将带有不同荧光基团的 PCR 产物两两混合进行测序。本试验将 10 个微卫星标记分为 5 组，标记组合及其所携带荧光基团见表 2。

表 2 微卫星标记分组情况

Tab. 2 The grouping of microsatellite markers in *A. sinensis*

分组	位点名称	荧光标记	片段大小/bp	分组	位点名称	荧光标记	片段大小/bp
A	Asi-75067	HEX	293	D	Asi-70421	FAM	309
	Asi-67648	FAM	302		Asi-56700	HEX	295
B	Asi-67123	FAM	276	E	Asi-77057	FAM	300
	Asi-73843	HEX	292		Asi-66034	HEX	309
C	Asi-72040	FAM	306				
	Asi-65194	HEX	296				

1.4 统计方法

应用 Excel 统计等位基因数；软件 ATetra1.2[9]分析计算期望杂合度、Shannon-Wiener 多样性指数；软件 GeneMarker V1.5 读取等位基因大小；软件 FaMoz[10]进行中华鲟亲子鉴定分析，计算排除率与累积排除率。

2 结果

2.1 微卫星标记的多态性

微卫星标记的等位基因、等位基因数和等位基因频率统计结果见表 3，10 个微卫星标记共检测到 124 个等位基因，其中等位基因数最多的位点为 Asi-77057，检测到 22 个等位基因；其次为 Asi-75067，检测到 17 个等位基因；等位基因数最少的位点是 Asi-66034，有 7 个等位基因。

表 3 中华鲟微卫星标记等位基因及其等位基因频率

Tab. 3 Number of alleles and allele frequencies of microsatellite markers in *A. sinensis*

位点	等位基因	等位基因频率	位点	等位基因	等位基因频率
	274	0.182		300	0.004
	276	0.050		302	0.004
	280	0.004		304	0.008
	284	0.070		308	0.066
Asi-75067	286	0.008	Asi-75067	314	0.004
	288	0.150		316	0.004
	290	0.128		320	0.008
	292	0.124		322	0.008
	296	0.178			

续表

位点	等位基因	等位基因频率
Asi-77057	238	0.004
	248	0.008
	250	0.012
	252	0.142
	256	0.012
	258	0.008
	260	0.004
	262	0.244
	264	0.020
	272	0.004
	276	0.008
	278	0.021
	280	0.004
	282	0.012
	284	0.200
	286	0.004
	294	0.004
	296	0.008
	298	0.008
	300	0.228
	304	0.041
	306	0.004
Asi-65194	281	0.016
	283	0.208
	288	0.074
	292	0.372
	294	0.005
	300	0.090
	302	0.085
	304	0.085
	314	0.065
Asi-67123	263	0.098
	269	0.053
	273	0.214
	274	0.004
	278	0.053
	282	0.019
	284	0.049
	287	0.192
	291	0.011
	294	0.144
	297	0.147
	303	0.008
	306	0.008
Asi-67648	300	0.011
	302	0.123
	305	0.062
	308	0.004
	312	0.202
	317	0.063
	320	0.140
	324	0.148
	327	0.109
	330	0.060
	333	0.014
	336	0.011
	339	0.004
	342	0.049
Asi-72040	281	0.077
	286	0.005
	298	0.005
	301	0.145
	304	0.010
	307	0.343
	311	0.222
	314	0.087
	317	0.005
	321	0.010
Asi-70421	280	0.004
	284	0.004
	286	0.004
	289	0.209
	300	0.044
	304	0.161
	307	0.229
	310	0.225
	313	0.116
	316	0.004

续表

位点	等位基因	等位基因频率	位点	等位基因	等位基因频率
	260	0.007		290	0.164
	287	0.007		293	0.118
	296	0.270		295	0.218
	298	0.292	Asi-73843	298	0.046
Asi-56700	300	0.109		300	0.060
	302	0.044		302	0.004
	304	0.044		305	0.029
	310	0.161		256	0.290
	312	0.066		291	0.283
	262	0.147		298	0.020
	274	0.004	Asi-66034	300	0.195
Asi-73843	278	0.004		303	0.058
	282	0.013		306	0.150
	284	0.017		312	0.004
	287	0.176			

2.2 遗传多样性分析

微卫星标记的遗传多样性检测结果显示(表4),10个微卫星位点期望杂合度变化范围为0.730 26~0.873 95,Shannon-Wiener多样性指数变化范围为1.561 61~2.219 60,说明这10个微卫星标记具有较高的多态性,可用于中华鲟亲子鉴定及系谱建立等遗传学分析。

表4　中华鲟微卫星各位点的遗传信息

Tab. 4　Genetic information of the microsatellite loci in *A. sinensis*

位点	等位基因数	期望杂合度	Shannon-Wiener 多样性指数
Asi-75067	17	0.864 46	2.143 82
Asi-67648	14	0.873 95	2.219 60
Asi-67123	13	0.850 79	2.059 02
Asi-73843	13	0.851 50	2.076 00
Asi-72040	10	0.766 69	1.618 31
Asi-70421	10	0.813 51	1.789 14
Asi-56700	9	0.778 06	1.719 69
Asi-65194	9	0.730 26	1.612 01
Asi-77057	22	0.811 23	1.987 92
Asi-66034	7	0.765 45	1.561 61

2.3 微卫星标记对已知亲本中华鲟家系的检测分析

通过对中华鲟繁殖亲本与子代基因型(表5和表6)比较发现,位点Asi-75067,Asi-67123,Asi-73843,Asi-72040,Asi-70421和Asi-65194对该家系成员扩增结果中,子代与亲本基因型保持一致，且 56 个子代基因型在该 6 个微卫星位点均呈杂合态。其余 4 个位点子代基因型与亲本基因型都存在不同程度的差异：位点 Asi-67648 有 13 个子代个体的部分等位基因 330，333 和 336 不是来自亲本，位点 Asi-56700 有 13 个子代个体的其中一个等位基因 310 不是来自亲本，位点 Asi-77057 有 4 个个体的 7 个等位基因不是来自亲本，位点 Asi-66034 有 2 个个体的一个等位基因 298 不是来自亲本，其余个体全部等位基因均来源于亲本（表 6）。

表 5　中华鲟 10 个微卫星标记的基因型

Tab. 5　Genotype of 10 microsatellite markers in *A. sinensis*

位点	父本基因型				母本基因型				子代 1 基因型			
Asi-75067	274	276	290	296	274	290	292	308	276	292	296	308
Asi-67648	302	312	317	327	312	320	327	342	302	312	327	342
Asi-67123	269	287	297		263	273	284	297	269	273	287	297
Asi-73843	287	293			262	290	295	298	262	287	293	298
Asi-72040	301	307	311		301	307	311		301	307	311	
Asi-70421	289	307	310	313	289	304	307		289	304	307	310
Asi-56700	298	300	312		296	298			296	298	312	
Asi-65194	292	304			283	288	292	283	292	292	304	
Asi-77057	262	284	300		252	262	284	300	262	284	300	
Asi-66034	256	291	300	306	256	291	300	303	256	291	300	

表 6　中华鲟子代与亲本等位基因比较结果

Tab. 6　Comparison of the alleles between parents and offspring of *A. sinensis*

位点	A	B	C	D						
Asi-75067	√									
Asi-67648		13	3	330	333	336				
Asi-67123	√									
Asi-73843	√									
Asi-72040	√									
Asi-70421	√									
Asi-56700		13	1	310						
Asi-65194	√									
Asi-77057		4	7	248	250	260	264	280	282	298
Asi-66034		2	1	298						

注：A 为子代全部等位基因来自亲本的位点；B 为非全部等位基因来自亲本的子代个体数；C 为来自非父母的等位基因个数；D 为非父母的等位基因。

2.4　排除率和累积排除率

由表 7 可知，位点 Asi-75067 的排除率最高，为 0. 555 491；Asi-65194，Asi-77057 和 Asi-66034三个位点的排除率较低，均低于 0.4。依据中华鲟微卫星标记排除率由高到低的顺序（表 7），当微卫星位点数目不小于 7 时，累积排除率均大于 0. 99，分别为 0. 992 799，0. 999 681和 0. 999 998。因此，确定 Asi-75067，Asi-67648，Asi-67123，Asi-73843，Asi-72040，

Asi-70421和 Asi-56700 这 7 个多态性较高的微卫星标记为中华鲟微卫星亲子鉴定体系。在某些标记不能顺利提供信息时,为达到排除率大于 0.99 的标准,把位点 Asi-65194,Asi-77057 和 Asi-66034 作为候选标记。

表 7　中华鲟微卫星标记的排除率及累积排除率

Tab. 7　Exclusion probability and cumulative exclusion probability of different microsatellite markers for parentage testing in *A. sinensis*

位点	PE1	PE2	PE3	CPE1	CPE2	CPE3
Asi-75067	0.555 491	0.717 114	0.880 012	0.555 491	0.717 114	0.880 012
Asi-67648	0.534 586	0.699 645	0.867 453	0.793 119	0.915 034	0.984 096
Asi-67123	0.526 405	0.692 778	0.862 250	0.902 022	0.973 896	0.997 809
Asi-73843	0.523 063	0.690 126	0.859 557	0.953 271	0.991 911	0.999 692
Asi-72040	0.442 998	0.620 577	0.806 304	0.973 972	0.996 931	0.999 940
Asi-70421	0.431 458	0.609 188	0.788 036	0.985 202	0.998 801	0.999 987
Asi-56700	0.407 490	0.585 600	0.771 956	0.991 232	0.999 503	0.999 997
Asi-65194	0.396 250	0.575 876	0.762 169	0.994 706	0.999 789	0.999 999
Asi-77057	0.399 454	0.577 398	0.758 192	0.996 821	0.999 911	1.000 000
Asi-66034	0.366 892	0.555 952	0.725 063	0.997 987	0.999 959	1.000 000

注:PE1 为单亲排除率;PE2 为父权排除率;PE3 为双亲排除率;CPE1 为单亲累积排除率;CPE2 为父权累积排除率;CPE3 为双亲累积排除率。

3　讨论

3.1　微卫星标记的多态性

微卫星标记分析中检测到遗传座位越多,得到正确鉴定结果的概率就越大;同时每个座位在群体中含有的等位基因数越多,得到正确鉴定结果的概率也会越高[11]。本研究初步选择 10 个微卫星标记,对 68 个中华鲟个体进行遗传多态性检测,该标记组合的平均等位基因数为 12.4,且每个座位上等位基因数均不低于 6。杂合度反应微卫星标记多态性的变化[12],本研究中 10 个微卫星标记平均期望杂合度为 0.810 59,平均 Shannon-Wiener 多样性指数为 1.878 712。综上所述,10 个微卫星标记均表现为高度多态性且能够提供丰富准确的遗传信息,可作为有效的遗传标记用于中华鲟亲子鉴定分析。

3.2　微卫星标记等位基因的分型

判定亲子关系的理论依据是孟德尔遗传定律,子代的基因组分别来自双亲。但微卫星标记在遗传过程中存在着突变的可能,且基因分型时,并非所有基因座都能获得正确的分型结果[13]。由表 6 可知位点 Asi-75067,Asi-67123,Asi-73843,Asi-72040,Asi-70421 和 Asi-65194 在中华鲟个体的扩增结果中,子代基因型均来源于亲本,符合孟德尔遗传规律。位点 Asi-67648,Asi-56700,Asi-77057 和 Asi-66034 在中华鲟个体的扩增结果中,子代部分基因型不存在于亲本基因型中,因此推测在遗传过程中,子代等位基因可能发生了基因突变或者错误分型。O'Reilly 等[14]发现平均每个基因座存在 2 %~3 %的基因分型错误而无法达到 100 %的准确。如果个体在多个基因座都出现错误的分型结果,就会产生错配现象,但如果在 1 ~ 2 个基因座出现错误的分型结果,仍会得到正确结果[15],因此位点 Asi-67648 和 Asi-56700 作为亲子鉴定核心体系不会影响亲子鉴定结果的准确性。此外,位点 Asi-65194,Asi-77057 和 Asi-66034 的排除率较低,故将其作为候选标记。

3.3 亲子鉴定分析

亲子鉴定中,常用的方法有两种:一种是基于排除概率的排除法;另一种是基于似然比的似然法。排除法是进行亲子鉴定最早使用且最为简单有效的方法[16]。本研究主要应用排除法分析多倍体中华鲟[6]的亲子鉴定体系。Ellegren 等[17]研究表明,组合 5 个微卫星座位(每个有 6 个以上的等位基因)可使排除率达 98 %以上,使用 10 个这样的座位排除率可达 99.99 %。张志和等[18]和高建伟[19]分别应用 18 对和 15 对微卫星座位对大熊猫和黑熊进行亲子鉴定研究,结果显示排除率均大于 99.99 %。本研究中,标记数达到 5 个时,其单亲、父权、双亲累积排除率分别为 97.40 %,99.69 %,99.99 %;标记数达到 7 个时,其单亲、父权、双亲累积排除率分别为 99.12 %,99.95 %,99.99 %;标记数为 10 时,其单亲、父权、双亲累积排除率分别为 99.80 %,99.99 %,100 %(表 7),其结果与 Ellegren 等[17]的研究基本一致。基于亲子鉴定累积排除率不低于 99 % 的标准[20]和尽量降低检测成本和减少工作量的原则,本研究确定微卫星标记 Asi-75067,Asi-67648,Asi-67123,Asi-73843,Asi-72040,Asi-70421 和 Asi-56700 为中华鲟亲子鉴定的核心体系。为进一步考虑亲子鉴定的准确性,下一步我们将在已有研究的基础上采用排除法和似然法相结合的方式,更为稳健的在多个中华鲟家系中开展亲子鉴定应用。

参 考 文 献

[1] Wang C,Kynard B,Wei Q W,et al. Spatial distribution and habitat suitability indices for non-spawning and spawning adult Chinese sturgeons below Gezhouba Dam,Yangtze River:effects of river alterations[J]. J Appl Ichthyol,2013,28.

[2] 危起伟,李罗新,杜浩,等. 中华鲟全人工繁殖技术研究 [J]. 中国水产科学,2013(1):1-11.

[3] 王绍先,王飞,刘成柏. DNA 分子标记技术在濒危物种保护中的应用[J]. 生态学杂志,2008,27(2):250-256.

[4] 成为为,汪登强,危起伟,等. 基于微卫星标记对长江中上 游胭脂鱼增殖放流效果的评估[J]. 中国水产科学,2014 (3):574-580.

[5] Rodzen J F, Famula T R, Bernie May. Estimation of parentage and relatedness in the polyploid white sturgeon (*Acipenser transmontanus*) using a dominant marker approach for duplicated microsatellite loci[J]. Aquaculture, 2004,232:165-182.

[6] Ludwig A,Belfiore N M,Pitra C,et al. Genome duplication events and functional reduction of ploidy levels in sturgeon (*Acipenser*, *Husoand Scaphirhynchus*)[J]. Genetics. 2001,158:1203-1215.

[7] 朱滨,常剑波,谭细畅,等. 湖鲟微卫星 DNA 引物应用于 中华鲟亲子关系分析的初步研究[J]. 水生生物学报,1999 (6):547-553.

[8] Salah M A,Iciar M. Universal and rapid salt-extraction of high quality genomic DNA for PCR-based techniques[J]. Nucleic Acids Res,1997,25:4692-4693.

[9] van Puyvelde K,van Geert A,Triest L. ATETRA,a new software program to analyze tetraploid microsatellite data: comparison with TETRA and TETRASAT[J]. Mol Ecol Resour,2010,10: 331-334.

[10] Gerber S, Chabrier P, Kremer A. FAMOZ: a software for parentage analysis using dominant, codominant and uniparentally inherited markers[J]. Mol Ecol Notes,2003,3:479-481.

[11] 钱林东,张自芳,田应华,等. 利用微卫星 DNA 标记进行黄牛的亲子鉴定[J]. 云南农业大学学报,2010,25 (1):69-74.

[12] 周磊,初芹,刘林,等. 利用微卫星和 SNP 标记信息 进行奶牛亲子鉴定的模拟研究[J]. 畜牧兽医学报,2011,42 (2):169-176.

[13] Lathrop G M, Hooper A B, Huntsman J W, et al. Evaluating pedigree data. 1. The estimation of pedigree error in the presence of marker mistyping[J]. Am J Hum Genet, 1983, 35: 241-262.
[14] O'Reilly P T, Herbinger C, Wright J M. Analysis of parentage determination in Atlantic salmon (*Salmo salar*) using microsatellites [J]. Anim Genet, 1998, 29: 363-370.
[15] 董世瑞,孔杰,张天时,等. 中国对虾微卫星家系鉴定的模拟分析与应用[J]. 水生生物学报,2008,32(1): 96-100.
[16] Jones A G, Ardren W R. Methods of parentage analysis in natural populations[J]. Mol Ecol, 2003, 12: 2511-2523.
[17] Ellegren H, Johansson M S, Berk K, et al. Cloning of highly polymorphic microsatellite in the horse[J]. Anim Genet, 1992, 23: 133-142.
[18] 张志和,沈富军,孙姗,等. 应用微卫星分型方法进行大熊猫父亲鉴定[J]. 遗传,2003,25(5): 504-510.
[19] 高建伟. 应用微卫 DNA 标记对延边圈养黑熊进行亲子鉴定和遗传多样性研究[D]. 延边:延边大学,2012: 1-97.
[20] 郭立平,徐丽,朱淼,等. 西门塔尔牛微卫星亲子鉴定 体系的优化[J]. 畜牧兽医学报,2013,44(6): 871-879.

Parentage Identification of Polyploidy *Acipenser sinensis* Based on Microsatellite Markers

Xin Miaomiao[1,2] Zhang Shuhuan[2] Wang Dengqiang[2] Wang Ke[2]
Zhang Lei[2] Huo Laijiang[2] Wei Qiwei[2,3]

(1. School of Life Science, Southwest University, Chongqing 400715, China;
2. Key Laboratory of Freshwater Biodiversity Conservation, Ministry of Agriculture of China, Yangtze River Fisheries Research Institute, Chinese Academy of Fishery Sciences, Wuhan 430223, China;
3. Freshwater Fisheries Research Center, Chinese Academy of Fishery Sciences, Wuxi 214081, Jiangsu, China)

Abstract 10 novel microsatellite markers were used for assessing genetic diversity of 68 *Acipenser sinensis* individuals. The results demonstrated that there were 124 alleles in 10 loci and the number of alleles, mean expected heterozygosities and Shannon-Wiener Diversity Indices per locus ranged from 7 to 22, from 0.730 26 to 0.873 95 and from 1.561 61 to 2.219 60, respectively. With the criterion that cumulative exclusion probabilities were not lower than 0.99, the 7 high polymorphic microsatellite markers (Asi-75067, Asi-67648, Asi-67123, Asi-73843, Asi-72040, Asi-70421 and Asi-56700) made up the system of paternity test for *A. sinensis*. The cumulative exclusion probabilities over the 7 loci of single parent, paternity and pair parent were 0.991 232, 0.999 503 and 0.999 997, respectively. These high polymorphic microsatellite markers provided a scientific basis for validating paternity testing efficiently with a lower cost in *A. sinensis*.

Key words Microsatellite *Acipenser sinensis* Parentage identification Polyploidy Cumulative exclusion probabilities

淡水人工培育中华鲟亚成体的海水驯化试验*

杨德国[1,2,**]　朱永久[1]　危起伟[1,2]　陈细华[1,2]　刘鉴毅[1,2]　王　凯[1]

(1. 中国水产科学研究院长江水产研究所 农业部淡水鱼类种质资源与生物技术重点开放实验室　湖北荆州　434000
2. 中国水产科学研究院淡水渔业研究中心　江苏无锡　214081)

摘　要　2000 年,采用阶段性递增水体盐度,即每 5 d 将水体的盐度逐渐提高 5,并不设稳定期的方法,对在淡水环境人工养殖近 3 年的中华鲟子一代亚成体(全长 105 ~ 149 cm,体重 5.8 ~ 15.0 kg)进行海水驯化试验。结果表明,经过 20 d 的驯化,较大规格的中华鲟亚成体即可适应盐度为 23 的海水环境,驯化的成活率平均达到 93.9 %(92.1 % ~ 95.7 %)。水体盐度增加时没有发现中华鲟有异常的行为表现,而驯化后的中华鲟在海水池塘中也能够正常生长。研究还表明,中华鲟对其栖息水体的盐度有较强的适应能力,但其耐盐性和适应能力可能与个体规格大小有一定的关系,即个体的规格越大,对水体盐度变化的适应能力就越强。

关键词　中华鲟　亚成体　盐度　驯化

中华鲟(*Acipenser sinensis* Gray)是属于鲟形目(Acipenseriformes)、鲟科(Acipenseridae)、鲟属(*Acipenser*)的一种大型鱼类。近代主要分布在黄海、东海、南海北部近海,以及长江、珠江、闽江、钱塘江、黄河等江河,朝鲜西南部和日本九州西部有过记录。目前,中华鲟主要见于黄海、东海近海和长江水域,闽江、钱塘江已基本绝迹,珠江数量也极少(张世义,2001 ;长江水系渔业资源调查协作组,1990;四川省长江水产资源调查组,1988)。为保护中华鲟自然资源,1988 年,我国已将其列为国家一级保护的野生动物(汪小炎,1994)。随着中华鲟自然资源的急剧衰退,中华鲟物种的保护受到越来越多的重视(杨德国等,2005;常剑波等,1999;危起伟等,1998;邓昕等,1997)。保护中华鲟物种的一个重要途径,就是实施易地保护或人工保护,既在中华鲟曾经栖息、分布的水域重新引进并逐渐建立中华鲟种群;或在全人工环境下养殖、保存一定规模的中华鲟群体。因此,在人工环境条件下将子一代中华鲟培育至性成熟,实现中华鲟的全人工繁殖,完成中华鲟的生活史,是实现上述目标的基础。

中华鲟是一种典型的过河口溯河产卵洄游的鱼类。在江河(主要是长江中上游江段)产卵、孵化的幼鲟,孵出后随江水向下游漂流,经 6 ~ 7 个月的降河洄游,体长达到约 20 cm 时幼鲟到达长江口,并逐渐进入东海、黄海等近海水域栖息和肥育。海区长成的中华鲟达到性成熟年龄后,开始从海洋进入江河,性腺在其向产卵场溯河洄游的过程中逐渐发育成熟,并在产卵季节到达产卵场产卵,产后的中华鲟又降河洄游至海洋栖息和肥育(杨德国等,2005;危起伟等,1994;长江水系渔业资源调查协作组,1990;四川省长江水产资源调查组,1988)。因此,在自然生态条件下,中华鲟的一生中有多次的过河口行为,其栖息环境也包括了海洋、河口和江河淡水等多种生态环境类型。2000 年,作者将部分淡水培育的中华鲟亚成体引入海水环境中进行养殖和培育,旨在探讨实现中华鲟全人工繁育的多种途径。为此,进行了淡水培育中华鲟亚成体的海水驯化试验。

---

* 基金项目:科技部社会公益研究专项(2000DIB50177);科技部科技基础性工作专项(2002DEA1004);中国水产科学研究院科研基金项目(2003-3-9)。

** 第一作者,男,1964 年出生,研究员,主要从事鱼类生态学和保护生物学研究,E-mail:yangdg@ yfi. ac. cn

1 材料和方法

1.1 试验鱼

本试验使用的中华鲟亚成体为3龄，是作者1997年秋季人工繁殖的子一代，其亲本来自长江。试验鱼规格为全长105～149 cm，体重5.8～15.0 kg（表1），共1 480尾。试验鱼的前期培育在福建省莆田市进行，培育时间约30个月（1997年10月底孵出至2000年4月进行试验），培育鱼池为鳗鱼养殖池改建的圆形和方形水泥池，水源为地下水及水库水，盐度不超过1。

表1 海水驯化试验中华鲟亚成体全长和体重

Table 1 Total length and body weight of sub-adult Chinese sturgeon

项目 Item	平均值±标准差（$n=20$） Mean±SD	幅度 Range
全长/cm Total length	127.6±13.1	105.0～149.0
体重/kg Body weight	10.7±2.88	5.8～15.0

1.2 试验地点和条件

驯化试验在厦门松浩实业有限公司南安石井镇东港养殖场养虾场进行。该养虾场位于厦门市东南部的沿海滩涂区，养虾场与海区有一道防浪堤相隔。养虾场建有10 000～20 000 m^2不同规格海水养殖池10个，砂质底质，池塘常年水深保持在1.5～3.0 m。池塘与海区由渠道和涵闸连通，并建有备用泵站，全年的大部分时间池塘水位通过海水涨潮保持，小潮汐时利用泵站抽提海水供水。该区域海水常年平均水温为19.9 ℃，水温变化范围在9.0～32.0 ℃，平均盐度24.8，盐度变化范围18.3～28.8。

在拟放养中华鲟的海水池塘附近修建面积为64 m^2（8 m×8 m）的水泥池8个作为海水驯化池。水泥池深0.8 m，接通淡水和海水水源，采用涡轮增氧机充气增氧。

1.3 试验方法

2000年4月中旬，首先将中华鲟亚成体从莆田养殖基地长途运输至长江水产研究所厦门中华鲟养殖基地（厦门松浩实业有限公司同安鲟鱼养殖场），暂养在直径为8～10 m的圆形水泥池内，水源为同安水库下排水。试验鱼暂养约7 d后基本稳定，开始进行海水驯化试验。

海水驯化开始前，预先将水泥驯化池内水体的盐度调节至3，然后按计划将中华鲟亚成体从同安养殖场运输至养虾场（运输时间约90 min，运输工具为活鱼运输车），并直接进入驯化池暂养和驯化。每5 d按照盐度5的幅度逐渐增加驯化池水体的盐度，当盐度达到23时将中华鲟放入池塘（池塘水体盐度为23.6），完成一组驯化试验需要的时间为20 d。试验中，受条件限制未设置专门的盐度调节池，试验水体的盐度直接在驯化池中调节。具体方法是，首先用拦网将试验鱼集中在驯化池一侧（约2/3），然后在另一侧设置比重计，沿池壁同时慢速加注海水和淡水，观察比重变化，调节海水和淡水注入量。驯化期间按照中华鲟成鱼养殖方式进行正常投饵、排污和管理，不间断充气保证水体溶氧量保持在5 mg/L以上。试验期间水温为17.1～21.2 ℃。采用RH-06型海水比重计（测量范围1.000～1.060，精度0.001）测量海水比重并换算成盐度。试验期间认真观察中华鲟的行为反应，做好记录。

2 结果和讨论

2.1 中华鲟亚成体海水驯化的成活率

8个驯化池分8组进行试验。根据试验鱼的规格，每组试验所使用的试验鱼尾数变动范围在 180 ~ 190 尾(表 2),但每组试验的试验鱼总生物量均控制在 1 900 kg 左右。驯化试验结束后中华鲟的成活率为 92.1 % ~ 95.7 %,总成活率为 93.9 %。由于试验条件限制,中华鲟在从同安养殖场运输至养虾场后,没有进行暂养和适应,而是直接开始海水驯化试验。这可能对试验结果尤其中华鲟成活率产生一定的影响,但因运输时间较短,影响的程度应较轻微。

表 2　中华鲟亚成体海水驯化成活率

Table 2　The survival rate of sub-adult Chinese sturgeon acclimatized

组别　Group	1	2	3	4	5	6	7	8	合计 Total
放鱼数/尾 No. of test fish	182	190	188	183	186	187	184	180	1 480
成活数/尾 Survival No.	170	175	180	171	176	173	176	169	1 390
成活率/% Survival rate	93.4	92.1	95.7	93.4	94.6	92.5	95.7	93.9	93.9

2.2　中华鲟在海水环境条件下的养殖和生长

中华鲟在经过驯化放入池塘后,能够较好地适应池塘环境。根据 2002 年 11 月对养殖个体进行抽样测定的结果(表 3)。完成海水驯化的中华鲟亚成体在海水池塘条件下,经过近 30 个月的生长,其平均全长从试验开始时的 127.6 cm(范围 105.0 ~ 149.0 cm),增长至 157.0 cm(范围 150.0 ~ 162.0 cm);平均体重也从 10.685 kg(范围 5.8 ~ 15.0 kg),增重至 20.7 kg(范围 18.0 ~ 24.0 kg)。全长和体重的日均增长值分别达到 0.031 cm/d 和 10.983 g/d。对不同淡水养殖方式下,中华鲟的生长速度进行比较的结果(杨德国等,2006),工厂化养殖、网箱养殖和池塘养殖中华鲟体重的日均增重分别为 11.89 g/d,6.14 g/d 和 4.48 g/d,海水池塘中养殖的中华鲟亚成体的日均增重也达到了 10.983/d,表明中华鲟在海水池塘中具有较好的生长速度。

表 3　中华鲟亚成体在海水池塘的生长

Table 3　The growth of sub-adult Chinese sturgeon in the seawater pond

项目 Item		平均全长/cm　Mean total length	平均体重/(g/尾) Mean body weight
测量时间　Date of test	2000-4-12	127.6(n = 20)	10 685(n = 20)
	2002-11-3	157.0(n = 6)	20 700(n = 6)
增加值　Increment		29.4	10 115
日均增长　Increment per day		0.032	10.983

为减少人为操作对中华鲟的刺激和影响,试验期间未进行中华鲟成活情况的专门检查,但在 2002 年底进行了拉网采样抽查。根据抽查结果以及日常管理记录的死亡鱼类数量计算,从完成驯化下塘至 2002 年底抽样期间,本试验条件下,中华鲟的成活率约为 50 %。海水池塘养殖中华鲟试验成活率较低的原因,主要是在养殖试验的前期阶段,因没有或仅对试验鱼投喂少量的人工饵料,试验鱼基本以池塘内混杂的野杂鱼、虾为饵料来源,饵料供应不足现象严重,造成试验鱼体质差。当夏季海水池塘水温升高(7 ~ 8 月超过 30 ℃)时,一些体质差的个体不能抵抗较长时间的高温,陆续出现死亡,降低了养殖的成活率。根据养殖试验期间日常管理和中华鲟死亡情况记录分析,试验中中华鲟的死亡时间,明显集中分布在每年的 7 ~ 10 月份,即养殖过程中死亡鱼是在放养数月后,在夏季高温期时出现的。由此推断,养殖成活率低应主要与偏冷水性的中华鲟对夏季池塘高水温环境不能较好适应有关,而与中华鲟对盐度的适应性没有直接的关系。

2.3 海水驯化期间中华鲟的行为反应及耐盐性分析

中华鲟是一种江海洄游的鱼类,其生物学属性决定了它具有一定的耐盐性。当中华鲟亚成体从淡水直接进入3盐度水体时,其行为与短途转运后的个体行为没有差异,一般在20~30 min内即进入正常的游动状态,未出现明显的呼吸频率加快、快速窜游、跳跃等应急反应现象。当水体盐度按照设计浓度梯度逐渐增加时,个体行为也无异常现象出现。在整个驯化期间,除了试验开始时因运输造成短时间(一般1 d)停食外,其他时间均能正常摄食,说明中华鲟亚成体对水体盐度的变化不敏感,具有一定的耐盐性。不过,由于本试验未在高于盐度23以上条件下进行,其耐盐性的盐度上限目前尚不清楚,今后有待进一步研究。

鱼类属于恒渗动物,主要通过肾脏和鳃对离子的转运来调节渗透压,海淡水或咸淡水洄游的鱼类一般都能适应较大的水体盐度变化(杨秀平,2002)。Krayushkina 等(2001)对海洋洄游性的短吻鲟(*Acipenser brevirostrum* LeSueur)和大西洋鲟(*Acipenser oxyrhynchus* Mitchill)亚成体进行盐化试验,测量其血清、尿及肠道液体的渗透压及 Na^+,K^+,Ca^{2+},Mg^{2+}离子浓度,观察肾脏及鳃上皮氯细胞结构,发现鲟类的肠道、肾脏和鳃上皮氯细胞在其渗透压调节中起到重要作用,认为不同种类鲟鱼(海淡水洄游型、咸水型、淡水型)肾脏结构的差异及鳃上皮氯细胞结构变化的程度与其对水体盐度的适应能力有一定关系,如海、淡水洄游性的短吻鲟和大西洋鲟肾单位的近端小管由两段组成,而淡水型(如匙吻鲟 *Acipenser transmontanus*)及咸水型(如鳇 *Huso huso*)种类则没有第2段近端小管;盐化后的短吻鲟和大西洋鲟鳃上皮氯细胞变化较咸水性的俄罗斯鲟(*Acipenser gueldenstaedtii*)更显著,因此海、淡水洄游性鲟类对水体盐度变化有更强的适应能力。本试验的结果对此有所证实。

2.4 不同海水驯化方式的比较及其保护生物学意义

我国对鲟类的海水驯化研究较少,常抗美等(2003)曾对杂交鲟(达氏鳇×史氏鲟)苗种进行过试验,平均全长11.7 cm,平均体重6.5 g的苗种,采用每2~3日调高水体盐度3的方法,经过36 d的驯化,可适应盐度25以上的海水,驯化成活率达到98 %,驯化期间杂交鲟可以正常生长。方彰胜等(2004)曾进行过中华鲟和史氏鲟的咸化试验,试验用中华鲟的规格为每尾体重1.0~1.3 kg,当采用连续递增盐度1的方法进行驯化时,中华鲟的平均成活率为77.6 %,开始出现死亡的盐度为13,出现最高死亡率的盐度则为23;当采用阶段性递增盐度驯化(每天递增1 ‰,每递增5 ‰后稳定5 d,然后继续下一阶段驯化)时,中华鲟的平均成活率为98.3 %,开始出现试验鱼死亡的盐度为29,最高死亡盐度为30,而无论采用何种驯化方式,史氏鲟的海水驯化成活率都低于中华鲟。本试验采用阶段性递增盐度但不设稳定期的方法驯化中华鲟亚成体,缩短了海水驯化的时间,但仍然取得93.9 %的平均成活率,进一步证实了中华鲟具有较广耐盐性的特有生物学属性。

通常,同种鱼类的较大个体比较小个体对盐度的变化有较强的适应能力,这可能与小个体有比较大的体表面积,需要付出比较多的能量才能调节水和离子的渗透压平衡有关(杨秀平,2002)。参照其他作者的试验结果进行比较分析表明,中华鲟对水体盐度的适应能力也与个体规格大小有一定的关系,即个体的规格越大,则其对水体盐度变化的适应能力就越强。这与 Altinok 等(1998)在对海湾鲟(*Acipenser oxyrinchus desotoi*)的耐盐性试验中得到的结论相似。这些结果,对确定中华鲟人工增殖放流的规格、不同规格中华鲟放流地点的选择以及中华鲟海水养殖的开展都具有一定的参考价值。而关于各种规格中华鲟在不同盐度的驯化时间和成活率,尚需要进行更多的试验研究才能确定。

参考文献

常剑波,曹文宣.1999.中华鲟物种保护的历史和前景.水生生物学报,23(6):712-720.

长江水系渔业资源调查协作组.1990.长江水系渔业资源.北京:海洋出版社:1-281.

常抗美,王日昕,刘晓丹,等.2003.杂交鲟的海水驯化试验.中国水产(1):39-40.

邓昕,邓中磷.1997.中华鲟的保护生物学研究进展.动物学研究,18(1):113-120.

方彰胜,王昀,李建军,等.2004.鲟鱼咸化试验.水利渔业(2):45-46.

四川省长江水产资源调查组.1998.长江鲟鱼类生物学及人工繁殖研究.成都:四川科学技术出版社:284.

汪小炎.1994.中国国家重点保护水生野生动物.北京:中国科学技术出版社:232.

危起伟,柯福恩,庄平,等.1994.论长江口中华鲟幼鱼的保护//第一届东亚地区国家公园与保护区会议暨 CNPPA/IUCN 第41届会议.北京:中国环境科学出版社:786-793.

危起伟,杨德国,柯福恩.1998.长江鲟鱼类的保护对策//21 世纪长江大型水利工程建设中的生态与环境保护(中国科协第十九次"青年科学家论坛").北京:中国环境科学出版社:208-216.

杨德国,危起伟,王凯,等.2005.人工标志放流中华鲟幼鱼的降河洄游.水生生物学报,29(1):26-30.

杨德国,危起伟,王凯,等.2006.淡水养殖中华鲟成鱼的3种方式及效果比较.南方水产,2(2):1-5.

杨秀平.2002.动物生理学.北京:高等教育出版社:417.

张世义.2001.中国动物志-硬骨鱼纲:鲟形目海鲢目鲱形目鼠鱚目.北京:科学出版社:209.

Altinok I, Galli S M, Chapman F A. 1998. Ionic and osmotic regulati on capabili ties of juvenile Gulf of Mexico sturgeon, *Acipenser oxyrinchus desotoi*. Comparative Bioch emis try and Physiology Part A: Molecular & Integrative Physiology, 120(4):609-616.

Krayushkina L S, Gerasimov A A, Smirnov A V. 2001. Hypoosmotic regulat ion in anadrcmous marine stu rgeon, with special reference to the structure and functi on of their kidneys and gill ch loride cells. Dok lady Biological Sciences, 378(3): 210-212.

Acclimation Experiment of Sub-adult Chinese Sturgeon (*Acipenser sinensis* Gray) from Fresh Water Culture in Sea Water

Yang Deguo[1,2]　Zhu Yongjiu[1]　Wei Qiwei[1,2]
Chen Xihua[1,2]　Liu Jianyi[1,2]　Wang Kai[1]

(1. Key Laboratory of Freshwater Fish Germplasm Resources and Biotechnolog y, Ministry of Agriculture, Yangtze River Fisheries Research Institute, Chinese Academy of Fishery Sciences, Jingzhou 434000;
2. Freshwater Fisheries Research Center, Chinese Academy of Fishery Sciences, Wuxi 214081)

Abstract Acclimation experiment of sub-adult Chinese sturgeon (*Acipenser sinensis*) in sea water was carried out in 2000. The fish used in the test was the first filial generation and had been bred in fresh water for about 3 years, its total length and body weight being (127.6 ± 13.1) cm (105 ~ 149 cm, n = 20) and (10.7 ± 2.88) kg (5.8 ~ 15.0 kg), respectively. The salinity in test water increased by 5 per 5 days in the test period. After 20 days, the salinity reached 23, and the average survival rate of test fish was 93.9 % (92.1 % ~ 95.7 %). Abnormal behaviors of Chinese sturgeons were not found as the salinity in test water became raised gradually. Obviously, the fish could endure the increased salinity and live normally in the seawater. The resultsalso showed that Chinese sturgeon could adapt well to the wide salinity range in water. In addition, the ability of Chinese sturgeon to adapt to salinity variation was possibly related to the body size, viz., the bigger the body of the fish, the stronger was its ability to adapt to the variation of salinity.

Key words *Acipenser sinensis*　Sub-adult　Salinity　Acclimation

中华鲟苗种工厂化养殖系统设计与应用*

杨德国[1,2,**]　危起伟[1,2]　朱永久[1]　杜　浩[1,2]　李罗新[1]

(1. 中国水产科学研究院长江水产研究所 农业部淡水鱼类种质资源与生物技术重点开放实验室　湖北荆州　434000
2. 中国水产科学研究院淡水渔业研究中心　江苏无锡　214081)

摘　要　根据中华鲟苗种的特殊生物学和生态学习性,设计出一套中华鲟苗种大规模批量培育的养殖系统。系统包括养殖容器、供水及水处理、水温调节、增氧、饵料供应等配套设施。通过多池型、多处理措施的综合协调处理方案,实现中华鲟苗种培育对养殖容器物理和水质环境的特殊要求,提高了系统的可靠性和生产效果,使中华鲟批量苗种培育成活率达到80%以上。

关键词　中华鲟　苗种培育　工厂化养殖系统　设计

中华鲟(*Acipenser sinensis* Gray)隶属于鲟形目鲟科鲟属,以个体硕大、生长快、寿命长而闻名,是国家一级保护野生动物。中华鲟曾是长江上游重要的渔获对象,由于水利工程建设、水质污染、过度捕捞等因素的影响,中华鲟的自然种群呈逐年衰退趋势[1]。20世纪70年代,我国首次在长江上游实现了中华鲟野生亲鱼的人工催产,并孵化出少量鱼苗[2]。葛洲坝水利枢纽修建后,长江水产研究所等单位组成的协作组于1983年在葛洲坝下也取得中华鲟人工繁殖成功[3],使中华鲟人工繁殖技术逐渐趋于完善,并开始进行较大规模的中华鲟人工放流工作。但由于该期间未能解决中华鲟幼鲟(10 cm以上大规格苗种)的培育技术难题,放流的中华鲟基本以刚孵出的仔鲟(卵黄囊苗,又称"水花",一般均不到10日龄)为主。如1984~1998年14年间(其中1994年因故未出苗),葛洲坝集团公司中华鲟研究所(原葛洲坝工程局水产处)共孵化出苗760.0万尾,最终放流仔鲟425.8万尾,放流幼鲟6.28万尾,放流幼鲟的比例仅为总放流量的1.45%,其主要原因就是幼鲟的培育技术未解决。在1995年以前,该单位仅极少数年份培育出幼鲟,但培育成活率一般都低于10%,每年生产的中华鲟幼鲟数量也极为有限,未能突破1万尾[4]。因此,解决中华鲟幼鲟培育尤其是批量培育的技术难题,建立适宜的中华鲟幼鲟批量生产技术和方法,就成为开展中华鲟幼鲟大规模放流的关键。

1　中华鲟苗种培育的主要方法

培育鲟鱼类苗种的方法主要有池塘培育和养殖工厂集约化培育,培育方式与所养殖的种类有较大的关系。池塘培育方式多应用在匙吻鲟、俄罗斯鲟等种类,其方法与其他经济鱼类苗种的培育方法相似,属于粗放型培育方式,在俄罗斯和一些独联体国家(包括苏联时期)以及伊朗等国得到部分应用[5]。而采用养殖工厂集约化培育则是目前鲟鱼苗种生产的主流方法,被多数鲟鱼养殖国家所采用[6]。发达国家建设的这类系统,多采用室内循环水养殖方式,系统自动化程度高,设备投资和运行的成本都较大。

---

* 基金项目:科技部社会公益研究专项(2000D IB50177);科技部科技基础性工作专项(2002DEA1004);中国水产科学研究院科研基金项目(2003-3-9)。

** 第一作者,男,1964年出生,研究员,主要从事鱼类生态学和保护生物学研究。

我国在进行中华鲟苗种培育时,较早也采用池塘培育的方式[2,7],但在苗种培育的早期阶段,其效果欠佳,因此后来较少采用。1983 年,在葛洲坝下获得中华鲟人工繁殖成功,但该阶段由于各方面条件的限制,关注的重点主要是中华鲟的人工催产技术和受精卵孵化技术,孵出的中华鲟苗种一般在 10 日龄以前(未开口摄食)即被放流,大规格苗种的培育技术研究未得到重视。在该阶段也进行了少量苗种集约化培育技术的试验,如曾有单位采用小型水泥池或玻璃钢圆形池培育中华鲟幼鱼,虽然有些年份也能培育出数千尾 10 cm 左右的幼鲟,但由于未能发现制约中华鲟苗种培育成活率的关键影响因素,尤其在中华鲟苗种培育设施的特殊要求方面重视不够,因而未能从根本上解决中华鲟苗种培育成活率低的问题,也未形成真正的中华鲟苗种工厂化养殖体系和技术。这也是我国中华鲟苗种培育成活率长期较低,不能形成大批量生产规模的主要原因。

2 中华鲟苗种工厂化养殖系统设计

2.1 制约成活率的主要因素

通过长期的试验和养殖实践观察发现,中华鲟苗种培育成活率与培育方式和养殖生态条件有非常密切的关系。主要体现在:

(1) 中华鲟的仔鱼期和幼鱼期主要在长江干流流水环境生活,对水域生态环境尤其是水体溶解氧具有较高要求。

(2) 中华鲟是一种底栖生活的鱼类,刚孵出的水花初期一般静卧水底,需要经历一段时间后才间歇性向水面串游;其卵黄囊硕大,且卵黄囊壁娇嫩,池塘底粗劣的环境和淤泥导致的低氧或负氧状态显然对其生活和生长极为不利。

(3) 中华鲟在秋季繁殖,苗种培育的生产季节是每年冬季,苗种生产的主要地点在长江中游的湖北地区,冬季时该地区池塘、湖泊、水库等自然水域的水温都较低;2000 年曾对太湖港水库(系养殖系统水源之一,面积约 233 hm^2)的水温进行周年监测,1,2,3,11,12 月的平均水温分别为 7.5 ℃,9.6 ℃,14.9 ℃,14.1 ℃,11.1 ℃,且水温的日变化幅度也较大,这种现象在池塘水体中则更为明显。

观察发现,在人工条件下,中华鲟苗种最适宜的水温是 18 ~ 22 ℃,而且水温的日变化幅度超过 3 ℃,对苗种的成活率有灾难性的影响[8]。此外,由于中华鲟苗(水花)具有特有的底栖群居生活习性,仔鱼在早期主要聚集在培育容器的底部和边角生活,而刚孵化的中华鲟苗有硕大的卵黄囊,在鱼体与容器壁的摩擦中,极易导致卵黄囊膜和鱼体腹部破损,故容器底部和四壁的物理特性(光滑度)对其成活率有非常明显的影响。

如上所述,中华鲟对栖息环境的水质条件具有较高的要求,养殖容器底部和边角往往又是最容易积淀污物的区域,仔鱼的群聚效应则对该区域的水环境提出了更高的要求。因此,相对恒定的水温,保持养殖容器光滑洁净的环境和稳定优良的水质就成为提高中华鲟苗种培育成活率的关键,池塘培育方式显然难以满足这些要求。

2.2 系统的构成及工艺设计

中华鲟苗种工厂化养殖系统由养殖容器和辅助设施构成。根据中华鲟苗种发育和生长的特点,将中华鲟苗种(规格 10 cm 以内)培育划分为两个阶段,使用不同的养殖容器。辅助设施包括供水、调温、增氧和生物饵料供应设施。其中养殖容器、增氧机和生物饵料池安排在一个 60 m ×60 m 的育苗车间内,其他辅助设施在车间外。养殖系统的总体构成和工艺流程如图 1。

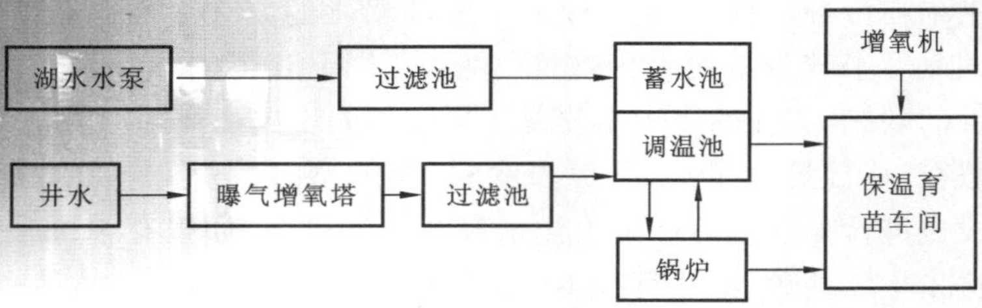

图1　中华鲟苗种工厂化养殖系统工艺流程

该系统诸单元的工艺设计则主要从解决上述中华鲟苗种培育的几个关键制约因素着手，通过多种措施综合运用，为中华鲟苗种营造良好的养殖生态环境。

2.2.1　养殖容器

根据中华鲟苗种不同发育阶段具有不同生活习性和行为的特点，采用分阶段使用不同养殖容器的方案。通过对养殖容器池型、规格、建造材料和生产工艺的不同设计，满足不同发育阶段苗种对容器底质环境的特殊要求。中华鲟卵黄囊苗身体娇嫩，游泳能力差，故采用直径为1 m的小型平底圆形容器，使仔鱼能够尽量分散在容器底部。容器制作时采用聚乙烯一次成模加工工艺，保证容器底部和内壁均具有良好的光洁度。容器的给水方向不固定，随着仔鱼生长其主动游泳能力逐渐增强，可逐渐调整给水角度，使容器内水体形成一定的流动状态，结合容器内壁光滑的特点和底部气石的共同作用，降低污物的沉积和附着效应，使容器底部具有良好水质环境。仔鱼全长到达5 cm时，已具有一定的游泳能力，此时则转移至直径为3.5 m的水泥池养殖。由于该期间中华鲟底栖性生活的特点仍未改变，对容器底质和底部水质环境仍有较严格要求，为保持底部一定光洁度和排污效果，采用圆形、加大底部坡度（坡比6 %）、底部和内壁贴瓷砖的方案，保证底部光滑和排污效果。

2.2.2　养殖水温

养殖系统水温的调控采用多种措施综合运用方案。首先是利用深井水水温恒定（18 ℃）的特点，在苗种生产期，以深井水作为主要的水源，天然湖水仅作为补充水源；其次是对育苗车间的建筑采用保温设计，为节省建设投资，育苗车间采用轻钢结构和简易塑料大棚屋面，冬季保温性能差，为此在大棚屋面下加入了一层厚度为2 cm的泡沫板保温层，加强车间保温效果；三是为养殖系统配备了1台2 t的热水锅炉，当苗种生产高峰期深井水水量不足时，利用锅炉对调温池的水进行辅助升温，该过程由一套水温监测和控制系统自动控制，同时在车间内布设供暖管道和散热器，当气温非常低时，将车间室内气温提高，并保持相对恒定，如此则有利于保持养殖容器的水温恒定。通过这些措施的综合运用，系统基本可保持冬季养殖容器水温恒定在18～22 ℃，水温日变化幅度一般不超过2 ℃，同时也相对节省能源消耗量。

2.2.3　养殖水质

养殖系统的水质通过以下措施予以保障。首先确保水源水质良好，对水源均进行过滤和增氧处理，其中深井水经过充分曝气增氧后，采用石英砂自流过滤，以去除氧化后的重金属盐；而湖水则直接泵入配套的过滤塔（石英砂）过滤。经过处理的深井水，其溶解氧一般可从2.8 mg/L提高到6.5 mg/L，总铁含量由1.25 mg/L下降至0.81 mg/L；而湖水经过滤后，其透明度则可从0.7 m提高到2.0 m以上。其次，在养殖容器结构和给排水工艺设计中，采用可提高养殖容器水体自净效果的设计方案。系统通过采用圆形、光滑的养殖容器，以及可随意调节给水角度的切角给水设计，在保障给水量的同时，使容器内部形成一定的、可随意调控流速大小的环流,这样既满足了苗种不同发育阶段对水体流速的需求,也可使容器内的污物随环流向设置于容器中部的排污口聚集,通过溢流式水位控制排污管,在水体交换过程中,大部分污物可及时、自动地排出容器,保证养殖容器内的水质始终处于较好状态。此外,系统设计采用螺旋增氧机和气石增氧方式,在为水体增氧的同时,可保持养殖容器底部形成局部的上升流,避免污物在容器底部沉积,提高排污效果。

3 系统特点

中华鲟苗种工厂化养殖系统对制约养殖效果的关键因素均采用多种措施协调作用的处理方案。当某个设施出现故障时,其他设施仍然可以为系统提供一定的支持,以提高系统整体运行的可靠性和稳定性。此外在整个系统设计中,实现了一种措施可在多个养殖环节中发挥作用的效果,提高了系统的整体运行效率。

系统的另外一个特点是易于建设,经济实用,投资节省。系统采用我国常见的建筑材料和设备,制作工艺简单,建设成本仅为国外同类系统投资的 1 /3;与国内类似系统比较,也节省投资 50 %以上[9],而养殖效果相当,并适用于多种鱼类苗种生产,具有通用性。

4 应用效果

中华鲟苗种工厂化养殖系统于 1999 年 10 月投入使用,当年即生产体长 10 cm 左右中华鲟幼鲟 5.73 万尾,苗种成活率达到 80.2 %。1999 ~ 2004 年,每年系统培育出体长 10 cm 左右幼鲟数量为 5.24 万 ~ 48.46 万尾,多年平均为 14.94 万尾,其中 2000 年生产出体长 10 cm 苗种 48.46万尾,培育成活率达到 80.2 % ~ 91.6 %,实现了中华鲟苗种的批量培育,且效果良好。在非中华鲟生产季节,如春季和夏季,系统也用于其他鱼类尤其珍稀濒危和名贵经济鱼类苗种的生产,包括史氏鲟、杂交鲟、胭脂鱼、美洲鲥、黄颡鱼、大口鲇、长吻鮠等,苗种培育成活率也稳定在 80 %以上,实现了系统的全年生产,提高了设施利用率。

受场地条件和资金的限制,系统目前没有配套建设相应的循环水水处理系统,也没有配套建设相应的亲鱼养殖池塘,使冬季生产的能源消耗相对较高,同时系统的全年生产能力也没有得到完全发挥。

参 考 文 献

[1] 汪松,乐佩琦,陈宜瑜. 中国濒危动物红皮书:鱼类[M]. 北京:科学出版社,1998:247.
[2] 四川省长江水产资源调查组. 长江鲟鱼类生物学及人工繁殖研究[M]. 成都:四川科学技术出版社,1988:284.
[3] 傅朝君,刘宪亭,鲁大椿,等. 葛洲坝下中华鲟人工繁殖[J]. 淡水渔业,1985(1):1-5.
[4] 肖慧,刘勇,常剑波. 中华鲟人工繁殖放流现状评价[J]. 水生生物学报,1999,23(6):572-576.
[5] Chebanov M, Billard R. The culture of sturgeons in Russia: production of juveniles for stocking and meat for human consumption[J]. Aquatic Living Resources, 2001, 14(6): 375-381.
[6] Williot P, Sabeau L, Gessner J, et al. Sturgeon farming in Western Europe: recent developments and perspectives[J]. Aquatic Living Resources, 2001, 14(6): 367-374.
[7] 赵云节,杜军. 中华鲟在成都地区池塘养殖试验报告[J]. 动物学杂志,1997(5):41-43.
[8] Zhu Y, Wei Q, Yang D, et al. Large-scale cultivation of fingerlings of the Chinese Sturgeon *Acipenser sinensis* for re-stocking: a description of current technology[J]. Journal of Applied Ichthyology, 2006, 22 (Suppl 1): 238-243.
[9] 刘艳红,罗国芝,孙佩英,等. 在闭合循环水产养殖系统中养殖高体革鯻的效果[J]. 上海水产大学学报,2003,12(2):130-134.

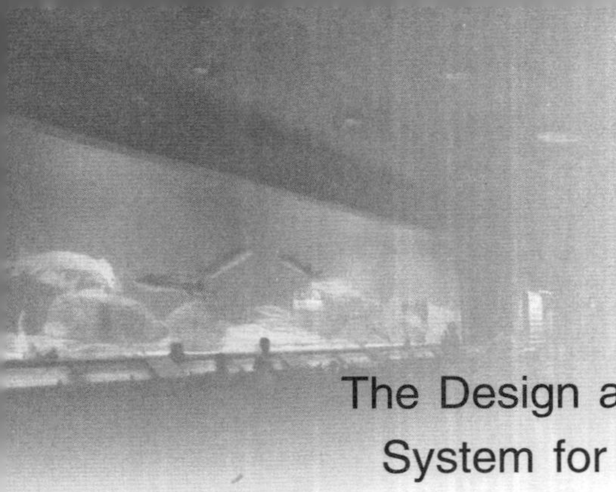

The Design and Running of a Recirculating Aquaculture System for Chinese Sturgeon Larvae and Fingerling

Yang Deguo[1,2] Wei Qiwei[1,2] Zhu Yongjiu[1] Du Hao[1,2] Li Luoxin[1]

(1. Key Laboratory of Freshwater Fish Germplasm Resources and Biotechnology, Ministry of Agriculture, Yangtze River Fisheries Research Institute, Chinese Academy of Fishery Sciences, Jingzhou, Hubei 434000;
2. Freshwater Fisheries Research Center, Chinese Academy of Fishery Sciences, Wuxi, Jiangsu 214081)

Abstract According to the special biological and ecological characteristics of Chinese sturgeon larvae, an aquaculture system which can mass batch cultivate Chinese sturgeon larvae has been designed. The system including a series of ancillary facility, such as breeding containers, water supplier, water treatment, water regular, aerator and feeder. It can realize the special requirement of Chinese sturgeon larvae for physical environment of breeding containers and water quality through more pool types and measures of comprehensive coordination treatments, and it also can increase the reliability of the system and production effect, to make the survival rate of batch cultivated Chinese sturgeon larvae more than 80 %.

Key words Chinese sturgeon Larval rearing Recirculating aquaculture system Design

中华鲟苗种的集约化生产系统*

杨德国[1,2,**]　危起伟[1,2,***]　朱永久[1]
刘鉴毅[1,2]　王　凯[1]　陈细华[1,2]　李罗新[1]

(1. 中国水产科学研究院长江水产研究所 农业部淡水鱼类种质资源与生物技术重点开放实验室　湖北荆州　434000
2. 中国水产科学研究院淡水渔业研究中心　江苏无锡　214081)

摘　要　以满足中华鲟苗种培育关键生态要素为依据，研制出中华鲟(*Acipenser sinensis* Gray)苗种集约化生产系统，实现了中华鲟幼鲟的批量生产。该系统由生产主设施系统和配套设施系统构成，其中主设施包括4种不同规格养殖池共2 534.8 m^2 及授精和孵化设施；配套设施系统包括亲鲟运输车、供水及水处理、增氧、保温升温、饵料供应及其他辅助设施。该系统鱼池采用不同的池型、建筑材料和制作工艺以适应苗种不同发育阶段生态要求，水温可调控，进排水方式可灵活调节，自流式排污，保证了池水的不同流态及良好稳定的水环境。该系统4个月可催产中华鲟亲鲟16尾以上，孵化166.4万粒中华鲟受精卵，抚育235.5万尾1～15日龄仔鱼，培育全长10～15 cm稚鱼31.31万尾。该系统的设施具有较好通用性，也适用于其他鱼类特别是珍稀名贵鱼类苗种的集约化批量生产。

关键词　中华鲟　人工繁殖　苗种培育　集约化系统

0　引言

中华鲟(*Acipenser sinensis* Gray)是国家一级保护野生动物。葛洲坝水利枢纽修建后，其生殖洄游通道被阻隔，在葛洲坝下江段形成的新产卵场范围有限，人工繁殖放流成为保护和增殖其自然资源的主要措施之一[1-6]。20世纪70年代，四川省水产资源调查组利用捕捞自金沙江的成熟亲鱼，栓养在江边进行人工催产获得受精卵，采用环道孵化、池塘育苗的方法获得少量鱼苗[7]。1983年，长江水产研究所等单位组成协作组在葛洲坝下取得中华鲟人工繁殖成功[8]；1984年，葛洲坝集团公司组成中华鲟研究所，实施中华鲟的人工放流工作。此后，中华鲟研究所和长江水产研究所几乎每年都进行中华鲟的人工繁殖和放流。但是，1995年以前，由于未能解决中华鲟幼鲟(体长10 cm左右，体重3～5 g的个体)培育的技术难题，人工培育幼鲟的成活率一般都在10 %以下，仅极少数年份达到20 %左右，每年培育幼鲟的数量基本上未能突破1万尾[9]。因此，放流的中华鲟个体绝大多数属于刚孵出的仔鲟(即“水花”)，放流苗种的成活率低，无法保证增殖效果。要实现大规格中华鲟苗种的规模放流，必需首先解决中华鲟幼鲟的培育技术难题。1996年开始，经过多次试验和比较，发现了影响中华鲟苗种培育成活率的一些关键生态因素[10]。然后，针对中华鲟人工繁殖和苗种培育不同阶段的特殊生态要求，对相应的设施进行了重新设计，改进了养殖工艺，对生产流程各环节进行了规范，使之配套和系统化，大幅度提高了培育人工繁殖中华鲟苗种的成活率，并于1999年在湖北省荆州市建

---

* 基金项目：中国长江三峡工程开发总公司三峡工程环境保护补偿经费科研类(施工区外)项目(SX96-1/HB，SX971-25/HB)；科技部社会公益研究专项(2000DIB50177)；科技部科技基础性工作专项(2002DEA1004)；中国水产科学研究院科研基金项目(2003-3-9)。

** 第一作者，男，1964年出生，湖北荆州人，研究员，从事鱼类生态学和保护生物学研究，Email：yangdg@ yfi. ac. cn

*** 通讯作者(Corresponding author)，男，1960年出生，江西南昌人，博士，研究员，从事保护生物学研究，Email：weiqw@ yfi. ac. cn

设了一个中华鲟人工繁殖和苗种培育生产基地。1999～2004 年期间，利用该系统共进行 5 次(2002 年因故未进行)中华鲟人工繁殖生产，每次培育的中华鲟幼鱼(体长 10～15 cm)均在 5 万尾以上，实现了中华鲟苗种的集约化批量生产。

1 生产流程及系统构成

中华鲟苗种生产流程包括亲鲟捕捞、运输、暂养、人工催产、授精、孵化和苗种培育 7 个环节，其生产系统由主设施分系统和配套设施分系统组成(图 1)。其中主设施包括各种养殖容器、授精和孵化设施；配套设施包括亲鲟运输、供水及水处理、增氧、保温升温、饵料供应及其他辅助系统。系统运行时，亲鲟暂养和催产在室外，人工授精、受精卵孵化、苗种培育、生物饵料培养及人工饵料加工等均在一座 60 m×60 m 的车间内进行。

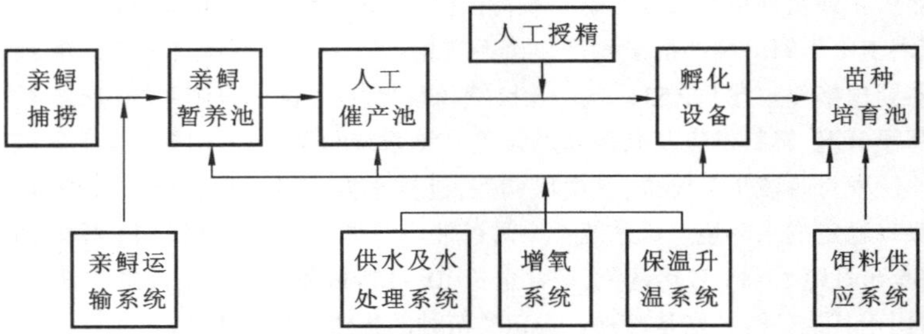

图 1　中华鲟苗种集约化培育系统

Fig. 1 Components of intensive breeding system of Chinese sturgeon larvae and fingering

2 设施及设计参数

2.1 养殖容器及孵化设施

根据中华鲟人工繁殖和苗种培育生产流程，该系统采用了三类不同规格和结构的养殖容器，以满足不同生产阶段功能的需要。其中一类为亲鲟池和催产池，主要用于野生亲鲟的暂养和催产操作，采用圆形砖混结构，因池型较大均建在室外，实际生产中可相互调剂使用。为满足中华鲟仔鲟和幼鲟对养殖环境的不同要求，设计、建造了两种苗种培育池，即仔鲟抚育桶[11]和幼鲟池，采用了不同的规格、池型和制作材料。容器详细设计参数和数量见表 1。

表 1　养殖容器设计参数及数量

Table 1　Designed parameters and number of the breeding ponds

类别	亲鲟池	催产池	仔鲟抚育桶	幼鲟池
形状	圆形锥底	圆形锥底	圆形平底	圆形锥底
材料	水泥池	水泥池	聚乙烯	水泥池
直径/m	12.0	16.0	1.0	3.0
高度/m	1.8	2.0	0.6	0.8
底坡比/%	6.0	6.0	0	10.0
面积/m^2	113.0	200.9	0.8	7.1
鱼池数/个	4	4	300	147
总面积/m^2	452.0	803.6	235.5	1 043.7

注：幼鲟池直径指鱼池内径，面积为有效面积。

该系统的孵化设施包括孵化池和孵化框两部分。孵化池为长方形，规格 4.0 m×1.0 m×0.5 m(长×宽×高)，共 12 个，砖混结构，采用 PVC 立管控制池内水位至 0.4 m，其上方纵向平行布置 2 排淋浴喷头，每排 10 个，喷头离池面距离为 0.35 m。每个孵化池可同时放置 14 个孵化框。孵化框采用杉木条作框架，其底部和四边的空隙处用腈纶窗纱布(40 目)封闭，孵化框规格为 55 cm×45 cm×20 cm(长×宽×高)，要求框架光滑，纱布绷紧无凹陷。该系统共配备孵化框 350 个。苗种培育池的主要工艺要求是池底和池壁光滑，无死角，水体流态可自由调整。

2.2 配套设施

2.2.1 亲鲟运输系统

采用东风 5 t 轻型货车改装的带吊臂多功能活鱼运输车 1 辆。活鱼箱采用不锈钢箱体并打磨光滑，箱体规格 3.4 m×1.9 m×0.8 m，箱体上部表面两侧及后部各安装宽 0.3 m 活动挡板一块，装卸亲鱼时开启，亲鱼入箱后闭合，保证运输时亲鱼和水体不溢出。随车吊臂动力由汽车引擎提供，最大起吊重量 2 300 kg，最大臂长 5 m，水平旋转角度 270°。活鱼箱配本田 2.2 kW一体式水泵及管道一套，另装备纯氧(氧气瓶和气石)增氧系统一套。运输亲鲟时，水泵和纯氧同时使用，使亲鲟有流水刺激，保持活鱼箱水体溶解氧 8 mg/L 以上[12]。

2.2.2 供水及水处理系统

水源两种，以紧邻基地的太湖港水库湖水为主水源，深井水为辅助水源，在冬、夏季协助调节水温。供水系统包括水泵房和深井潜水泵、储水塔、过滤塔及管道。水泵房配 2 套 11 kW 和 1 套 7.5 kW 低扬程水泵，可单独或同时运行，根据储水塔水位自动控制，总供水能力 600 m^3/h。深井 2 口，15 kW 潜水泵提水，手动控制，供水能力 120 m^3/h。储水塔 1 座，储水量400 m^3，净水位1.8 m 以维持供水压力。水源经不同的管道(PVC 材料)系统输送至暂养池、催产池和车间(图 2)。

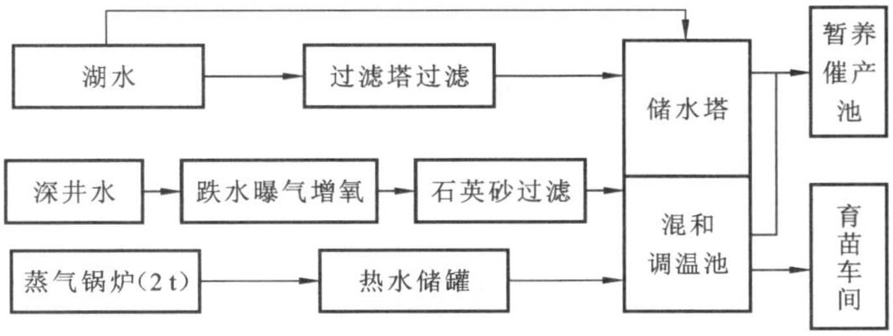

图 2　供水系统流程

Fig. 2　Flow chart of water supply in the intensive breeding system of Chinese sturgeon larvae and fingerling

供水系统配有 2 套水处理设施。其中湖水采用压力式过滤塔过滤、紫外线消毒，水处理能力 100～120 m^3/h。井水则采用跌水盘曝气增氧、石英砂过滤塔自流过滤，处理能力 80 m^3/h。湖水处理后透明度显著升高，总铁含量有所下降；井水处理后溶解氧显著升高，总铁由 1.25 mg/L下降至 0.81 mg/L，能满足中华鲟苗种培育的水质要求。井水处理后透明度反而有所下降，主要是因为原井水铁含量高，须充分曝气才能使所有铁离子氧化为絮状化合物被过滤，本系统采用简易跌水曝气增氧方式，加上滤前储水池容积小，故有部分铁离子在滤后才被氧化，降低了水体透明度(表 2)。进入蓄水池的这些絮状物通过其自然沉淀作用，大部分不会进入供水管道。

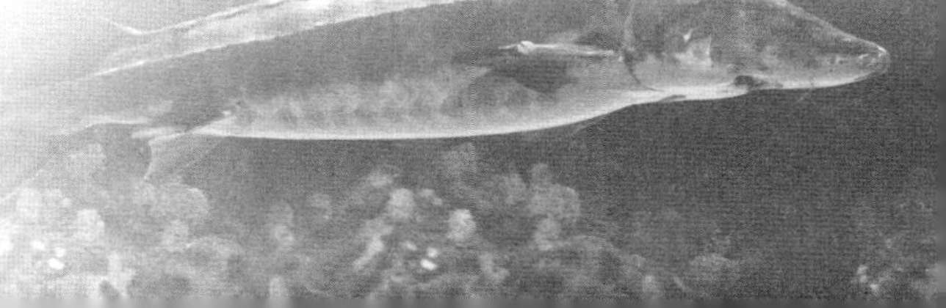

表 2　水处理前(A)后(B)湖水和井水 4 理化因子水平

Table 2　Values of four factors in lake water and groundwaterbeing disposed fore-and-aft

	pH		DO/(mg/L)		总铁/(mg/L)		透明度/m	
	A	B	A	B	A	B	A	B
湖水	7.8	7.8	8.2	8.2	0.68	0.55	0.7	2.0
井水	7.0	7.2	2.8	6.5	1.25	0.81	1.0	0.8

2.2.3　增氧系统

采用涡流式增氧机通过散气石对苗种培育池的水体充气增氧。每个仔鲟抚育桶安装 2 个散气石，幼鲟培育池安装 4 个，共使用 3 台 1.1 kW 和 2 台 2.2 kW 增氧机。

2.2.4　保温升温系统

配备 2 t 蒸气锅炉 1 台，50 m^3 钢制热水罐一座。当冬季水温和气温过低，井水不能调节水温至 18 ℃时，使用锅炉提升水温和车间室温。水温调节采用电子控温仪自动控制。为节省能源，培育车间采用保温设计，安装有暖气管道和散热器，车间顶棚加盖厚 2.5 cm 泡沫板保温。

2.2.5　饵料供应系统

车间内修建有 4 个条状水蚯蚓暂养、培育池，规格 55.0 m × 0.5 m × 0.3 m，总面积 110 m^2，以孵化池和苗种池的排水为水源。另外，配切肉机和小型饲料成型机各一台，用于水蚯蚓和软颗粒配合饲料(150 kg/h)加工。

该系统配套 150 kW 变压器提供电力供应，同时配 90 kW 发电机组一套为备用电源。

3　系统的生产能力及实例分析

3.1　系统生产能力

该系统主要针对中华鲟人工繁殖和苗种培育要求设计和建造，一个生产季节(约 4 个月)的生产能力见表 3。实际运行时，通过合理调度，系统部分阶段生产能力可大幅提升，如亲鲟暂养、催产、受精卵孵化和仔鲟抚育的规模可提高到设计能力的 2 倍。该系统全年均可生产，也适用于其他鲟鱼类、胭脂鱼、美洲鲥、黄颡鱼、大口鲶、长吻鮠等珍稀名贵鱼类的人工繁殖和苗种培育。

表 3　系统设计生产能力

Table 3　Capacity of the intensive breeding system of Chinese sturgeon larvae and fingerling

设施	阶段或鱼长	数量	容纳量/(尾/m^2)或(粒/m^2)	总生产能力/尾或粒
暂养池	雌鲟	4	2/池	8
催产池	雌鲟	4	2/池	8
孵化框	受精卵[a]	168	40 000	1 664 000
抚育桶	1 ~ 15 日龄[b]	300	10 000	2 355 000
幼鲟池	3 ~ 5 cm	147	2 000	2 087 400
	5 ~ 8 cm		800	834 960
	8 ~ 10 cm		500	521 850
	10 ~ 15 cm		300	313 110

注：a：孵化框孵化受精卵数量 1 万粒/个；

b：聚乙烯抚育桶放养仔鲟数量 8 000 尾/个。

3.2 系统应用实例

系统于1999年9月底初步建成，当年10月投入使用。当年因建成初期，系统保温升温设备未配套，影响了苗种培育效果；但仍培育出体长10 cm左右中华鲟幼鲟5.73万尾，苗种成活率达到80.2%。1999年以后，除2002年外（该年度未进行中华鲟人工繁殖生产），每年均利用该系统进行中华鲟人工繁殖和苗种生产，5年共催产中华鲟亲鲟36尾，其中34尾成功产卵，年均获卵187.0万粒，多年平均受精率达到61.84%；年均孵化出卵黄囊苗92.69万尾，多年平均孵化率为45.1%。1999～2004年期间，每年在该系统培育出体长10 cm左右幼鲟5.24万～48.46万尾，多年平均达14.94万尾，苗种培育成活率80.2%～91.6%，平均86.4%，其中2000年生产出体长10 cm苗种48.46万尾，超过了设计生产能力。

4 讨论

本系统的技术和工艺有以下特点。

4.1 同时控制水源和车间室内温度保持水温恒定

研究表明，中华鲟稚鱼生长的最适水温是18～22 ℃，水温长期低于15 ℃时则很难成活，而水温的剧烈变化（每小时变化幅度超过2 ℃）也会显著降低苗种成活率[10]。中华鲟属秋季繁殖鱼类，苗种生产季节在冬季（主要是11月至次年3月），此时湖北地区同期池塘、湖泊、水库等天然水体的水温一般都低于18 ℃，且日变幅大。本系统运行时，将水源水温控制在18～20 ℃，车间室温控制在25～30 ℃，实现了养殖水体水温稳定在20 ℃左右（18～22 ℃）的效果。实践证明，同时调控水源水温和车间室温，对保持养殖水体水温的相对恒定非常重要，在气温极低时效果尤其明显。同时，采用这种调温方式还可降低系统运行能耗。

4.2 使用内壁光滑的养殖容器保持水质优良

培育鲟鱼类苗种主要有两种方法：一种是池塘培育，前苏联及现独联体国家较常见[13]；另一种是美国、德国等采用的室内循环水集约化培育[14]。池塘培育属粗放型生产，产量和成活率都较低，而集约化培育自动化程度高，投资和运行成本较大。在自然环境中，体长10～20 cm的中华鲟主要在长江干流流水环境中生活，对水质条件要求较高。中华鲟是一种底栖型鱼类，尤其刚孵化的中华鲟苗（水花）腹部有一个大的卵黄囊，且具有群居习性。内壁粗糙的养殖容器，其底部易积聚污物而形成局部污染，池塘的淤泥常使池底处于低氧或负氧状态，不利中华鲟成活；同时，粗糙的养殖容器底部或池塘底的杂物也会对鱼苗的卵黄囊造成直接损坏，最终均导致苗种成活率下降。本集约化养殖系统采用内壁光滑的养殖容器，污物（残饵和粪便）不易在容器壁（底）黏附和沉积，在较微弱的水流下也可达到较好的排污效果，使容器底部始终有较洁净的环境和养殖水体良好的水质。光滑的内壁也大大减少了容器壁对鱼苗卵黄囊的直接损伤，提高了存活率。

4.3 使用不同的池型和建材为不同发育期的苗种提供适宜的生态环境需求

不同发育阶段的中华鲟苗种，具有不同的生活习性和行为，本系统通过使用不同的池型、建材和制作工艺，为不同发育阶段的苗种营造不同的生态环境。刚孵出的中华鲟卵黄囊苗身体娇嫩，游泳能力差，故采用直径为1 m的小型平底圆形容器，使仔鱼尽量分散在容器底部；容器采用聚乙烯一次成模工艺制作,保证其底部和内壁有良好光洁度;采用可调方向的进水方式,以便根据仔鱼游泳能力的变化,通过调整进水角度在容器内营造不同的水流状态。仔鱼全长生长到5 cm以上后,游泳能力加强,摄食强度加大,则采用直径3.5 m的圆形水泥池养殖;通过加大底部坡度(坡比6 %)、内壁贴瓷砖保证池底光滑,提高排污效果。

该系统投资较节省。据测算,建成本系统共投资约280万元,仅为国外同类系统投资的1/3,与国内类似系统比较,节省投资50 %以上[15]。1999年以来,以该系统较好的可控性为基础,通过对生产流程各环节管理措施和生产技术的优化和规范化,使多年中华鲟苗种培育成活率稳定达到80 %以上,实现了中华鲟苗种培育的精益化生产。当然,该系统的部分功能尚待完善。如若能在水处理系统中配套建设水循环利用系统,进行养殖用水尤其冬季温水(一般18 ℃左右)的重复利用,则可减少苗种生产的水资源和能源消耗,降低成本。另外,如系统配备一定面积的池塘用于培育一些经济水生动物亲本,则可大大拓宽系统使用范围。

5 结论

一些鱼类,如中华鲟,具有特殊的生态习性,其苗种在不同发育阶段对生态环境有不同的需求。本系统根据中华鲟生态习性、发育特点及生态环境需求,分阶段设计养殖设施,包括亲鲟暂养池、催产池、仔鱼抚育桶和幼鲟培育池,并使用不同的池型、建材和制作工艺,营造出符合苗种各发育期生态要求的养殖环境。在分阶段养殖设施基础上,配套完善的、具有高可控性的供水、水处理、调温及增氧系统,通过系统集成提高了系统整体运行的稳定性和通用性,实现了中华鲟亲鲟催产到苗种培育的一站式批量生产。同时,系统的整体性和高可控性,使其也适用于其他鱼类尤其珍稀名贵鱼类苗种的批量生产,并可全年使用,提高了设施利用率。

参考文献

[1] 余志堂,许蕴玕,邓中粦.葛洲坝水利枢纽下游中华鲟繁殖生态的研究[A]//鱼类学论文集[C].北京:科学出版社,1986: 1-14.

[2] 胡德高,柯福恩,张国良.葛洲坝下中华鲟产卵情况初步调查及探讨[J].淡水渔业,1983(3):15-18.

[3] 胡德高,柯福恩,张国良,等.葛洲坝下中华鲟产卵场第二次调查[J].淡水渔业,1985(3):22-24.

[4] 胡德高,柯福恩,张国良,等.葛洲坝下中华鲟产卵场调查研究[J].淡水渔业,1992(5):6-10.

[5] Wei Q, Zhang J, Ke F, et al. Biology, fisheries and conservation of sturgeons and paddlefish in China[J]. Environmental Biology of Fishes,1997,48(4/6):241-255.

[6] 危起伟,杨德国,柯福恩.长江鲟鱼类的保护对策[A]//黄真理,傅伯杰.21世纪长江大型水利工程建设中的生态与环境保护(中国科协第十九次"青年科学家论坛")[C].北京:中国环境科学出版社,1998:208-216.

[7] 四川省长江水产资源调查组.长江鲟鱼类生物学及人工繁殖研究[M].成都:四川科学技术出版社,1988:284.

[8] 傅朝君,刘宪亭,鲁大椿,等.葛洲坝下中华鲟人工繁殖[J].淡水渔业,1985(1):1-5.

[9] 肖慧,刘勇,常剑波.中华鲟人工繁殖放流现状评价[J].水生生物学报,1999,23(6):572-576.

[10] Zhu Y, Wei Q, Yang D, et al. Large-scale cultivation of fingerlings of the Chinese Sturgeon *Acipenser sinensis* forre-stocking:a description of current technology[J]. Journal of Applied Ichthyology,2006,22(Suppl. 1):238-243.

[11] 危起伟,杨德国,朱永久,等.中华鲟幼鱼培育装置[P].中国专利:ZL200520098089.6,中华人民共和国国家知识产权局,北京,2006.

[12] 朱永久,刘鉴毅,危起伟,等.中华鲟活鱼运输箱[P].中国专利: ZL200520098090.9,中华人民共和国国家知识产权局,北京,2006.

[13] Chebanov M, Billard R. The culture of sturgeons in Russia:production of juveniles for stocking and meat for human consumption[J]. Aquatic Living Resources,2001,14(6):375-381.

[14] Williot P, Sabeau L, Gessner J, et al. Sturgeon farming in Western Europe: recent developments and perspectives[J]. Aquatic Living Resources, 2001, 14(6): 367-374.
[15] 刘艳红,罗国芝,孙佩英,等. 在闭合循环水产养殖系统中养殖高体革鯻的效果[J]. 上海水产大学学报,2003,12(2):130-134.

Intensive Reproduction System for Chinese Sturgeon (*Acipenser sinensis*) Larvae and Fingering

Yang Deguo[1,2] Wei Qiwei[1,2] Zhu Yongjiu[1] Liu Jianyi[1,2]
Wang Kai[1] Chen Xihua[1,2] Li Luoxin[1]

(1. Yangtze River Fisheries Research Institute, Chinese Academy of Fishery Sciences, Key Laboratory of Freshwater Fish Germplasm Resources and Biotechnology, the ministry of Agriculture of PRC. Jinzhou 434000, China;
2. Freshwater Fisheries Research Center, Chinese Academy of Fishery Sciences, Wuxi 214081, China)

Abstract Chinese sturgeon (*Acipenser sinensis*) is an adromous fish. In order to protect species, artificial reproduction of it has been caught out, and by now the intensive reproduction has been achieved. Intensive reproduction system for Chinese sturgeon fry is composed of primary and assistant system. The former includes four breeding ponds of different sizes with a total area of 2902.3 m^2, fertilization and hatch equipment. The later consists of transportation vehicle for parent fish, system of providing and disposing water, offering oxygen, maintaining and raising temperature, offering feed, and so on. This system needs little investment, has good harmonization and is convenient to operate. Fish ponds are made of different construction materials and techniques with different shapes to meet ecological demands for larvae and fingerlings at different development stages. The water temperature and the pattern of infall and discharge can be adjusted conveniently with automatic draining contamination, which provides the culture ponds with different flow patterns, good and stable water environment. The system can induce more than 16 parent fish to spawn in single one season, hatch 1.664×10^6 fertilized eggs and culture 2.355×10^6 larvae of 1 ~ 15 days old and 0.4233×10^6 fingerlings with 10 ~ 15 cm in total length, which can achieve the complete process of reproducing Chinese sturgeon. The system is the same with intensive reproduction and larger scale breeding of fingerling for other fishes.

Key words *Acipenser sinensis* Artificial propagation Fingering and fry rearing Intensive system